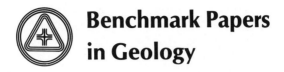

Benchmark Papers
in Geology

Series Editor: Rhodes W. Fairbridge
Columbia University

A selection from the published volumes in this series

A complete listing of volumes published in this series begins on p. 379.

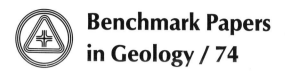

**Benchmark Papers
in Geology / 74**

A BENCHMARK® Books Series

MODERN CARBONATE
ENVIRONMENTS

Edited by

AJIT BHATTACHARYYA
Rensselaer Polytechnic Institute

and

GERALD M. FRIEDMAN
Rensselaer Polytechnic Institute

Hutchinson Ross Publishing Company

Stroudsburg, Pennsylvania

LIBRARY OF CONGRESS CATALOGING IN PUBLICATION DATA
Main entry under title:
Modern carbonate environments.
    (Benchmark papers in geology ; 74)
    Includes index.
    1. Rocks, Carbonate—Addresses, essays, lectures.
I. Bhattacharyya, Ajit Kumar,   II. Friedman, Gerald M.
1917-   III. Series.
QE471.15.C3M6   1982   552'.5   82-11816
ISBN 0-87933-436-3

Distributed worldwide by Van Nostrand Reinhold Company Inc.,
135 W. 50th Street, New York, NY 10020.

CONTENTS

Contents

SERIES EDITOR'S FOREWORD

The philosophy behind the Benchmark Papers in Geology is one of collection, sifting, and rediffusion. Scientific literature today is so vast, so dispersed, and, in the case of old papers, so inaccessible for readers not in the immediate neighborhood of major libraries that much valuable information has been ignored by default. It has become just so difficult, or so time consuming, to search out the key papers in any basic area of research that one can hardly blame a busy person for skimping on some of his or her "homework."

This series of volumes has been devised, therefore, as a practical solution to this critical problem. The geologist, perhaps even more than any other scientist, often suffers from twin difficulties—isolation from central library resources and immensely diffused sources of material. New colleges and industrial libraries simply cannot afford to purchase complete runs of all the world's earth science literature. Specialists simply cannot locate reprints or copies of all their principal reference materials. So it is that we are now making a concerted effort to gather into single volumes the critical materials needed to reconstruct the background of any and every major topic of our discipline.

We are interpreting "geology" in its broadest sense: the fundamental science of the planet Earth, its materials, its history, and its dynamics. Because of training in "earthy" materials, we also take in astrogeology, the corresponding aspect of the planetary sciences. Besides the classical core disciplines such as mineralogy, petrology, structure, geomorphology, paleontology, and stratigraphy, we embrace the newer fields of geophysics and geochemistry, applied also to oceanography, geochronology, and paleoecology. We recognize the work of the mining geologists, the petroleum geologists, the hydrologists, and the engineering and environmental geologists. Each specialist needs a working library. We are endeavoring to make the task of compiling such a library a little easier.

Each volume in the series contains an introduction prepared by a specialist (the volume editor)—a "state of the art" opening or a summary of the object and content of the volume. The articles, usually some twenty to fifty reproduced either in their entirety or in significant extracts, are selected in an attempt to cover the field, from the key papers of the last century to fairly recent work. Where the original works are in foreign languages, we have endeavored to locate or commission translations. Geologists, because of their global subject, are often acutely aware of the oneness of our world. The selections cannot therefore be restricted to any one country, and whenever possible an attempt is made to scan the world literature.

To each article, or group of kindred articles, some sort of "highlight commentary" is usually supplied by the volume editor. This commentary should serve to bring that article into historical perspective and to emphasize its particular role in the growth of the field. References, or citations, wherever possible, will be reproduced in their entirety—for by this means the observant reader can assess the background material available to that particular author, or, if desired, he or she too can double check the earlier sources.

A "benchmark," in surveyor's terminology, is an established point on the ground that is recorded on our maps. It is usually anything that is a vantage point, from a modest hill to a mountain peak. From the historical viewpoint, these benchmarks are the bricks of our scientific edifice.

RHODES W. FAIRBRIDGE

PREFACE

A spurt in oil exploration and the consequent discovery of large oil and gas reservoirs in carbonate rocks in the Middle East, the United States, Mexico, and other places in the 1950s prompted oil companies to take a fresh look at carbonate rocks and their environmental relationships.

The companies directed their attention to modern, tropical, shallow-water carbonate sediments especially in the Bahamas. This effort resulted in the publication of excellent key papers on the depositional environments of various modern carbonate-forming areas. The initial investigation on the Bahama Banks yielded data on modern carbonate-forming environments in what was thought to be an easily accessible area. But soon the limitation of the Bahamas model was realized and investigations were undertaken in other more distant areas representing a broader spectrum of carbonate environments. A thorough understanding of facies relationships in modern carbonate sediments provides a process-response model that allows a much more precise definition of facies relationships in the ancient carbonate rock record. Noel P. James (1978, p. 105) correctly writes that carbonate sediments are born, not made, and each depositional subenvironment is dominated by a set of distinct processes that results in a distinct set of primary textures, structures, lithology, and consequent distribution of primary porosity. It is this set of distinctive characteristics that is commonly used as a template in deciphering facies relationships in ancient carbonate rocks and helps locate the presence and distribution of porous potential reservoir facies. Carbonate rocks form more than 50 percent of known reservoirs of oil and gas throughout the world.

The project of compiling key papers on modern carbonate-forming environments was started to help researchers and exploration geologists studying facies relations and searching for oil and/or gas in carbonate rocks. The task of choosing papers for this volume was not an easy one; page restrictions forced us to omit excellent papers that have added much to the understanding of the subject. Many papers included in this volume were shortened. However in shortening, care was taken not to hurt the main theme of the papers. The editors provided comments on fifteen papers published in this volume.

The papers in this volume include classical studies of selected carbonate-forming regions, carbonate production, and cold-water carbonates. Part I includes papers from selected carbonate regions — the Bahamas, Florida, the Persian Gulf, and Shark Bay in Australia. These regions represent both the moist, humid, tropical and the dry, arid, to semiarid environments. Part II includes papers on carbonate production. Three out of four papers in this part discuss the origin of carbonate mud from different view points. The fourth paper focuses on the role of reefs in the generation of carbonate sediments. Part III includes one translated excerpt of a German paper and an abridged version of a paper evaluating the role of cold-water environments in carbonate production. These papers draw the attention of geologists to the role of cold versus hot tropical climates in the generation of carbonate rocks.

We are grateful to Professor Rhodes W. Fairbridge, the series editor, who read the original manuscript and offered constructive suggestions. We are also grateful to the authors who not only gave permission to reproduce their papers, but also supplied the necessary reprints and original negatives or prints of their illustrations, wherever available. Roy D. Nurmi, Anne Reeckmann, and Steven Chisick, all of Rensselaer Polytechnic Institute, helped us considerably in obtaining the material for this volume. Vin Buffaline and Luanne Wheeler typed the manuscript. We acknowledge their help gratefully.

AJIT BHATTACHARYYA
GERALD M. FRIEDMAN

REFERENCE

James, N. P., 1978, Introduction to Carbonate Facies Models, in *Facies Models,* Roger G. Walker, ed., Geoscience Canada, Reprint series 1, 211p.

CONTENTS BY AUTHOR

MODERN CARBONATE ENVIRONMENTS

INTRODUCTION

Starting in the 1950s the enormous petroleum reservoirs discovered in carbonate rocks in the Devonian of Canada, in Pennsylvanian and Permian strata of New Mexico and Texas, and in various Mesozoic and Cenozoic deposits in the Middle East prompted oil companies to increase their efforts in carbonate research. A most important part of this effort related to the study of depositional models from modern carbonate settings, particularly in shallow marine environments. Such models have served as an important link between Holocene settings and analogous ancient counterparts, especially in the subsurface in which petroleum geologists analyze and predict carbonate reservoirs. A geologist must be familiar with modern sediments where close relationships between geological processes and products have been established. A successful geologist, especially a subsurface geologist, must be process-oriented. He must match observed features in a rock with experience from modern sediments. His reasoning is by analogy, the geologist's stock-in-trade. The ultimate goal is the reconstruction of the pattern of ancient depositional environments and the interpretation of facies changes. Modern oil exploration for stratigraphic traps relies on the prediction of such lateral facies changes.

The search for potential reservoirs in carbonate sequences must consider depositional processes and products. Facies models, based on a study of modern settings, are useful aids in interpreting carbonate sequences. These models are developed in large part from experience in modern depositional environments, including vertical and lateral relationships, internal textures, structures, and compositions, and an indication of the diagenetic changes that may be expected. Most sequences show some variation from idealized facies models, and the distribution of primary porosity in carbonate rocks is a function of such models. Primary porosity and hydrocarbon-reservoir characteristics are commonly related to the energy in the environment of deposition. In general, carbonate sediments deposited in high-energy environments, such as reefs and shoals, are prone to have good primary and secondary porosity.

The papers of this volume include classical studies on selected carbonate regions, carbonate production, and cold-water carbonates. Although a mix between industrial effort and academic interest, the papers in this volume were largely sponsored by the petroleum industry. Paper 1, Black's classical paper, resulted from academic interest. However those of Illing (Paper 2) and Stockman, Ginsburg, and Shinn (Paper 12) are the outcome of industrial sponsorship. In the late 1950s and early to mid-1960s Shell Oil Company had an active laboratory in Florida that monitored carbonate processes and depositional environments in the Bahamas and Florida. Stockman, Ginsburg, and Shinn were part of the Shell carbonate team. Friedman was at that same time responsible for carbonate research for the Pan American Petroleum Corporation, now the Amoco Production Company (a Standard Oil of Indiana subsidiary), which supported the work of John Imbrie then of Columbia University. Purdy's classical paper (Paper 3) based on his doctoral thesis under Imbrie was an outgrowth of this sponsored research. Co-compiler of this volume Friedman nostalgically recalls field work in the Bahamas with J. Imbrie, E. G. Purdy, N. D. Newell, R. G. C. Bathurst, and Richard Scalan of the Phillipps Petroleum Corporation that funded the Columbia University team. Loss of a ship to the depth of the Bahamas was an inevitable part of this experience. Matthews's paper (Paper 11) resulted from a doctoral thesis under Purdy after the latter graduated from Columbia University and joined the faculty of Rice University. This work again was oil-company sponsored. Funding of the study of Logan and Cebulski, (Paper 9) was by the American Petroleum Institute, and their paper was published as part of a memòir of the American Association of Petroleum Geologists. Even where authors provide academic addresses the ultimate destinations of these authors was industry, thus Kendall's (Paper 8) address is stated as Department of Geology, University of Sydney, Australia, but his current affiliation is the Gulf Oil Corporation. Leonard's credited address on his paper (Paper 15) is Rensselaer Polytechnic Institute, but he is now with Amoco Production Company.

Carbonate sediments in clear, warm, sunlit waters of tropical shallow marine seas, unlike their terrigenous counterparts, originate as organically secreted solids formed within the basin of deposition and remain at or very near to their original sites of formation. They occur either as skeletal debris, as organically secreted carbonate mud, or in various combinations of the two. Skeletal materials are secreted by the metabolic activities of living organisms and included as whole skeletons or broken fragments. Ordinarily most of this skeletal debris is accumulated following the death of these organisms, but some skeletal materials are generated and shed during the life processes, such as the molting of ostracodes and trilobites. Coccoliths shed plates during life, many one-celled foraminifers make tests throughout their life. These discarded tests and shells often constitute significant parts of marine sediments.

Living reefs are another important source of skeletal sediment. Fishes that thrive in millions around reefs graze and rasp off the reefs producing skeletal sand; boring organisms convert the solid skeleton of reefs into skeletal sand. Although the total volume of skeletal material of modern and ancient reef-building organisms preserved intact is small, the reef builders supply copious amounts of skeletal sand that is distributed around the reef (Friedman and Sanders, 1978). In studying Jamaican reefs Goreau (1963) stated that "over 70 percent of the total $CaCO_3$ contained in the large reef systems is in the form of fine, unconsolidated sand deposited in thick beds over large areas adjacent to the living reef frame" (p. 127). There are few quantitative studies on carbonate production by reefs but a beginning in this direction has been made by Goreau (1961, 1963) and Chave et al. (Paper 13). Study of interactions of different borings and other organisms vis-a-vis reef and consequent sedimentation and cementation are of prime importance in the understanding of distribution of porosity and permeability in reef and associated facies. Zankl and Schroeder's paper (Paper 5) should act as a primer for those desiring a deep insight into the various processes that operate in a reef system.

These skeletal and other nonskeletal sands, in a current-agitated environment, respond similarly to the terrigenous sands and are molded into several different types of smaller and larger bedforms, as has been demonstrated from the Bahamas and Florida (Ball, 1967).

Sorting and particle-size analysis of these skeletal sands need caution and prudence. The size-frequency distribution of particles in Bahama sediment (Paper 3) and the Alacran reef complex, Yucatan (Folk and Robles, 1964) indicate that coarseness or fineness of carbonate sediment does not necessarily correlate with wave energy or water depth. However, it has been demonstrated by Ginsburg (Paper 6) that study of particle size and nature of constituent particles can aid in the interpretation of sedimentary environments (see also Friedman, 1961, on this subject).

In addition to these skeletal particles nonskeletal carbonate particles are important, for example ooids, pellets, pelloids, intraclasts, and grapestones that produce thick carbonate deposits in the rock record. Many of these constituents were indentified over a century ago. Sorby (1879) was the first to notice the occurrences of grapestone; but it was Illing (Paper 2) who made a most comprehensive study into

3

the significance and quantitative importance of these materials from the Bahamas, marking the beginning of a modern scientific approach to the study of carbonate sediments.

In addition to skeletal and nonskeletal particles, modern carbonate sediments and ancient carbonate rocks contain substantial amounts of 3 to 4 micron-size needles and platelets of carbonate crystals too small to resolve under normal petrographic microscopes. The lack of skeletal remains in these fine-grained limestones and laboratory production of lime mud inorganically led many to believe that these are inorganically produced sediments from supersaturated sea water. It was again Sorby (1879) who visualized that "fine calcareous mud, probably formed from various organic bodies by decomposition and mechanical wearing,"(p. 76) but who was ignored until 1955 when Lowenstam made a direct approach to the problem and conclusively demonstrated that postmortem disintegration of fragile codiacean (green) algae produces prismatic aragonite needles of 2 to 10 microns in length. Electron microscopy has further revealed that many fine-grained, apparently nonfossiliferous deep-sea limestones consist almost entirely of the remains of coccoliths (Fisher et al., 1967), a rediscovery dating back to Thomas Henry Huxley who, in his classical nineteenth-century essay "On a piece of chalk," stated unequivocally "I termed these bodies coccoliths" (Eisler, 1967, p. 42). Since then many geologists have noted that lime mud is polygenetic in origin. Papers 10, 11, and 12 have been included in this volume to illustrate the different viewpoints.

Other important biologic factors in the accumulation of carbonate sediment are the trapping and baffling action carried out by filamentous blue-green algae, now known as Cyanophyte bacteria, and thereby generating different morphological varieties of stromatolites showing millimeter-thin laminae. Such laminated structures are also common in ancient limestones. After the organic matter of algae has decomposed, the only indication of their former presence is the finely laminated structure. That stromatolites are algal-generated structures was first pointed out by Black (Paper 1). Various other aspects of algae and algal-generated structures have been discussed by Monty (Paper 4) among others.

Other papers that have been included in this volume discuss the various aspects of carbonate-forming environments in the Bahamas, Florida, Persian Gulf, and Shark Bay region of Australia. We hope that the few papers that have been included here on carbonate environments will help readers to understand facies relationships in carbonate rocks of the past.

Papers 14 and 15 have been included simply to highlight the

occurrences of carbonates in cold waters and to focus on the possibility that not all ancient carbonates are of tropical shallow-water origin. It is hoped that researchers will feel encouraged to have a fresh look at ancient limestones in this perspective.

REFERENCES

Ball, M. M., 1967, Carbonate Sand Bodies of Florida and the Bahamas *Jour. Sed. Petrology* **37:**556–591.

Eisler, L., ed., 1967, *On a Piece of Chalk by Thomas Henry Huxley,* Scribner's, New York, 90p.

Fisher, A. G., S. Honjo, and R. E. Garrison, 1967, Electron Micrographs of Limestones and their Nanofossils, Princeton University Press, Princeton, N. J., 141p.

Folk, R. L., and R. Robles, 1964, Carbonate Sands of Isla Perez, Alacran Reef Complex, Yucatan, *Jour. Geology* **72:**255–292.

Friedman, G. M., 1961, Distinction between Dune, Beach, and River Sands from their Textural Characteristics, *Jour. Sed. Petrology* **31:**514–529.

Friedman, G. M., and J. E. Sanders 1978, *Principles of Sedimentology,* Wiley, New York, 792p.

Goreau, T. F., 1961, On the Relation of Calcification to Primary Productivity in Reef Building Organisms, in *The Biology of Hydra and of some other Coelenterates,* H. M. Lenhoff and W. F. Loomis, eds., University of Miami Press, Coral Gables, Florida, pp. 269–285.

Goreau, T. F., 1963, Calcium Carbonate Deposition by Coralline Algae and Corals in Relation to Their Roles as Reef Builders, *New York Acad. Sci. Annals* **109:**127–167.

Lowenstam, H. A., 1955, Aragonite Needles Secreted by Algae and some Sedimentary Implications, *Jour. Sed. Petrology* **25:**270–272.

Sorby, H. C., 1879, Structure and Origin of Limestones: Anniversary Address of the President, *Geol. Soc. London Quart. Jour.* **35:**56–95.

Part I

SELECTED CARBONATE REGIONS

Editors' Comments
on Papers 1 Through 5

BAHAMAS

Algal Laminae

Among sedimentary structures found on modern carbonate tidal-flat sediments, algal layering is one of the most conspicuous but least understood. It has been described from different modern carbonate tidal flats, for example, Bahamas, Florida, Red Sea, Persian Gulf, and Shark Bay, Australia (Davies, 1970; Friedman et al. , 1973; Friedman, 1980a, 1980b; Ginsburg, 1955; Ginsburg and Hardie, 1975; Illing et al., 1965; Kendall and Skipwith, 1968; Logan et al., 1974; and Shinn et al., 1969). But questions about specific aspects of depositional environments or sedimentary processes that result in these laminae still remain unanswered. Recently Hardie and Ginsburg (1977) attempted to answer the mechanism and implications of such uniform millimeter-thin layering in the geologic record.

Maurice Black, in a pioneering study of sediments of Andros Island, Bahamas, first pointed out the role of blue-green algae, now considered to be Cyanophyte bacteria, in the binding of fine-grained sediments developing into characteristically laminated structures. The results of his study are found in Paper 1. He describes different kinds of algal laminae and postulates some recurrent or rhythmic change in the environment to cause such laminated deposits. The modern analogues of ancient stromatolites that he describes here are predominantly of a low-salinity lacustrine type, but supratidal and intertidal algal mats of normal salinity may also build such structures. Black observed that characteristic laminae that have a distinct geographic zonation are produced not by a single algal species but by a community of organisms; he also noted that particular species of algae may build up more than one kind of structure. Black's early studies in stratigraphy and paleobotany prepared him to recognize the importance of Bahamian algal-laminated sediments in the study of Precambrian and early Paleozoic stromatolites. It is unfortunate that Black's intention to complete a more detailed comparison of ancient and modern stromatolites was never realized.

In addition to his interest in modern carbonate sediments, Black investigated the Cretaceous coccoliths in the English chalk (Black, 1953; Black and Barnes, 1959) and modern coccoliths in deep-sea deposits (1965). His work on chalk established the stratigraphical value of coccoliths, opening up a whole new field of paleontological and stratigraphical research through the use of electron microscopes in their studies.

Calcareous Sands

L. V. Illing's long interest in the interpretation of carbonate sediments, specifically the sediments forming on the Bahaman Banks and their role in petroleum geology, began with his research at Cambridge, England, in 1948. He made a thorough study of skeletal and nonskeletal sands that mantle the major part of the vast platform (100,000 km^2 or 40,000 square miles) of the Bahama Banks and disproved the then prevailing idea regarding the nature and origin of some of the nonskeletal sands. He demonstrated that the nonskeletal grains are not erosional products of preexisting limestones but are formed penecontemporaneously within the basin mostly through biochemical processes. He also postulated a physicochemical origin for ooids that has since been disputed (Bathurst, 1971). In his study, Illing also discussed the origin of other nonskeletal grains, measured their sizes and sorting, and interpreted the controlling factors in their distribution on the Bahama Banks.

This classical study, excerpts from which are presented as Paper 2, generated much interest among geologists to study other modern carbonate environments and many publications have appeared since then. Still Illing stands out as the torch bearer and his work marks the beginning of a modern scientific approach to the study of Holocene carbonate environments and their sediments. Certain concepts have changed since the publication of his article. For example, it has been recognized that many pelletlike structures (peloids) may generate through recrystallization (Beales, 1958; Bathurst, 1971; McKee et al., 1969). "Intraclast," a term coined by Folk (1959), includes penecontemporaneously eroded carbonate sediments, ranging in size from silt to boulders. The "derived grains" of Illing are now known as "lithoclasts" (Folk, 1959). Again, "friable aggregates" of Illing have been replaced by Purdy (1963) with "mud aggregates." Purdy has recognized some of the "lumps" of Illing as "grapestones," which Purdy has defined as "cemented aggregates of skeletal and nonskeletal grains in which the carbonate cement is restricted largely to the aggregate's periphery, the interior of aggregates tending to remain porous" (p. 344). The sediments identified by the term "oolite" by Illing are now known to be aggregates of individual ooids having similar characteristics to Illing's oolite. Nevertheless, one is struck by the keen observations and thorough nature of his work as reflected in Paper 2.

Sedimentary Facies

Intricate variations in the lithology and faunal contents usually associated with ancient limestones have forced geologists to look into their modern counterparts and study in detail the distribution and lateral variation in the lithology and faunal assemblages. Of the limited number of marine environments where carbonate sediments are forming today, the Bahama Banks have attracted the most attention of geologists. Purdy's doctoral dissertation on the carbonate sediments of the northwestern part of the Great Bahama Banks has been one of the most outstanding in this aspect. His dissertation was published in the *Journal of Geology* in 1963 in two parts. (The second part, on sedimentary facies, is Paper 3. The first part is not reprinted here.) His study forms a standard reference work for ecologic and sedimentologic interpretation both of modern carbonate sediments and their ancient analogues. In the first part, he describes the petrography of the constituent carbonate grains of the Bahama Banks and attempts to depict the relationship between these constituent grains as a reaction group hierarchy. In the second part (Paper 3),

Purdy attempts to delineate sedimentary facies on the basis of quantitatively important grain types and traces the various factors controlling origin and lateral distribution of them.

Stromatolitic Deposits

After Black's historical paper in 1933, (Paper 1) many geologists became impressed by the similarity of modern and ancient stromatolites, but unfortunately no one made detailed investigations on their mechanism of formation or of the microstructures of modern algal mats occurring in various depositional environments. Moreover, very few geologists were aware of the enormous strides made by botanists in their analysis of the biological, physiological, and symbiotic organization of these fascinating algae. As a result, misunderstanding generated false notions and concepts about the mechanism, environmental setting, or ecology of both modern and fossil stromatolites. Monty (Paper 4) tried to set the record straight and emphasized the need to integrate the data on ecology, physiology, and metabolism of blue-green algae for a correct understanding and interpretation of modern and ancient stromatolites.

Monty suggested that soft algal biscuits and balls that are found in freshwater are the modern analogues of lithified calcareous oncolites in the rock record and not the ones described by Ginsburg (1960) and Gebelein (1969) from the shallow marine setting of Florida and Bermuda. Oncolites furthermore appear to be a product of low-energy environments. He further cautions geologists about the overemphasis of the concept of inter- and supratidal settings as habitat of stromatolites or the overemphasis of the trapping and binding activity of algae in the formation of stromatolites, for (1) inter- and supratidal flats are but one of the many settings of stromatolite formation, and (2) many freshwater algae exist that can secrete calcium carbonate to form coherent algal structures and seem to lose this characteristic in marine environments. Monty analyzed the distribution of blue-green algal structures on Andros Island and adjacent marine banks and provided a detailed account of macro- and micro-structures of freshwater stromatolites and their zonation. He emphasized that the environment shapes stromatolites whereas algae themselves remain a passive spectator (see also Logan et al., 1964). Moreover, the observed abundance of stromatolites in the shallow marine realm is not only because of predation and boring as suggested by Garrett (1970), but Monty believed this to be the result of "direct

11

intercommunity competition in which the competitive weakness of blue-green algae has driven them to settle in habitats where their competitors could not grow" (p. 776).

Besides the generation of algal laminae and distinctive fabrics of algal origin, Bathurst (1967) described another kind of algal mat that plays an important role in sediment binding and stabilization. These algal mats are gelatinous, transparent, colorless, and elastic and do not leave any imprint in the fossil record. These mats are distinct from algal stromatolites. Bathurst thus cautions the carbonate sedimentologists

> to bear in mind that, where ancient limestones have textures which would conventionally be interpreted as indicating low-energy deposition, they may in fact have been laid in a high-energy environment where the floor was immobilized by a mat. (1967, p. 737)

It is interesting to note the recent findings of a luxurious growth of blue-green algal mats and stromatolites below a huge thickness of ice in all four lakes of the Dry Valley region of Antarctica (Young, 1981). Algae in these lakes occur at depths of 27–60 m covering the whole lake bottom. Waters of these four lakes have different salinities, one of which is so salty that it never freezes. This observation indicates that blue-green algae are hardy organisms, adept at growing in extreme and hostile environments. Occurrence of algae under such harsh environments has raised some intriguing questions. How could these algae, which depend on sunlight for photosynthesis and life, survive and thrive in a place where no sun shines at least four months of the year? Scientists are still in search of an answer.

Modern Reefs

Reefs are wave-resistant bioherms of colonial, skeletal metazoans forming rigid exoskeletal structures that flourish under a particular tectono-sedimentary environment. Abundant colonial and noncolonial organisms thrive in and around modern reefs and are natural laboratories for the study of benthic marine ecology. As a result, modern reefs have become the pleasant hunting ground of geologists, especially as large oil and gas reserves occur in many fossil reefs. For proper evaluation of ancient reefs and interfingering lithic facies, an unprecedented surge in studies of modern reefs has provided data on their morphology, ecology, and mineralogy of the skeletal framework and their diagenesis (Friedman, 1968; Friedman et al., 1974; Friedman et al., 1976; Glynn et al., 1972; Goreau, 1959; Goreau and Wells, 1967; Goreau and Land, 1974; Hartman and Goreau, 1970; Heckel, 1972; Jones and Endean, 1967; Land and Epstein, 1970; Land and Goreau, 1970; Logan, 1969; Maiklem, 1968; Matthews, 1974; Maxwell, 1962;

Macintyre, 1977; Mesolella et al., 1970; Newell et al., 1959; Purdy, 1974; Schroeder, 1972; Shinn, 1966; Smith, 1948; Stoddart, 1969; Tracey et al., 1948; Umbgrove, 1947; Vaughan, 1915*a*, 1915*b;* and many others).

But identifying reef reservoirs in the subsurface poses problems and frustration because of the apparent lack of framework-building and the dominance of cryptocrystalline cement, which has the appearance of lime mud or micrite. There are many such buildups in the Phanerozoic rock record that lack framework builders in position of growth, and look like limemud buildups (Crowley, 1969; Friedman and Sanders, 1978, p. 175; Heckel, 1972; Heckel and Cocke, 1969; Wilson, 1957, 1962; and Wray, 1964, 1968). Anatomic dismemberment (Hubbard, 1976) and bioerosion may have been the cause of degraded framework. Endolithic borers may even convert skeletal fragments into peloids and a reefrock composed of peloids may puzzle unwary geologists and may be explained incorrectly (Friedman, 1981; Lees, 1964, and Schwarzacher, 1961). Furthermore, calichification and "chalkification" of reefs, emergent at some time in their history, are quite capable of completely erasing their depositional texture (Friedman, 1981). To an unsuspecting geologist, these degraded reefs, made up apparently of lime mud, will be mistaken as low-energy environments; yet they formed at the highest energy possible. Considering the extensive bioerosion that takes place in a reef, it is rather surprising that any reef material is preserved in the rock record.

Moreover, no two bioherms are identical in size and configuration because of their inherent complex manner of growth and associated facies, which in turn may determine to a large extent the quality and distribution of porosity so vital for an oil reservoir (Stanton, 1967). Thus a proper understanding of ancient reef facies demands detailed information on the interaction of complex biologic and sedimentologic processes in and around an active modern reef.

Paper 5 is unique. It describes the reef morphology and the different organic and inorganic processes that operate in the Holocene reefs, interactions and interrelationships that go together in the final makeup of the reefs and associated facies and their ultimate preservation in the rock record.

REFERENCES

Bathurst, R. G. C., 1967, Subtidal Gelatinous Mat, Sand Stabilizer and Food, Great Bahama Bank, *Jour. Geology* **75:**736–738.

Bathurst, R. G. C., 1971, *Carbonate Sediments and Their Diagenesis,* Elsevier, Amsterdam, 620p.

Beales, F. W., 1958, Ancient Sediments of Bahamian Type, *Am. Assoc. Petroleum Geologists Bull.* **42:**1845–1880.

Black, M., 1953, Constitution of the Chalk, *Geol. Soc. London Proc.* **1499:**81–82.

Black, M., 1965, Coccoliths, *Endeavour* **24:**131–137.

Black, M., and B. Barnes, 1959, The Structure of Coccoliths from the English Chalk, *Geol. Mag.* **96:**321–328.

Crowley, D. J., 1969, Algal-bank Complex in Wyandotte Limestone (Late Pennsylvanian) in Eastern Kansas, *Kansas Geol. Survey Bull. 198,* 52p.

Davies, G. R., 1970, Algal-laminated Sediments, Gladstone Embayment, Shark Bay, Western Australia, in *Carbonate Sedimentation and Environments, Shark Bay, Western Australia,* B. W. Logan, G. R. Davies, J. F. Read and D. E. Cebulski, eds., Am. Assoc. Petroleum Geologists Mem. 13, pp. 169–205.

Folk, R. L., 1959, Practical Petrographic Classification of Limestones, *Am. Assoc. Petroleum Geologists Bull.* **43:**1–38.

Friedman, G. M., 1968, Geology and Geochemistry of Reefs, Carbonate Sediments, and Waters, Gulf of Aqaba (Elat), Red Sea, *Jour. Sed. Petrology* **38:**895–919.

Friedman, G. M., 1980*a*, Reefs and Evaporites at Ras Muhammad, Sinai Peninsula: A Modern Analog for One Kind of Stratigraphic Trap, *Israel Jour. Earth-Sci.* **29:**166–170.

Friedman, G. M., 1980*b*, Dolomite is an Evaporite Mineral: Evidence from the Rock Record and from Sea-Marginal Ponds of the Red Sea, in *Concepts and Models of Dolomitization,* D. H. Zenger, J. B. Dunham and R. L. Ethington, eds., Soc. Econ. Paleontologists and Mineralogists Spec. Pub. 28, pp. 69–80.

Friedman, G. M., 1981, Recognition of Post-Paleozoic Reefs: An Experience in Frustration, *Internat. Coral Reef Symposium, 4th, Manila, Proc.* p. 21.

Friedman, G. M., and J. Sanders, 1978, *Principles of Sedimentology,* Wiley New York-Chichester-Brisbane-Toronto, 792p.

Friedman, G. M., A. J. Amiel, M. Braun, and D. C. Miller, 1973, Generation of Carbonate Particles and Laminites in Algal Mats—Examples from Sea-Marginal Hypersaline Pool, Gulf of Aqaba, Red Sea, *Am. Assoc. Petroleum Geologists Bull.* **57:**541–557.

Friedman, G. M., A. J. Amiel, and N. Schneidermann, 1974, Submarine Cementation in Reefs: Examples from the Red Sea, *Jour. Sed. Petrology* **44:**816–825.

Friedman, G. M., S. A. Ali, and D. H. Krinsley, 1976, Dissolution of Quartz Accompanying Precipitation and Cementation in Reefs: Example from the Red Sea, *Jour. Sed. Petrology* **46:**970–973.

Garrett, P., 1970, Phanerozoic Stromatolites: Noncompetitive Ecologic Restriction by Grazing and Burrowing Animals, *Science* **169:**171–173.

Gebelein, C. D., 1969, Distribution, Morphology and Accretion Rate of Recent Subtidal Algal Stromatolites, Bermuda, *Jour. Sed. Petrology* **39:**32–49.

Ginsburg, R. N., 1955, Recent Stromatolitic Sediments from South Florida (abstract), *Jour. Paleontology* **29:**723.

Ginsburg, R. N., 1960, Ancient Analogues of Recent Stromatolites: *Internat. Geol. Cong., 21st, Norden, Reports, Part XXII,* pp. 26–35.

Ginsburg, R. N., and L. A. Hardie, 1975, Tidal and Storm Deposits, Northwestern Andros Island, Bahamas, in *Tidal Deposits. A Case Book of Recent Examples and Fossil Counterparts,* R. N. Ginsburg, ed., Springer-Verlag, New York-Heidelberg-Berlin, pp. 201–208.

Glynn, P. W., R. H. Stewart, and J. E. McCosker, 1972, Pacific Coral Reefs of Panama: Distribution and Predators, *Geol. Rundschau* **61:**483–519.

Goreau, T. F., 1959, The Ecology of Jamaican Coral Reefs. 1. Species Composition and Zonation, *Ecology* **40:**67–90.

Goreau, T. F., and L. S. Land, 1974, Fore-Reef Morphology and Depositional Processes, North Jamaica, in *Reefs in Time and Space: Selected Examples from the Recent and Ancient,* Leo F. Laporte, ed., Soc. Econ. Paleon. Mineralogists Spec. Pub. 18, pp. 77-89.

Goreau, T. F. and J. W. Wells, 1967, The Shallow-Water Scleractinia of Jamaica: Revised List of Species and Their Vertical Distribution Range, *Bull. Marine Sci.* **17:**442–453.

Hardie, L. A. and R. N. Ginsburg, 1977, Layering: The Origin and Environmental Significance of Lamination and Thin Bedding, in *Sedimentation on the Modern Carbonate Tidal Flats of Northwest Andros Island, Bahamas,* L. A. Hardie, ed., The Johns Hopkins University Press, Baltimore-London, pp. 50–123.

Hartman, W. D., and T. F. Goreau, 1970, Jamaican Coralline Sponges: Their Morphology, Ecology, and Fossil Relatives, *Zoological Soc. London, Symp.* **25:**205–243.

Heckel, P. H., 1972, Pennsylvanian Stratigraphic Reefs in Kansas, Some Modern Comparisons and Implications, *Geol. Rundschau* **61:**584–598.

Heckel, P. H., and J. M. Cocke, 1969, Phylloid Algal-Mound Complexes Outcropping in Upper Pennsylvanian Rocks of Mid-Continent, *Amer. Assoc. Petroleum Geologists Bull.* **53:**1058–1074.

Hubbard, J. A. E. B., 1976, Sediment Budgeting: A Timed Study of Scelractinian Necrology, *Internat. Geol. Cong., 25th, Australia, Proc.* **3:**845.

Illing, L. V., A. J. Wells, and J. C. M. Taylor, 1965, Penecontemporary Dolomite in the Persian Gulf, in *Dolomitization and Limestone Diagenesis — A Symposium,* L. C. Pray and R. C. Murray, eds., Soc. Econ. Paleontologists and Mineralogists Spec. Pub. 13, pp. 89–111.

Jones, D. A., and R. Endean, 1967, The Great Barrier Reefs, *Sci. Jour.* **3:**44–51.

Kendall, C. G. St. C., and P. A. d'E. Skipwith, 1968, Recent Algal Mats of a Persian Gulf Lagoon, *Jour. Sed. Petrology* **38:**1040–1058.

Land, L. S., and S. Epstein, 1970, Late Pleistocene Diagenesis and Dolomitization, North Jamaica, *Sedimentology* **14:**187–200.

Land, L. S., and T. F. Goreau, 1970, Submarine Lithification of Jamaican Reefs, *Jour. Sed. Petrology* **40:**457–462.

Lees, A., 1964, The Structure and Origin of the Waulsortian (Lower Carboniferous) "Reefs," West-Central Eire, *Royal Soc. London Philos. Trans.,* ser. B, **247:**484–531.

Logan, B. W., 1969, Carbonate Sediments and Reefs, Yucatan Shelf, Mexico. Part 2. Coral Reefs and Banks, Yucatan Shelf (Yucatan Reef Unit), *Amer. Assoc. Petroleum Geologists Mem. 11,* pp. 129–198.

Logan, B. W., P. Hoffman, and C. D. Gebelein, 1974, Algal Mats, Cryptalgal Fabrics and Structures, Hamelin Pool, Western Australia, *Am. Assoc. Petroleum Geologists Mem. 22,* pp. 140–194.

Logan, B. W., R. Rezak, and R. N. Ginsburg, 1964, Classification and Environmental Significance of Algal Stromatolites, *Jour. Geology* **72:**68–83.

McKee, E. D., and R. C. Gutschick, 1969, History of Redwall Limestone of Northern Arizona, *Geol. Soc. America Mem. 114,* 726p.

Macintyre, I. G., 1977, Distribution of Submarine Cements in a Modern Caribbean Fringing Reef, Galeta Point, Panama, *Jour. Sed. Petrology* **47:**503–516.

Maiklem, W. R., 1968, The Capricorn Reef Complex, Great Barrier Reef, Australia, *Jour. Sed. Petrology* **38:**785–798.

Matthews, R. K., 1974, A Process Approach to Diagenesis of Reefs and Reef-Associated Limestones, in *Reefs in Time and Space: Selected Examples from the Recent and Ancient,* Leo F. Laporte, ed., Soc. Econ. Paleontologists and Mineralogists Spec. Pub. 18, pp. 234–256.

Maxwell, W. G. H., 1962, Lithification of Carbonate Sediments in the Heron Island Reef, Great Barrier Reef, *Geol. Soc. Australia Jour.* **8:**217–238.

Mesolella, K. J., H. A. Sealy, and R. K. Matthews, 1970, Facies Geometries within Pleistocene Reefs of Barbados, West Indies, *Amer. Assoc. Petroleum Geologists Bull.* **54:**1899–1917.

Newell, N. D., J. Imbrie, E. G. Purdy, and D. L. Thurber, 1959, Organism Communities and Bottom Facies, Great Bahama Bank, *Am. Mus. Nat. History Bull.* **117:**177–228.

Purdy, E. G., 1963, Recent Calcium Carbonate Facies of the Great Bahama Bank. I. Petrography and Reaction Groups, *Jour. Geology* **71:**334–355.

Purdy, E. G., 1974, Reef Configurations: Cause and Effect, in *Reefs in Time and Space: Selected Examples from the Recent and Ancient,* Leo F. Laporte ed., Soc. Econ. Paleontologists and Mineralogists Spec. Pub. 18, pp. 9–76.

Schwarzacher, W., 1961, Petrology and Structure of Some Lower Carboniferous Reefs in Northwestern Ireland, *Am. Assoc. Petroleum Geologists Bull.* **45:**1481–1503.

Schroeder, J. H., 1972, Fabrics and Sequences of Submarine Carbonate Cements in Holocene Bermuda Cup Reefs, *Geol. Rundschau* **61:**708–730.

Shinn, E. A., 1966, Coral Growth Rate, an Environmental Indicator, *Jour. Paleontology* **40:**233–240.

Shinn, E. A., R. M. Lloyd, and R. N. Ginsburg, 1969, Anatomy of a Modern Carbonate Tidal Flat, Andros Island, Bahamas, *Jour. Sed. Petrology* **39:**1202–1228.

Smith, F. W. W., 1948, Atlantic Coral Reefs, University of Miami Press, Coral Gables, Florida, 112p.

Stanton, R. J., 1967, Factors Controlling Shape and Internal Facies Distribution of Organic Carbonate Buildups, *Am. Assoc. Petroleum Geologists Bull.* **51:**2462–2467.

Stoddart, D. R., 1969, Ecology and Morphology of Recent Coral Reefs, *Biol. Review* **44:**433–498.

Tracey, J. I., H. S. Ladd, and J. E. Hoffmeister, 1948, Reefs of Bikini, Marshall Islands, *Geol. Soc. America Bull.* **59:**861–878.

Umbgrove, J. H. F., 1947, Coral Reefs of the East Indies *Geol. Soc. America Bull.* **58:**729–778.

Vaughan, T. W., 1915a, On the Recent Madreporaria of Florida and West Indies, *Carnegie Institution of Washington Year Book* **14:**220–231.

Vaughan, T. W., 1915*b*, The Geologic Significance of the Growth Rate of the Floridian and Bahamian Shoal Water Corals, *Wash. Acad. Science Jour.* **5**:591–600.

Wilson, F. W., 1957, Barrier Reefs of the Stanton Formation (Missourian) in Southwest Kansas, *Kansas Acad. Science Trans.* **60**:429–436.

Wilson F. W., 1962, A Discussion of the Origin of the Reeflike Limestone Lenses of the Lansing Group (Upper Pennsylvanian) of Southeast Kansas—Geoeconomics of the Pennsylvanian Marine Banks in Southeast Kansas, *27th Field Conference Guidebook,* Kansas Geol. Society, pp. 101–105.

Wray, J. L., 1964, Archaeolithophyllum and Abundant Calcareous Algae in Limestones of the Lansing Group (Pennsylvanian), Southeastern Kansas, *Kansas Geol. Survey Bull.* **170**:1–13.

Wray, J. L., 1968, Late Paleozoic Phylloid Algal Limestones in the United States, *Internat. Geol. Cong., 23rd, Proc.* **8**:113–119.

Young, P., 1981, Thick Layers of Life Blanket Lake Bottoms in Antarctica Valleys, *Smithsonian* **12**:52–61.

1

Reprinted by permission from pages 165-170 and 175-191 of *Royal Soc. London Philos. Trans., ser. B,* **222:**165-192 (1933)

The Algal Sediments of Andros Island, Bahamas.

By MAURICE BLACK, *Trinity College, Cambridge.*

CONTENTS.

I.—INTRODUCTION.

In the course of a geological reconnaissance of Andros Island, in the Bahamas, it was found that the lower forms of plant life, especially the Blue-green Algæ, play an important part in the process of sedimentation. In addition to those forms which actively contribute calcium carbonate to the sediment, there are other species which function primarily as sediment binders, without necessarily precipitating any lime themselves. Such sediment-binding algæ usually impart characteristic structures to the medium in which they grow; and in the interior of Andros, where such deposits are

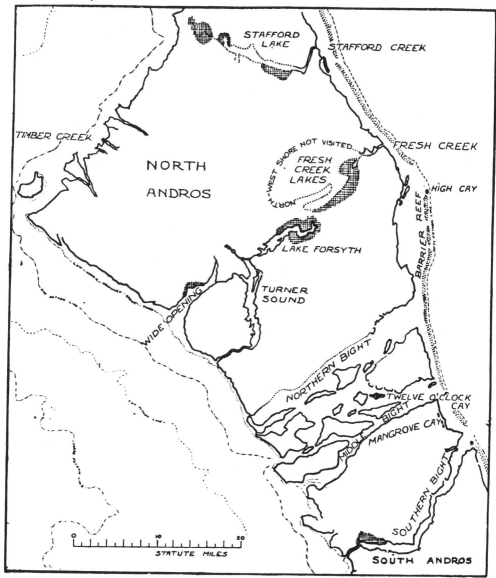

Fig. 1.—Map of North Andros Island. The localities from which algal sediments are described in this paper are shaded.

now accumulating over large areas, structures are being produced which are reminiscent of those found in some of the great limestone formations of the Lower Palæozoic and Upper Precambrian. In view of this, and of the supposed algal origin of certain of these limestone structures, it is felt that a detailed description of the Bahaman sediments will provide an example of a modern Cyanophyceous deposit, which may prove useful for comparison with older limestones of similar structure.

The field work on which this paper is based was undertaken during the spring of 1930, in the course of an expedition to Andros Island, made possible through the assistance of the Rouse Ball Fund of Trinity College, Cambridge, and of the Percy Sladen Fund of London. It is the author's pleasant duty to acknowledge his gratitude to the trustees of these two funds for their generous support. This expedition to the interior of Andros Island formed part of the programme of the International Expedition to the Bahamas, under the general direction of Dr. RICHARD M. FIELD, of Princeton, U.S.A.

The Bahama Bank is a region of shallow water limestone sedimentation. It consists of a great submarine plateau, standing almost at sea level, and thus forming a great stretch of shallow sea, surrounded on all sides by deep water. The rocks which form the islands are very pure limestones of Pleistocene Age, and the modern sediments on the bank are also entirely of calcite and aragonite, without any admixture of siliceous or argillaceous material. Thus the region is one of exclusively limestone sedimentation, and in this respect is closely comparable with some of the seas of Palæozoic and later Precambrian times, in which carbonate sediment prevailed over considerable areas.

Andros Island, in which the algal beds are found, has an area of some two thousand square miles, and is the largest island in the Bahamas. Along the east coast there is a well developed barrier reef, backed by a narrow ridge of limestone hills, but the interior and the western part of the island are of entirely different character. Low-lying, marshy plains, interspersed with shallow lakes and outcrops of limestone, are found in the interior, whilst in the westernmost part of the island, locally known as "The Marl," the limestone outcrops disappear, and the country consists of white, unconsolidated limestone-mud, known as *Drewite*. This is a desolate region, with bare white drewite flats, very sparsely covered with halophytic vegetation. The whole island, but especially the western part, is dissected by an intricate system of tidal creeks and mangrove swamps, which render it liable to heavy flooding under favourable circumstances. As this flooding has an important effect upon the growth of the algal deposits, by providing an intermittent supply of sediment, and possibly by causing an alteration in the salinity of the surface water, we shall return to this question in more detail later.

II.—ECOLOGY OF THE ALGAL FLORA.

(a) Distribution of the Genera.

The western edge of the Great Bahama Bank lies some eighty miles off the coast of Andros, and the intervening shallow water never reaches a depth greater than about four fathoms. The sea floor is mantled with bottom deposits of white, mostly fine-grained, calcium carbonate sediments, with occasional outcrops of Pleistocene limestone. Owing to the shallowness and clarity of the water, and to the whiteness of the bottom deposits, the light intensity on the sea floor is very great, and the bottom flora consists very largely of Chlorophyceæ and marine angiosperms. Dredgings between the edge of the bank and the west coast of Andros brought up mainly specimens of *Penicillus, Rhipocephalus, Udotea, Halimeda, Batophora* and *Caulerpa*. It was evident, in examining the dredgings, that the basal filaments of the algæ, in permeating the surface layers of the unconsolidated muds and sands upon which they grow, have a considerable binding effect upon the sediment. This is even more conspicuous in the zone between tides, where the purely filamentous forms, such as *Derbesia*, assist in producing the same effect. Although the sediment is by no means hard, it becomes tenaciously cohesive, and resists the erosive action of ordinary tidal currents.

Above the intertidal belt, the Chlorophyceæ are no longer dominant, and are almost completely replaced by the Cyanophyceæ. Wherever the surface sediments were examined, they were found to be permeated by algal filaments and cells, or to be covered by a skin consisting of algal colonies growing on the surface. In regions of intermittent sedimentation, such as the coastal belt, and some of the flat, low-lying areas in the interior of the island, these algæ were found to play an active part in the process of sediment accumulation. The colonization of newly deposited sediment by filamentous algæ first of all binds together the sediment, preventing its being easily washed away again, and then produces a felt of algal filaments, which is sometimes quite thick and dense. In nearly all the species involved, the filament is enclosed in a mucilaginous sheath, to which mineral particles very readily adhere. Thus any fresh sediment brought into the region is at once trapped amongst the filaments.*

The most important species in the Bahaman sediments are :—

Gloeocapsa atrata (TURPIN) KÜTZING.
G. granosa (BERKELEY) KÜTZING var. chlora nov.†
G. fusco-lutea (NAEGLI) KÜTZ.
G. gelatinosa KÜTZ.
G. magma (BREBÍSSON) KÜTZ.
G. rupestris KÜTZ.
G. viridis sp. nov.

Aphanocapsa grevillei (HASSAL) RABENHORST.
A. marina HANSGING.
Symploca læte-viridis GOMONT.
Phormidium tenue (MENEGHINI) GOMONT.
Schizothrix braunii GOMONT.
Plectonema atroviride sp. nov.
Scytonema androsense sp. nov.
S. crustaceum AGARDH. var. *catenula* nov.

It will be noticed that the Rivulariaceæ, so frequently associated with the formation of lake balls and water biscuits, are of no importance here.

These algæ were found to be growing, not in pure colonies, but in rather complex communities, which are still under investigation, and will not be described in detail

---

* This sediment-binding action of filamentous algæ is also of common occurrence in regions where the sediment consists of non-calcareous mud, and colonization of fresh sediment by such species as *Microcoleus chthonoplastes* and *Vaucheria thuretii* has been shown to be important in tidal salt marshes. For example, see CAREY and OLIVER, 1918, p. 173, and Plate XV, upper figure.

† The new species and varieties of algæ are described in section VII, p. 186.

until more of the species involved have been grown in artificial culture. Roughly speaking, it may be said that the communities in areas affected by the tides are dominated by *Symploca læte-viridis* and several non-filamentous forms; that as one goes from the tidal belt inland towards the less saline water, species of *Scytonema* appear, and grow intimately with a large number of other species, important amongst which are *Schizothrix* and *Aphanocapsa*; and finally, that in places where the ground water is not saline, or which are affected by rain-water alone, almost pure colonies of *Scytonema* are found, without admixture of any species belonging to the Oscillatorieæ. Thus we have the distribution shown in Table I.

TABLE I.

Maximum Salinity : parts per 1000.	Nature of Habitat.	Flora.
30–40	Shoal water below tides	Chlorophyceæ : *Udotea, Penicillus, Halimeda.*
15–36	Between tides, near H.W.M.	Cyanophyceæ : *Phormidium, Symploca,* and unicellular forms.
? ca. 2	Above H.W.M. Inland Marl flats	Cyanophyceæ : *Scytonema, Plectonema, Schizothrix,* and unicellular forms.
0	Above H.W.M. Limestone outcrops	*Scytonema* alone.

III.—DESCRIPTION OF THE ALGAL SEDIMENTS.

Where definite algal structures are developed in the sediments, their growth form depends to a large extent upon local conditions, and a different type of algal head is found in each of the geographical belts mentioned above.

Under the shoal waters of the Great Bahama Bank, no complex algal heads were observed, and the Cyanophyceæ, which are elsewhere responsible for these structures, were not found to be present in large numbers. The Chlorophyceæ, which make up the bulk of the flora here, were found to permeate the sediment and bind the particles together, without producing any banded or radial structure. It was only above low-water mark, where the Cyanophyceæ are the prevalent algæ, that alga-controlled lamination and algal heads with characteristic internal structures were found.

The simplest type, fig. 2, and figs. 21 and 22, Plate 22, was found in localities which were frequently flooded by sea water. Round the western entrance to Southern Bight, filaments of *Symploca læte-viridis* permeate the sediment as it is deposited, and play an important part in binding it, thus preventing re-erosion. During intervals of non-deposition the alga continues to grow in company with several other species, and a thin organic layer is produced. Beyond this, however, the alga does little to modify the structure of the sediment, except by accentuating the mechanically formed laminations.

Somewhat more distinctively alga-controlled sediments were found about high-water mark at Twelve O'clock Cay and in the Wide Opening, figs. 17 and 18, Plate 21, where, in addition to this kind of lamination, simple algal heads are developed, fig. 3, and fig. 24, Plate 22. These take the form of irregularly scattered domes usually an inch or two high, and four or five inches in diameter, each possessing a crude concentric lamination.

It was in the interior of Andros Island that the most highly developed type of algal head was found, and where the most extensive algal deposits were discovered. Bordering the fresh-water lakes, and forming the flat ground between the limestone ridges, are large stretches of drewite marshes, which are covered with circular algal heads possessing a peculiar structure of approximately concentric laminæ, fig. 5, and figs. 19 and 20, Plate 21.

In all these forms, parallel or concentric lamination is the dominant element in the structure of the algal heads, and the mineral matter is soft and uncemented sediment, mechanically entrapped by the algæ, without any perceptible addition of secondary crystals. On slightly more elevated ground, however, the algal heads assume an entirely different structure, fig. 28, Plate 22. The colonies consist of radiating filaments, without much interstitial sediment, and the whole algal head is often strongly cemented with carbonate crystals precipitated round the filaments. Two main factors are probably responsible for this difference in structure : firstly, the absence of sediment, which leaves the algal head porous, and secondly, the different properties of the water, which is rain-water with a certain amount of freshly dissolved calcium carbonate in solution, but practically no other dissolved salts.

The types of algal structures described below are restricted to the sedimentary forms ; those algal heads in which the clastic sediment does not play an essential part are still under investigation, and will be dealt with in a later publication. In the paragraphs immediately following, only the positions of the localities are indicated, and the local conditions are described in more detail in Section V, p. 179, after the significance of the environment has been discussed.

[Editors' Note: Material has been omitted at this point.]

IV.—The Significance of the Lamination.

Each of the four types of algal-head described above is characterized by the possession of a laminated structure, which may arise in one or more of three ways :—

(1) Rhythmic variation in the quantity of filaments of one species or group of species, with relation to the quantity of sediment.

(2) Alternation of two species or of two groups of species.

(3) Sedimentary lamination in mineral particles enclosed between the filaments.

In each development of concentrically laminated sediment investigated on Andros Island, it was found that the presence of two of these laminating processes could be traced. Thus :—

The lamination in Type A arises from a combination of 1 and 3.

     ,,   ,,   Type B   ,,   ,,   2 and 3.

     ,,   ,,   Types C and D   ,,   ,,   1 and 2.

There can be little doubt that each of these laminating processes represents a response to some definite recurrent or rhythmic change in the environment, and it is a matter of considerable stratigraphical interest to know what these changes are, and to what extent we can rely upon the evidence of these laminated bodies to reconstruct the external conditions under which such sediments were formed. It is impossible to enter into the question exhaustively until more results of experimental work on the algæ in culture are available, but one or two of the more important environmental conditions as observed in the field may be discussed here.

(a) Deposition of the Sediment.

The simplest rhythm is that expressed in the sediment itself, but even here there are several factors involved. In the lower part of the tidal belt, shifting and redeposition of sediment takes place as a result of tidal scour, and as a result of storms. Our observations went to show that on the marl flats where the algal deposits are found, tidal action by itself was not responsible for any appreciable sedimentation, but may become quite important during violent storms, when large quantities of sediment are stirred up into the water. If the sediment in suspension is very fine grained, the mucilaginous sheaths of the algæ entrap quite an appreciable quantity, and appear to do this selectively for the smallest grains in suspension. Thus, if a mixture of mud, silt, and sand is drifted over the algal heads at a velocity sufficient to keep the sand grains moving, the mud particles cling to the mucilage of the filaments, but very little sand is retained. Laminæ which have the characteristics of ordinary gravity sedimentation are also formed in most algal deposits in the tidal belt of the island ; in these, each lamina begins with comparatively coarse grains, often between 0·1 mm. and 1·0 mm. in diameter, and passes upwards into fine mud. Such laminæ are frequently about one millimetre thick, but may vary considerably from this. They appear to effect a complete smothering of the algal head, probably by sediment settling out after a heavy storm ; we were not fortunate enough to observe the actual deposition of this kind of lamina in progress, but abundant examples were found where algal heads had quite recently been smothered by fresh deposits of mechanical sediment.

In the widespread algal deposits of the interior of Northern Andros Island, no trace of mechanical lamination was discernible, and the sediment is all extremely fine in texture. The interior lakes are not tidal, but here again the sediment is supplied to the algal marshes by a process of flooding. This takes place in two entirely different ways : by heavy rainfall, and by invasion by sea water from the banks to the west. In the first case the salinity of the ground water is reduced, and the transportation of sediment is very small, being a slight washing of material from the land into the lakes and creeks. There is possibly also a little transference of sediment from the lake floors on to submerged parts of the algal flats, when the bottom deposits are stirred up by waves, and drifted by wind-blown surface currents on to flooded areas bordering the lakes. The solution effects are probably more important.

In the second case conditions are quite reversed ; vast quantities of sea water are piled up on the shoals, and are swept over the low-lying parts of Andros as a result of violent westerly winds. During such storms, the water over the banks is laden with churned-up sediment and the whole of the flooded area is liable to be smothered under a film of fine white mud, which is left behind when the flood water retreats. The prevailing winds are easterly in direction, and do not cause this kind of flooding. It is probably produced on a grand scale only by hurricanes, which are liable to visit this region during the autumn months. These are circular storms, revolving in an anti-clockwise direction, and travelling along a north-westerly course across the archipelago. Consequently, during the earlier part of a storm, the shoal water on the banks is violently agitated, and becomes strongly charged with suspended sediment. When the wind reverses, and blows from the west, a great wave of drewite-laden water is forced over the island. If the storm is particularly violent, this flood water crosses the island completely, and finds its way through the eastern creeks into the lagoon behind the barrier reef—a cross-country journey of some forty miles.

The sediment brought into the centre of the island by such flood water is extremely fine grained, since all the coarser constituents are deposited before the water has travelled far in from the west coast. This uniformity of grain size prevents any mechanical lamination, or any noticeable difference in size between the particles agglutinated round the mucilage of the algæ, and those particles which settled out under gravity.

(b) Growth of the Algæ.

The growth rate of any particular colony of algæ is probably controlled by a complex set of conditions, important variables amongst which are :—humidity, supply of carbon dioxide, temperature, salinity of the ground water, and deposition of sediment. These conditions are now being investigated experimentally with the algæ in culture but owing to the extremely slow growth rate of most of the species involved, and the difficulty of obtaining cultures free from bacteria and fungi, it will be some time before a satisfactory series of measurements can be made. Field observations were made on the variations of salinity, sedimentation, and water level, and the effects of rhythmic fluctuations of these upon the growth of the algal heads was qualitatively examined.

(i) *The Effect of Salinity.*—In this region, *Symploca læte-viridis* and *Phormidium tenue* seem to be characteristic of salt water, in the case of *Symploca* ranging up to 36 parts per thousand. Experimental work, however, indicates that these two species can tolerate considerable ranges of salt concentration, and can flourish both in nearly fresh water and in highly saline water. With *Scytonema* and *Plectonema*, on the other hand, the salinity has a strong controlling influence. Very little *Scytonema* was found in places liable to be washed by undiluted sea water. With salinities between 1·0 and 10·0 per thousand, there tends to be an alternation between *Scytonema* and the *Schizothrix-Aphanocapsa* association. Where the salt content of the water is appreciably less than one part per thousand, flourishing colonies of pure *Scytonema* were found. The growth habit assumed in these circumstances is quite unlike the forms already described in this paper; the filaments assume a radial arrangement, and deposit a finely crystalline cement of calcium carbonate between the filaments. In North Andros, this kind of structure is best developed in places where the main supply of water is fresh rain-water, and where there is only a negligible influx of water-borne sediment.

It is thus possible that an intermittent growth of *Scytonema* might be controlled by fluctuations in salinity; for instance, the layers with *Scytonema* in the high level mud flats at Southern Bight may represent periods during which the mud flats were free from flooding by sea water, and the salt leached out of the surface layer by rain.

(ii) *The Effect of Sedimentation.*—Experimental work on the two most important groups of species—those belonging to the Scytonemataceæ, and those belonging to the Oscillatoriaceæ—has already shown that whereas the former are extremely slow-growing, the Oscillatoriaceæ are capable of surprisingly rapid growth in a fresh culture. Indeed, an ordinary lamina of sediment, a few millimetres in thickness, could be permeated and bound fast by filaments of *Schizothrix, Symploca* or *Phormidium* in a few days, whereas it would be a matter of months before a skin of *Scytonema* could form over the new surface. Thus intermittent access of fresh sediment to a region equally advantageous to each of the two genera, would undoubtedly cause laminæ with a prepondernace of *Scytonema* to alternate with laminæ containing a preponderance of *Schizothrix*, which is, indeed, what occurs in the Lake Forsyth region.

The same kind of process is probably responsible for the dark organic layers in the specimens from Twelve O'clock Cay. Here the sedimentary layers contain chiefly a filamentous form (probably *Symploca* again), and the more organic layers consist of a complex colony of deeply coloured unicellular forms, which make a dense film over the whole surface of the structure. The filamentous forms probably grow comparatively quickly through the newly deposited sediment; the unicellular forms, however, do not do this, but only colonize the exposed outer surface, where they form a dark coloured, mucilaginous layer which protects the sediment underneath from re-erosion.

(iii) *Effect of Fluctuations in Water Level.*—The seasonal rise and fall of water level in the interior of Andros has the effect of alternately gently flooding and draining the algal flats, but desiccation is probably quite rare. The effect of partial drying is to suspend the growth of the algæ, and possibly induce the formation of resting spores. On re-wetting, *Scytonema* takes a very long time to begin a fresh growth, whilst *Schizothrix* produces an abundance of filaments comparatively quickly.

V.—Relationship to the Environment.

(a) *Lake Forsyth.*

The Lake Forsyth country forms part of the interior belt of low salinity which runs through the centre of Andros Island. Wide expanses of drewite flats surround the north-easterly part of the lake, and stretch for considerable distances north and south. They are colonized by algal heads of Type C, which form a closely set pavement of raised discs, and give a very characteristic appearance to the ground, figs. 19 and 20, Plate 21. To the north-east the algal flats are interrupted by a low outcrop of limestone, fig. 6, which separates this region from a series of brackish lakes draining into Fresh Creek. Towards this limestone ridge, upon which drewite deposits rest unconformably, the spermophyte vegetation becomes more important, and conditions become unsuitable for the formation of well-shaped algal heads. There are also regions within the marl flats where algal heads are not formed, and the algal beds give place to undifferentiated sediments, as, for instance, in local developments of mangrove swamp, where the ground is permanently under water. In these localities a greyish coloured ooze, consisting of drewite with a large proportion of organic matter, is being deposited.

The lake is not tidal, and appears to derive enough water from the drainage of the surrounding country to keep its outlet, the White River, in constant flow. For the same reason, although the surface of the lake is only a few feet above sea level, the salinity of its water is normally quite low, and our observations showed it to vary but little from 1·25 parts per thousand. The territory surrounding the lake is subject to a gentle seasonal flooding, caused directly by fluctuations in the rainfall. During the time when we were in camp on the shore of the lake, the water level was high enough to swamp parts of the algal flats, but a large proportion of the marshes still stood above water. The floor of the lake is covered with a very fine-grained deposit of calcium carbonate, and it is possible that storms such as ordinary Northers, which stir up sediment into the water, may also be responsible for depositing it over the marshes during these seasonal floods. However, this may be, the hurricane floods appear to be the most important agents in spreading subaqueous sediments over the land.

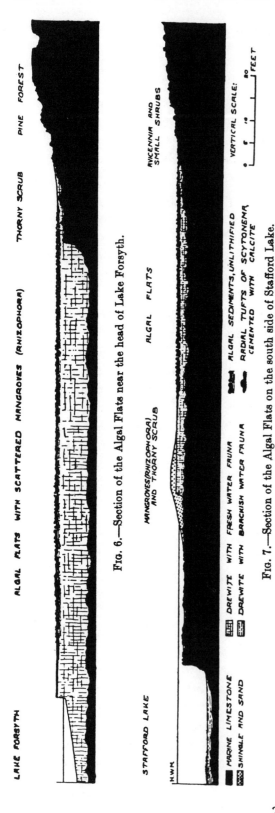

FIG. 6.—Section of the Algal Flats near the head of Lake Forsyth.

FIG. 7.—Section of the Algal Flats on the south side of Stafford Lake.

(b) Stafford Lake.

The eastern end of Stafford Lake is bordered at water level by a limestone platform which is locally covered by a thin deposit of peaty drewite, see fig. 7. At the junction between the drewite and the limestone, several distinct growth forms of algal colonies are to be found, but where the drewite is thick enough to mask the limestone completely, algal heads exactly like those of Lake Forsyth appear, with the same species of *Schizothrix* and *Scytonema*. At the eastern end of the lake, such deposits are restricted in area by the limestone outcrops, but further west, the thickness of unlithified sediment becomes greater, and the drewite flats increase in extent until they stretch completely across the lake. The ground water conditions appear to be more closely related to the fresh water areas in the south and west, rather than to Stafford Lake itself, which is in part tidal, and distinctly saline (14–20 parts per thousand at our camp, lat. 24° 50′ N., long. 78° 0′ W.). This difference is also borne out by the fossil fauna of the surface layers of drewite, which show a fresh-water (Amnicolid) assemblage, quite distinct from the brackish water assemblage in the sediments now accumulating on the floor of Stafford Lake. Consequently, although the drewite flats on which the algal beds are now being formed border a saline lake, it appears that the local conditions of growth are similar to those at the headwaters of Lake Forsyth. The algal flats are separated from the shores of Stafford

Lake by a ridge of partly cemented gravel, and where this is absent, it is found that there is an area with no algal deposits at all comparable with those of Lake Forsyth.

On the rugged outcrops of limestone which rise through the drewite marshes at the eastern end of the lake, the conditions for algal growth are quite different from those which prevail on the marshes themselves. The supply of sediment is negligible, the available moisture is restricted to rain-water, and the rock surface drains fairly rapidly. The algal heads found here are of the radial type mentioned on p. 170.

(c) The Southern Bight.

The north shore of Southern Bight is bordered by drewite mud flats which rise with an almost imperceptible slope from the bight to higher ground beyond the reach of flooding by ordinary spring tides. The lowest belt, which remains covered by a few inches of water at low tide, supports a vegetation dominated by green algæ; *Caulerpa paspaloides* on the soft mud, and *Batophora oestedi* where there is a solid substratum. Separated from this lower mud flat by a slight step in the surface of the ground, is a higher mud flat which is not reached by ordinary neap tides, but is slightly flooded by spring and other heavy tides. This forms an area of slow deposition, for if such flooding happens to coincide with rough weather, during which sediment is stirred up into the sea, this belt of land is covered by a thin film of mud which settles out before the water retreats again. This mud flat is colonized by *Symploca*, and the sediment takes on the conspicuously laminated structure (Type A) described on p. 171. Going further beyond the reach of ordinary tides, the surface of the drewite is found to be slightly lithified, presumably through the action of rain-water, and outcrops of Pleistocene limestones also appear; conditions here become favourable for the uninterrupted growth of *Scytonema*, and the sedimentary structures give place to partly calcified algal heads, similar to those found at Stafford Lake and near the headwaters of Fresh Creek.

(d) Twelve O'clock Cay.

Twelve O'clock Cay is a small island lying in the land-locked sea between Middle and Northern Bights. The algal heads, which are here all of Type B, are of particular interest because they occur on a mud flat which is washed at every tide by water of essentially the same salinity as the sea, 36–38 parts per thousand on the Great Bahama Bank. Although this region is normally sheltered from heavy wave action, the sediment is subjected to constantly shifting currents, and a considerable amount of the finer material is consequently winnowed away, leaving a sediment of a distinctly sandy texture. The tide rises over the algal flats quite gently, and with normal weather conditions there is no transport of sediment. It is probably only during quite severe storms that sediment is swept over the algal flats and deposited there. Towards low-water mark, the algal deposits pass laterally, with no abrupt transition, into loose submarine sediments with no organic matrix or structure.

(e) *The Wide Opening.*

The algal beds which occur on the north side of the Wide Opening are similar to those of Twelve O'clock Cay, except in their being colonized to a greater extent by mangroves. They are probably washed at every tide unless the wind sets strongly off shore, when the water is blown out of the Wide Opening. The salinity varies considerably, since there is a substantial amount of brackish water drainage entering this gulf. Our only determination gave a salinity of 17·87 parts per 1000, but as this sample was taken at lowest ebb tide, the figure is undoubtedly lower than the normal ; with the rising tide, the Wide Opening fills up with sea water—37·8 per 1000—and by the time that the water level is high enough to flood the algal beds, the salinity must have risen to between 36 and 37 parts per 1000.

(f) *The Fresh Creek Lakes.*

The lakes at the head of Fresh Creek present an interesting series of algal deposits, which are still under investigation. In the localities examined, which are all along the southern shore of this lake system, there are extensive outcrops of limestone. The ground stands several feet above the surface of the lakes, and appears to receive sediment at very rare intervals. Although the lake water is slightly saline, the ground water is fresh, and the recent sediments have a hard, cemented crust. Most of the algal deposits are non-sedimentary, and resemble the cemented forms from Stafford Lake and Southern Bight. There is, however, one sedimentary form (Type D), which is remarkable because the algal heads become completely detached, and are transported by the wind until they are deposited finally in one of the lakes.

(g) *Review of Relationship to the Environment.*

Reviewing the geographical relationships of these algal deposits, we see that they mark a borderline facies between land and water—and, indeed, they often occupy territory which the cartographer hesitates to assign to either. Their structure depends upon several rhythmic processes, foremost amongst which is the alternate flooding and draining of the areas in which the algæ grow. Since they are essentially mechanical sediments, as opposed to the organic accumulations of coenoplase and calcareous algæ, they can only develop in regions where calcium carbonate is the dominant sediment. Thus, unlike the lime extracting algæ, they could not produce their characteristic forms in streams or lakes, unless there were also a sediment of solid carbonate in suspension.

(h) *Early Development of the Algal Heads, and Relation with the Sediment below.*

In all the localities examined, the surface of the sediment between the algal heads, no matter what its former irregularities, is completely covered by a mucilaginous algal skin, and there can be little doubt that the algal sediments described in this paper originated as algal films on mechanically deposited sediments. In the early stages of development, the form which an algal head may take seems to depend to some extent upon the nature of the sediment. In regions of sandy sediment, such as Twelve O'clock Cay, the characteristic, dome-shaped structure probably originates round some irregularity on the surface of the sediment; this becomes colonized by algæ, which begin to accumulate sediment around them. Once the domed structure has been initiated, the algal head will tend to increase its size by the addition of fresh layers deposited concentrically round the earlier ones, producing algal heads of Type B.

In the interior of Andros, round Lake Forsyth and Stafford Lake, the algal heads are initiated, most probably, in a different way. The sediment is of a fine enough

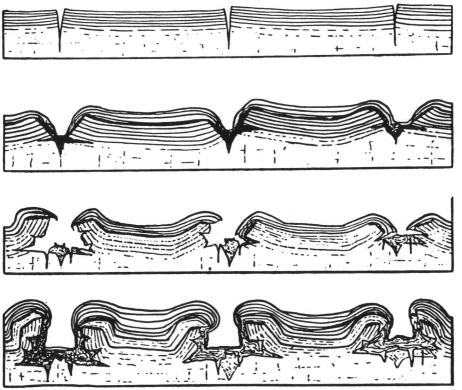

Fig. 8.—Early Stages in the Development of Algal Heads of Type C. × ¼. For explanation, see the text.

texture to produce desiccation cracks on a slight drying out of the surface layers, fig. 8. Once this has happened, each desiccation polygon becomes sheathed in a felt of algal filaments, and with further growth, the polygonal outline is lost, and a circular disc is produced. Further growth tends to fill up the gaps between the disks, and to cause a doming of the upper surface. A repetition of desiccation at this stage causes the upper layers of the algal head to shrink; the peripheral part then tears away from the material filling the interspaces between neighbouring algal heads, and a slight upward curling of the outer edge gives a concave or saucer-shaped form to the upper surface (see fig. 4). A renewal of growth then gives rise to the characteristic structure

31

of Type C. The development of Type D probably follows the same plan, but does not appear to go so far as in that of Type C.

When algal heads of Type C attain a considerable thickness, it is often found that only the top inch or inch-and-a-half retains an appreciable quantity of organic matter, and although the earlier formed layers still retain their laminated appearance, the laminæ are now marked out as alternating grey and white bands, with relatively little plant material left. That the disappearance of this organic matter is brought about by bacterial decomposition, is made probable by the recent discovery of very active cellulose and hemi-cellulose destroying bacteria, and also of agar-liquefying bacteria, in the mangrove swamps of western Andros. (BAVENDAMM, 1931 and 1932.)

VI.—COMPARISON WITH OTHER RECENT CYANOPHYCEOUS DEPOSITS.

Amongst recent calcareous algal deposits, the Bahaman sediments stand apart from all other recorded examples in being completely unlithified, and in consisting of clastic material collected by, but not precipitated by, the algæ concerned. This tendency of the Cyanophyceæ to collect clastic material and retain it amongst the mucilaginous sheaths of the thallus, is no recent discovery, and the importance of these plants as pioneer colonizers of newly deposited mudflats has long been recognized. These phenomena have, however, mainly been observed in regions where argillaceous (and to a less extent, arenaceous) sediments are predominant, as for example on the shores of estuaries and deltas (CAREY and OLIVER, 1918, pp. 170–173), and the possibility of the development of this kind of algal colonization on a vaster and more elaborate scale in regions of limestone sedimentation has not fully been realized.

Numerous occurrences of calcareous concretions have been described in which the lime is believed to have been precipitated through the activity of blue-green algæ. In all these, the calcium carbonate is thrown out of solution because the physiological activities of the plant cells disturb the chemical equilibrium of the surrounding water ; the solid carbonate, as a consequence, is completely external to the plant cells, and may either come down as a loose precipitate, or as a hard stony encrustation round the algal colonies. All the limestones which are known to originate in this way are being precipitated in fresh-water lakes and streams, in contrast with the Bahaman sedimentary deposits, which range from fresh lake water to water of ocean salinity. A further point of importance is that all the previously described cyanophyceous limestones either require a nucleus or hard substratum upon which to grow, or else they take the form of nodules with a structure of hard, concentric envelopes. The stratigraphical relationships of the Bahaman deposits are somewhat different, for here the algal beds themselves form an incident in the deposition of a series of stratified sediments, and they develop on the surface of completely unconsolidated material.

An example of a cyanophyceous deposit which shows some interesting points in common with the Bahaman material is described by W. WETZEL (1926) from the valley of the Rio Loa, where it flows through the Atacama Desert in Chile. The unlithified algal structures form a tough leathery coating (*Lederhäute*), on the granite blocks of the river bed. They are found to contain *Oscillatoria*, *Nostoc*, and a filamentous bacterium, *Crenothrix*. The river water is slightly saline (3·49 parts per 1000), which is a suggestive point when it is remembered that these *Lederhäute*, and the Bahaman deposits, which are also mostly in somewhat saline water, have in common the peculiarity of remaining unlithified. Another point of interest is that the Atacama *Lederhäute* also enclose clastic sediment, which, however, is non-calcareous, but consists of fine sand and mud, insoluble in hydrochloric acid. Consequently, if detrital calcite is assumed to be absent, it is possible to estimate the relative proportions of precipitated carbonate and clastic sediment. An analysis gave the following results, showing the clastic material and carbonate to be present in approximately equal amounts :—

Organic Matter	9·5
Clastic Sediment	45·0
Calcium Carbonate	37·8
Magnesium Carbonate	4·2

In the Bahaman deposits there is no such ready method of distinguishing between clastic and precipitated material, since both consist entirely of soluble carbonates. Nevertheless, in Type A and Type B the texture of the sediment is sufficiently coarse for many of the individual grains to be easily recognized as fragments of molluscan or foraminiferal shells, or as clastic grains derived from older limestones. Type B can thus be seen at a glance to consist almost entirely of clastic material, and Type A very largely so.

The hard, stony cyanophyceous limestones have much less in common with the Bahaman deposits. They generally occur as discoidal or spherical bodies, up to a foot or more in diameter ; internally they are often cavernous, and the general structure may be roughly radial, or, occasionally, concentric. " Water Biscuits " and " Lake Balls " have been recorded and described from many localities, amongst which are the Finger Lakes of New York State (CLARKE 1900), various lakes in Michigan and Minnesota (POWELL 1903), and streams in Pennsylvania (RODDY 1915). The concentric arrangement becomes more noticeable where a seasonal or other change in water-level causes alternate drying and moistening ; a roughly concentric structure is found in specimens from the flat gravel banks on the shore of the Rhine near Konstanz, where the algæ are covered only when the river is high (PIA 1926, pp. 45–47).

In an extremely interesting occurrence described by Sir DOUGLAS MAWSON from the Robe district, South Australia, the water biscuits have a well-developed concentric structure, owing to alternate desiccation and flooding (MAWSON 1929). The region in which the deposits are found is flat and almost at sea level, and is subjected to seasonal flooding ; the conditions thus appear to have much in common with those prevailing in parts of Andros Island. The algal deposits, nevertheless, show interesting differences, for whereas the Bahaman algal heads are unlithified, and merely form a surface modification of the sediment upon which they grow, the water biscuits are formed each as a separate, lithified individual, completely detached from its substratum, and deriving its carbonate largely or entirely by precipitation. Thus each water biscuit consists of a series of concentric shells, each one entirely enclosing the previously-formed part of the structure—an arrangement in strong contrast with the construction of the Bahaman deposits.

The author knows of no records of modern cyanophyceous limestones in process of formation in the open sea. Høeg has recently described a series of post-glacial stromatolitic plates, discovered on the cliffs of Malmø and Svenør, in the Oslo Fjord, and has shown that they are probably of marine origin. Unfortunately, their growth appears to have ceased some considerable time ago, and it is difficult to reconstruct the conditions which favoured the deposition of calcite in this particular from. Høeg (1929) suggests that the carbonate was precipitated from sea water through the agency of blue-green algæ or of bacteria, and points out that the structure and arrangement of the stromatolitic plates is strongly in agreement with an origin by precipitation. In this they differ entirely from the Bahaman deposits, for they are thoroughly lithified and have a compact crystalline structure.

A comparison with the Precambrian and early Palæozoic stromatolites, to which the Bahaman sediments show a certain resemblance, is beyond the scope of this paper, and will be considered separately in a later communication.

VII.—Description of Species.

Gloeocapsa granosa (Berkeley) Kützing, fig. 9.—A plant resembling *G. granosa* in its habit was found in the marshes near Lake Forsyth. It differs from typical

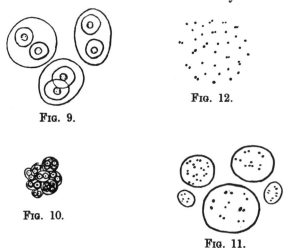

Fig. 9.

Fig. 12.

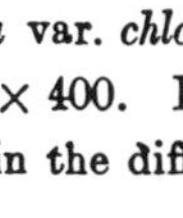

Fig. 10.

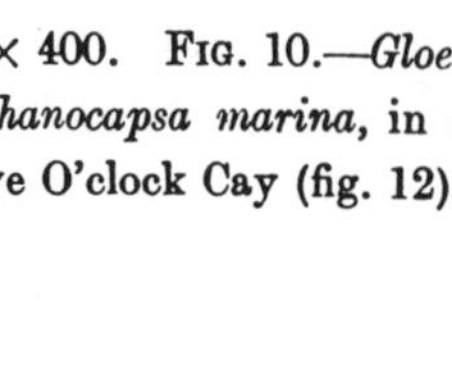

Fig. 11.

Fig. 9.—*Gloeocapsa granosa* var. *chlora*, Lake Forsyth. × 400. Fig. 10.—*Gloeocapsa viridis*, sp. nov., Twelve O'clock Cay. × 400. Figs. 11 and 12.—*Aphanocapsa marina*, in oval colonies from Lake Forsyth (fig. 11), and in the diffuse form, from Twelve O'clock Cay (fig. 12). × 1000.

material in its larger size, and also in the conspicuous jade-green pigmentation of the sheaths, especially in the case of families of four or more cells, in which the innermost sheaths are often deeply coloured. This form is therefore best regarded as a variety of *G. granosa*, for which the name *chlora* is proposed.

G. granosa (BERKELEY) var. *chlora* nov. Cells apple-green, 4–6 microns in diameter; individual sheaths olive-green to jade-green in colour, more strongly pigmented in the older colonies, non-lamellose, 12–14 microns in diameter. Colonies up to 75 microns in size, the inner common sheaths pale green, the outer ones colourless.

Habitat.—Amongst other Cyanophyceæ on the moist surface of chalk marshes; at the type locality, the water is almost fresh.

Locality.—Lake Forsyth, Andros Island, Bahamas.

Gloeocapsa viridis sp. nov., fig. 10.—Cells 0·75–1·0 microns in diameter; sheaths 2·5–5·0 microns, bright grass-green, usually strongly lamellose; plants in clusters from a few cells to groups of about twenty microns.

Habitat.—On sandy ground, liable to be flooded by sea water; at the type locality, intimately associated with *G. magma*.

Locality.—Twelve O'clock Cay, Middle Bight, Andros Island, Bahamas.

Aphanocapsa marina HANSGING, figs. 11 and 12.—The small *Aphanocapsa* which is found in great abundance at Lake Forsyth, and at Twelve O'clock Cay, only differs from the typical *A. marina* in the rather greater size of its cells, which range from 0·5 to 0·75 microns, as compared with 0·4 to 0·5 microns in the material described by HANSGING and by FOSLIE (1890, p. 169). The cells of the Bahaman material are pale green; near Lake Forsyth, where the water is almost fresh, the plant is sometimes found in well-defined, oval colonies, up to 12 microns in diameter, but it is usually in the form of a gelatinous, shapeless, mass, filling the spaces between other blue-green algæ. At Twelve O'clock Cay, the plant was found abundantly near high-water mark, as a colourless extended mass, pervading the spaces between sand grains, or amongst other algæ.

Plectonema atroviride sp. nov., fig. 13.—Filaments 4–5 microns in diameter, straight, rarely with a slight spiral tendency, branched freely at first, but later sparingly; false branches usually solitary. Sheaths firm, the outermost one hyaline, the inner deeply coloured, dark jade-green, almost opaque. Trichome one micron broad, with cells 6–8 microns in length.

Habitat.—Growing amongst other Cyanophyceæ, especially in tufts of *Scytonema*, in chalk marshes which are liable to be flooded by fresh water.

Locality.—At the head of Lake Forsyth, and near the southern shore of Stafford Lake, Andros Island, Bahamas.

Remarks.—This species resembles *P. nostocorum* in its form and habit, but differs in its thick and deeply coloured sheaths, and also in the greater length of its cells. In its unusually long cells, it may be compared with *P. terebrans*, but differs again in its stout sheaths. It also shows some similarity to *P. radiosum*, but is a distinctly smaller plant. From all other species of *Plectonema* with a similar size or habit, it differs in the elongate shape of its cells.

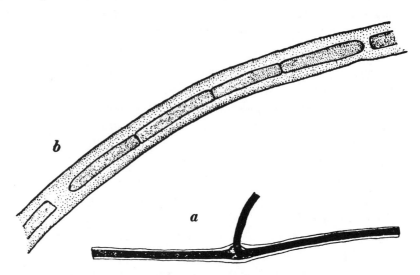

Fig. 13.—*Plectonema atroviride* sp. nov., Lake Forsyth, Andros. (*a*) Part of a filament, to show the two sheaths, the inner one deeply pigmented, and also the method of branching. × 700. (*b*) Part of a filament with the inner sheath bleached to show the trichome. × 3000.

Scytonema androsense sp. nov., fig. 14.—Plants forming matted strata and small woolly tufts, dark lead-grey to deep green in colour. Filaments with a very sharp outer edge, 15–30 microns in diameter, and up to a centimetre in length. Sheaths strongly lamellose; the innermost lamella is deep sage-green in colour, and is sometimes dark enough to obscure the trichome inside; outer sheaths, very pale green to colourless; at certain levels, the lamellæ are strongly divergent, and spread out to form broad, pale green or colourless funnel-shaped ocreae. Trichome 6–10 microns in diameter, with cells quadrate or slightly longer than broad; cell contents pale blue-green. Heterocysts the same width as the trichome, or a little wider, one-and-a-half times or twice as long as broad; outer wall of the heterocyst hyaline, enclosing an inner, golden yellow, granular centre. False branches free, and usually divergent at the base.

Habitat.—In chalk marshes, moistened by water with a salinity of 1·25 parts per thousand, or less.

Locality.—Lake Forsyth, Stafford Lake, and near the Fresh Creek Lakes, in Northern Andros Island, Bahamas.

Remarks.—This species falls into the division *Petalonema* of Bornet and Flahaut, but differs from other ocreate species of similar dimensions and structure in its intense green colour. It differs further from most of the species in this group, such as *S. densum,* *S. velutinum,* and *S. alatum,* by its characteristic heterocysts; from *S. crustaceum,* which it resembles in size and shape of heterocysts, it differs strongly in the lack of any trace of cohesion between the false branches.

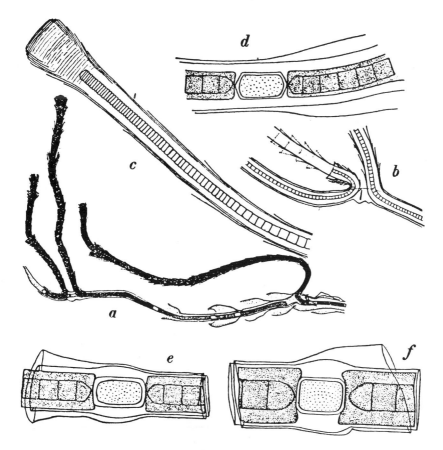

Fig. 14.—*Scytonema androsense* sp. nov., Lake Forsyth. (*a*) Part of the plant to show the spreading habit of the ocreae. × 120. (*b*) Method of branching. × 240. (*c*) Characteristic growing apex of a filament. × 480. (*d*) Part of a filament to show the deeply pigmented inner sheath broken at a heterocyst. × 800. (*e*) and (*f*) Typical heterocysts. × 800.

Scytonema crustaceum AGARDH, figs. 15 and 16.—Associated with *S. androsense* in the interior of Andros Island, is another species of *Scytonema*, which approaches *S. crustaceum* var. *incrustans* in many of its characteristics, but differs in the behaviour of its false branches. Unlike the typical *S. crustaceum*, and its variety *incrustans*, which give rise to branches in pairs, this form most commonly produces its false branches singly, and a paired arrangement is less frequently seen. Where the false branches do arise in pairs, they are usually immediately ascending, and are free for their entire length, fig. 16a. False branches arising singly, however, frequently pierce only the innermost sheath of the parent filament, and run within the outer sheaths for some distance before they break free, fig. 16b. Where the false branches entirely fail to break through the sheaths of the parent filament, fig. 16c, a peculiar disorganization of the trichome is sometimes observed to take place, fig. 16d; the filament produces a bulbous swelling, within which the trichome develops into a tangled mass of *Nostoc*-like chains, or may even break down into separate spherical cells.

In view of these peculiarities, this plant is best regarded as a variety of *S. crustaceum*, for which the name *catenula* is proposed.

S. crustaceum AGARDH. var. *catenula* nov.—Plants forming matted strata and woolly tufts, dark grey to deep green in colour. Filaments 20–45 microns in diameter, often with an irregular outer edge. Sheaths of strongly divergent lamellæ, olive-brown in colour. Spreading ocreæ absent. Trichome 5·0–12·5 microns in breadth, with spherical

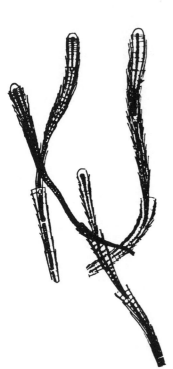

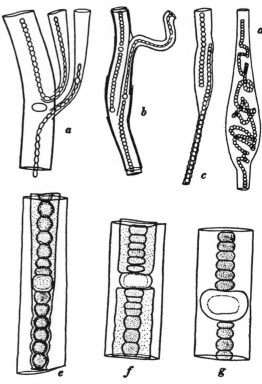

FIG. 15.—*Scytonema crustaceum* var. *catenula*, Lake Forsyth. Part of plant to show the divergent sheaths. × 150.

FIG. 16.—*Scytonema crustaceum* var. *catenula*. (a) Normal method of branching. × 200. (b)–(d) Branches with only partial rupture or non-rupture of the sheaths. × 150. (e)–(g) Typical heterocysts. × 500.

or compressed barrel-shaped cells; cell contents pale blue-green. Heterocysts compressed, wider than the trichome, and broader than their own length, colourless or olive-green, homogeneous. False branches usually arising singly, less commonly in pairs; solitary false branches frequently failing to pierce one or more of the outer sheaths, so that the branch runs within the parent filament, between two of the sheaths, or produces a bulbous swelling, within which the trichome becomes loosely aggregated or even disorganized into separate spherical cells.

Habitat.—In chalk marshes, associated with *S. androsense*.

Locality.—Marshes round the head of Lake Forsyth, Stafford Lake, and the Fresh Creek Lakes, Andros Island, Bahamas.

VIII.—SUMMARY.

(1) Algal deposits, with characteristic structures, in some respects resembling certain Palæozoic stromatolites, are developing over a considerable area in Northern Andros Island.

(2) Several geographical belts can be recognized in the island, and the algal deposits in each belt possess a growth form distinct from that shown by algal deposits growing in other belts.

(3) The calcium carbonate which goes to make up these deposits is not necessarily precipitated by the algæ responsible for the structures, but consists of essentially mechanically entrapped sediment.

(4) The algal deposits are entirely unlithified, and do not require a solid substratum upon which to grow. They originate normally as a surface modification of unconsolidated sediments, and well-developed algal sediments are found to pass laterally and also downwards, into ordinarily bedded sediments.

(5) Generally the characteristic structure is produced, not by a single species, but by a community of organisms ; this is especially true of the structures with a dominantly concentric arrangement, as opposed to a radial one.

(6) Any particular species of alga may enter into, and help to build up, more than one type of rock structure.

(7) Although this kind of structure is best developed in regions of low salinity, it is also found in localities where the salinity is essentially that of the open ocean.

IX.—REFERENCES.

BAVENDAMM, W. (1931). ' Ber. deuts. bot. Ges.,' vol. 49, p. 288.

   (1932). ' Arch. Mikrobiol.,' p. 205.

CAREY and OLIVER (1918). "Tidal Lands, a Study in Foreshore Problems," Blackie & Sons.

CLARKE, J. M. (1900). ' Bull. N.Y. State Museum,' No. 39, vol. 8, p. 195.

FOSLIE, M. (1890). "Marine Algæ of Norway."

HØEG, O. A. (1929). ' K. norske Vidensk. Selsk. Skr.,' No. 1.

MAWSON, Sir DOUGLAS (1929). ' Quart. J. Geol. Soc.,' vol. 85, p. 613.

PIA, J. (1926). "Pflanzen als Gesteinsbildner." Berlin, Gebr. Borntraeger.

POWELL, H. (1903). ' Minn. Bot. Stud.,' p. 75.

RODDY, H. J. (1915). ' Proc. Amer. Phil. Soc.,' vol. 54, p. 246.

WETZEL, W. (1926). ' Centralbl. Min.,' B, 354–361.

2

Reprinted from pages 1-3, 7-8, 10, 11, 12-13, 16-45, 47, 54, 55, 67, 74, 75, 76, 77, 78-81, 85, 86, and 91-95 of *Am. Assoc. Petroleum Geologists Bull.* **38**:1-95 (1954)

BAHAMAN CALCAREOUS SANDS[1]

LESLIE V. ILLING[2]
London, England

ABSTRACT

The major part of the Bahama Banks is covered with a mantle of sands composed of calcium carbonate. They rest on a basement of similar sediment which has been consolidated into limestone by the subaerial deposition of calcite cement. Along the extreme edges of the Banks, the sand consists of the débris of neritic organic skeletons; elsewhere it is predominantly non-skeletal, and composed of grains of cryptocrystalline aragonite.

These non-skeletal sand grains have hitherto been regarded as a break-down product of the limestones, but the present study, based on the Ragged Island area in the southeast part of the Great Bahama Bank, disproves this, and shows that such derived grains are rare in the recent bottom sediment. The sand grains are instead considered to be primary: they have formed, and are now forming, by the physico-chemical and bio-chemical extraction of aragonite from the sea water, which is saturated or supersaturated with calcium carbonate.

The grains develop by the progressive aragonitic cementation of friable aggregates of calcareous silt particles. Further cementation tends to join the grains into lumps, in which an outer dense cryptocrystalline layer commonly allows the slower growth of larger aragonite prisms in the interior parts. The habit and surface texture of the lumps undergo a sequence of changes when traced from the ocean edge toward the interior of the Banks. Corrasion prevents excessive lump growth, and the typical bottom sediment is a well sorted medium-grade sand. The traces of calcareous mud in the sands contain no evidence of "drewite" needles, which are restricted to the protected "shelf lagoons" west of Andros and Abaco Islands.

The controlling factor in the formation and distribution of the sands is the tidal currents of cooler oceanic water which sweep onto the Banks with velocities diminishing toward the interior, precipitating calcium carbonate as the water warms and evaporates. Sedimentation is therefore most rapid along the borders of the Banks. Aided by sea-level changes, this has led to the building of lines of elongate cays, parallel, and close to the ocean edge, making each Bank a gigantic atoll.

Superficial aragonite oöliths develop on the protected beaches of the cays, and in the track of currents through the inter-cay channels. East of the "Tongue of the Ocean" and elsewhere close to

[1] Manuscript received, September 3, 1953.

[2] Petroleum geologist, London. The writer is indebted to the officers of British Bahamian Oil Development Ltd. for the use of their facilities in collecting the material for the investigations summarized in this paper, which is based on the writer's doctorate thesis at Cambridge University, England. Thanks are extended to C. S. Lee and M. Black for their help in the field and in the laboratory, and to Margaret A. Illing, the writer's sister, who studied the foraminifera found in the sediments. The investigations were carried out in 1948–1950 at the Sedgwick Museum, Cambridge, under a grant from the Department of Scientific and Industrial Research. Chemical analyses of sediments and sea waters were covered by a grant from The Royal Society, London. Thanks are due to G. L. Hobson for his critical reading of the script, and to A. Barlow, J. A. Gee, and A. L. Greig for help with the preparation of the illustrations.

the borders of the Banks, there are extensive areas where the sand is swept into a curvilinear pattern-work of submarine dunes, which occasionally adopt barchan shape. These have not been sampled, but are considered favorable for the formation of true, as opposed to superficial, oöliths.

Evidence from the well drilled to a depth of 14,587 feet on Andros Island suggests that calcareous sands, similar to the recent deposits, have played a large part in building the Bahama Banks.

INTRODUCTION

The spectacular variations in the lithological and faunal assemblages associated with fossil limestone reefs have emphasized the importance of detailed studies of their recent counterparts. The finding of large oil reservoirs in similar and related environments has focussed further research into the problems of "reefing," yet this same "reef-mindedness" of petroleum geologists and others has tended to obscure the fact that such deposits include but a few of the host of different types of calcareous sediment.

Of the limited number of marine environments in which the formation of recent limestones may be studied, the Bahama Banks have attracted increasing attention. Previous writers have dwelt chiefly on the formation of the precipitated calcareous muds formed in certain restricted localities, and on the coral-algal reefs scattered along the margin of the Banks, notably east of Andros Island. The present paper is concerned with the nature and origin of the mantle of calcium carbonate sands which covers by far the largest part of the Bahama Banks. It is shown that, contrary to earlier views, these calcareous sands are neither dominantly bioclastic, nor are they derived from earlier deposits. Instead, the sand grains, oölitic and otherwise, are now in process of formation on the sea bed as a result of the extraction of calcium carbonate from the supersaturated shallow warm sea water; the stages in their growth, the processes involved and the variations in the sediment with environment are discussed, in the hope that they will shed some light on the origin of similar calcareous deposits in the geological column elsewhere in the world.

The Bahama Banks are a huge submarine plateau, just below sea-level, covering more than 40,000 square miles. They are remarkably flat, with only a few marginal areas in which the depth of water exceeds 6 fathoms; low islands, whose only elevations are long ridges of dune sands, cover about a tenth of the Banks (Fig. 1).

This vast platform is surrounded by precipitous slopes, dropping to oceanic depths. The Straits of Florida and the Santaren Channel at the west are 300–500 fathoms deep. The Old Bahama Channel separates the Banks from Cuba, and deepens eastward from 1,000 to 1,500 fathoms. Soundings of more than 2,500 fathoms have been recorded in the Atlantic Ocean along and close to the northeast edge of the plateau (Ewing *et al.*, 1946; Schalk, 1946).

The Banks are dissected by similar steep-sided channels of deep water. Providence Channel, in the north, separates the Little Bahama Bank from the Great Bahama Bank, and, at the southeast, several smaller isolated Banks are entirely surrounded by ocean. The Great Bahama Bank is itself almost split in two by the blind channel of the "Tongue of the Ocean," 1,000 fathoms deep.

[*Editors' Note:* Material has been omitted at this point.]

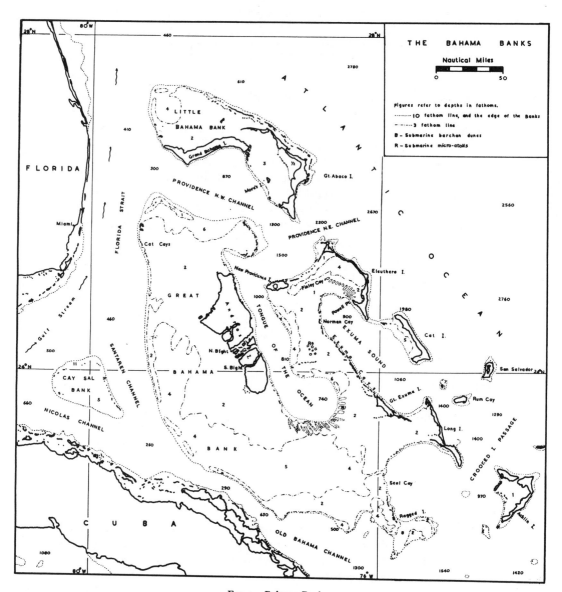

FIG. 1.—Bahama Banks.

PRESENT INVESTIGATIONS

The present work is based on observations and sampling made mainly during seismic operations in 1948 in the "Ragged Island Concession" of British Bahamian Oil Development Ltd., covering the southeastern extremity of the Great Bahama Bank (Figs. 1 and 3). Several journeys by air between Nassau and Ragged Island afforded excellent opportunities to study the distribution of marine vegetation and reefs, the pattern of underwater sand accumulations, the general morphology of the Banks, the form of the arcs of cays, and their relation to the abrupt oceanic margins. The water is so clear that every detail of the sea bottom is visible in varying shades of color. Along the edges of the Banks the light blue-green of the shallow water and the channels between the vivid green cays changes suddenly to deep ultramarine where the Banks plunge steeply to oceanic depths; Plate 7.1 conveys better than words the beauty and fascination of such scenes. This wealth of detail is not visible from on board ship; unless the bottom of the boat is equipped with a glass window, the nature of the sea bed can only be interpreted from the changing hue of the water. The air view is thus of immense value in the study of sedimentation on the Banks. In no other way can one see the whole pattern of underwater sand accumulations. By directing attention to areas of special interest, it serves further as a control in the collecting of bottom samples. It was not possible to fly over all the parts of the Concession from which bottom samples were collected. Conversely there was no opportunity to collect from the many underwater dunes and reefs seen in such spectacular form from the air. Yet the uniformity of marine environment over wide areas of the Banks makes it possible to apply the evidence gained in one area to the interpretation of data collected in the other.

Much valuable information of the accumulation of sediment in the critical areas on the margins of the Banks has been obtained from overlapping air-photographs covering the arcs of cays and their continuations as submarine reefs.

Besides the Ragged Island material, a few small and scattered samples were dredged during a voyage along the edge of the Banks off Eleuthera and the Exuma Islands. Numerous bottom samples from the Little Bahama Bank, including some of the marly sediment from the brackish swamps of Abaco, were collected and kindly placed at the writer's disposal by C. S. Lee. Preliminary reports of his expedition have already been published, together with many interesting observations on the structure of the Banks as a whole that emerged from the various programs of geophysical exploration carried out by B.B.O.D. in its concessions (Lee, 1951).

[*Editors' Note:* Material, including Figures 2, 5, 6, and 7, has been omitted at this point.]

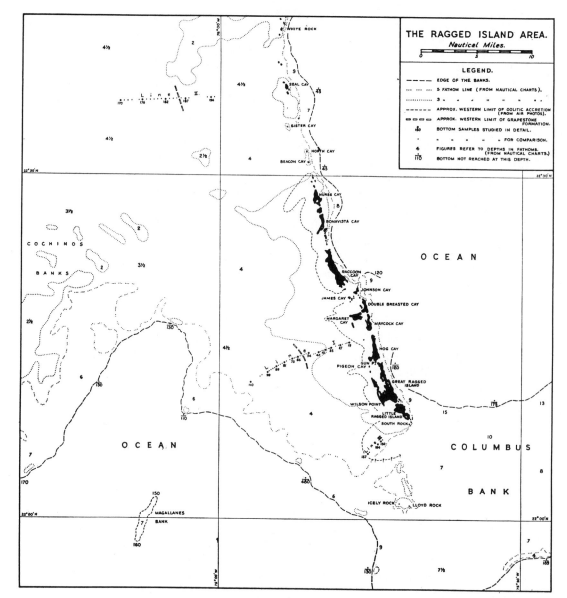

FIG. 3.—Ragged Island area.

LESLIE V. ILLING

TABLE I. PERCENTAGE MECHANICAL ANALYSIS OF SEDIMENT SAMPLES (ON WEIGHT BASIS)

Grades		Gravel		Coarse Sand		Medium Sand		Fine Sand		Silt	Mud	Quartile Parameters				
Sub-Grades+		(a)	(b)	(c)	(d)	(e)	(f)	(g)	(h)	(j)		MD	Q_1	Q_3	So	Sk
Sample Nos.	Depth Ft.	%	%	%	%	%	%	%	%	%	%	mm.	mm.	mm.		
LINE I																
19	13	—	0.2	3.9	20.4	43.1	19.4	6.7	3.4	2.9	tr	0.43	0.30	0.60	1.41	0.97
27	16	1.3	1.7	14.7	31.8	32.0	11.1	3.0	2.0	2.4	tr	0.50	0.40	0.86	1.47	0.99
35	17	0.7	1.2	9.2	25.4	35.8	15.6	6.5	3.0	2.6	tr	0.50	0.36	0.73	1.42	1.05
41	18	0.9	1.0	5.7	23.4	41.7	17.8	4.7	2.4	2.4	tr	0.47	0.37	0.66	1.34	1.10
47	19	4.3	1.9	11.0	27.0	37.8	12.1	2.2	1.3	2.4	tr	0.56	0.39	0.84	1.47	1.04
54	19	0.7	1.9	13.2	29.3	34.9	12.4	3.3	2.1	2.2	tr	0.56	0.39	0.84	1.48	1.03
64	20	1.2	1.8	20.0	32.4	29.6	9.4	3.0	1.5	1.1	tr	0.65	0.44	0.96	1.48	1.00
73	21	1.7	3.3	18.6	27.8	27.1	11.4	4.9	3.2	2.0	tr	0.62	0.38	0.96	1.59	0.95
81	22	1.8	1.8	18.3	28.4	29.8	12.5	4.3	1.8	1.3	tr	0.60	0.39	0.94	1.55	1.02
89	25	3.2	3.7	17.9	24.4	26.4	15.1	6.0	2.0	1.3	tr	0.59	0.36	1.00	1.67	1.03
99	26	1.5	2.7	16.0	26.3	29.7	16.0	5.3	1.6	0.9	tr	0.56	0.36	0.90	1.58	1.03
110	27.	0.2	1.9	12.8	22.6	31.8	19.9	8.0	2.1	0.7	tr	0.49	0.32	0.78	1.57	1.02
LINE II																
170	30	1.2	6.4	24.1	27.5	24.0	11.1	3.6	1.4	0.8	tr	0.73	0.44	1.13	1.61	0.92
176	32	2.4	11.7	29.5	30.0	18.5	4.6	1.6	0.8	0.9	tr	0.90	0.58	1.38	1.54	0.99
182	26	2.7	14.6	28.6	31.7	14.8	3.9	2.1	1.0	0.6	tr	0.94	0.63	1.47	1.53	1.04
187	28	8.2	5.7	13.1	22.3	22.7	18.7	7.2	1.3	0.8	tr	0.59	0.33	1.05	1.78	0.99
194	31	0.4	1.1	9.8	24.9	33.3	19.0	7.7	3.0	0.8	tr	0.49	0.31	0.75	1.55	0.97
LINE III																
150	28	tr	0.1	1.7	19.8	51.3	25.6	1.3	0.1	0.1	tr	0.44	0.35	0.58	1.30	1.03
154	28	0.4	0.9	5.5	24.1	36.7	27.6	4.3	0.4	0.1	tr	0.44	0.32	0.67	1.46	1.09
157	28	1.4	2.2	9.9	27.4	30.7	19.4	7.7	0.9	0.3	tr	0.52	0.33	0.80	1.55	0.98
LINE IV																
B-52	18	—	0.3	4.3	13.7	32.2	33.5	10.9	4.7	0.4	tr	0.36	0.26	0.52	1.41	1.04
B-58	15	0.2	2.4	11.9	17.7	20.3	20.2	20.7	6.3	0.3	tr	0.38	0.20	0.73	1.91	1.01
LINE V																
B-34	1	0.2	0.2	0.4	1.3	4.2	8.4	33.8	46.8	4.7	tr	0.12	0.09	0.17	1.36	1.08
B-37	13	0.7	0.7	2.4	6.9	17.4	17.3	23.7	19.3	11.5	0.1	0.19	0.11	0.39	1.91	1.15
B-38	6	0.8	1.3	12.0	48.1	32.8	3.8	0.8	0.2	0.2	tr	0.67	0.53	0.86	1.27	1.01
B-39	5	0.1	0.3	3.4	19.8	54.9	18.2	2.6	0.4	0.3	tr	0.46	0.37	0.58	1.27	1.01
B-40	5	2.5	1.1	7.5	25.4	34.1	20.9	7.2	1.0	0.3	tr	0.49	0.33	0.67	1.43	1.11
B-41	5	0.6	0.8	6.1	20.7	32.9	28.2	9.0	1.5	0.2	tr	0.42	0.29	0.64	1.49	1.05
B-42	6	0.2	0.6	5.9	24.2	39.6	20.7	7.5	1.3	0.2	tr	0.47	0.33	0.66	1.43	0.97
B-43	7	1.7	1.6	11.4	33.4	31.0	11.9	5.7	1.8	1.5	tr	0.58	0.39	0.84	1.38	0.86
B-44	8	1.8	2.7	11.8	27.4	29.6	13.9	7.4	3.8	1.6	tr	0.55	0.34	0.84	1.57	0.94
B-45	10	tr	0.1	1.5	6.1	17.5	21.6	28.2	21.1	3.9	tr	0.20	0.12	0.36	1.70	1.11
B-46	10	tr	0.1	1.8	10.6	39.4	30.9	10.7	5.4	1.1	tr	0.37	0.26	0.50	1.40	0.96
B-47	39	tr	0.3	1.3	19.1	76.2	2.3	0.2	0.2	0.4	tr	0.51	0.47	0.58	1.12	1.04
B-48	65	14.1	9.2	13.2	27.0	30.2	4.7	0.9	0.3	0.4	tr	0.75	0.52	1.55	1.73	1.43
LINE VI																
B-13	6	20.0	6.2	5.5	14.1	27.1	14.6	6.0	3.6	2.9	tr	0.55	0.34	1.90	2.36	2.13
B-14	12	1.3	1.7	4.5	15.8	44.2	21.0	7.6	2.8	1.1	tr	0.44	0.31	0.59	1.38	0.95
B-15	17	3.3	2.3	4.3	9.6	20.1	28.6	17.1	9.9	4.8	tr	0.30	0.18	0.50	1.69	1.00
B-16	12	0.8	1.5	2.7	5.3	30.0	34.6	16.9	6.1	2.1	tr	0.32	0.21	0.43	1.43	0.91
B-17	10	9.3	5.2	15.7	32.8	22.3	7.7	3.7	2.0	1.3	tr	0.74	0.48	1.13	1.54	0.98
B-18	8	1.3	2.5	4.1	6.7	15.6	31.8	29.5	7.1	1.4	tr	0.26	0.18	0.41	1.51	1.13
VARIOUS																
B-7	0	0.5	6.8	36.0	17.2	27.0	6.0	4.1	2.2	0.2	tr	0.83	0.46	1.25	1.65	0.83
B-8	0	—	0.2	1.5	6.8	66.7	17.4	5.6	1.6	0.2	tr	0.44	0.36	0.49	1.17	0.92
B-9	0	—		0.2	11.9	54.8	29.7	2.9	0.5	0.2	tr	0.27	0.23	0.32	1.17	1.00
B-11	0	6.0	6.5	11.7	25.9	26.8	10.0	5.0	4.1	3.8	0.2	0.60	0.37	0.98	1.63	1.00
B-19	0	—		0.1	11.5	79.4	8.2	0.5	0.1	0.2	tr	0.47	0.41	0.52	1.13	0.95
B-22	42	20.5	3.1	8.7	33.0	31.6	1.6	0.7	0.3	0.5	tr	0.72	0.53	1.50	1.68	1.53
B-35	—	0.4	0.6	0.4	0.9	4.3	20.0	59.0	13.5	0.8	0.1	0.17	0.14	0.22	1.28	1.03
ABACO																
L-20	2	—	—	0.9	3.9	9.9	7.6	8.0	12.2	54.8	2.7	0.060	0.03*	0.17	2.6*	1.2*
L-21	2½	5.5	5.4	7.6	16.4	15.6	9.2	7.3	8.3	23.7	1.1E	0.35	0.076	0.80	3.24	0.50
L-23	3	0.3	1.8	8.1	19.7	22.0	15.5	10.8	11.0	10.3	0.5E	0.37	0.15	0.68	2.13	0.75
L-24	2	—	0.7	0.3	1.7	8.1	13.0	11.8	15.0	47.2	2.2E	0.076	0.04*	0.20	2.2*	1.4*
L-25	2	1.6	2.3	5.4	15.0	18.0	10.7	8.8	12.1	25.0	1.1E	0.25	0.072	0.52	2.84	0.67
L-28	2½	0.9	1.1	4.0	10.6	12.9	9.1	8.4	14.7	38.0	0.3E	0.11	0.05*	0.43	2.8*	1.9*
L-30	25	0.4	1.3	4.7	11.0	20.3	17.3	13.1	19.9	11.8	0.2E	0.25	0.10	0.48	2.19	0.77
L-32	18	0.1	0.4	1.3	9.0	33.2	32.6	15.8	5.9	1.7	tr	0.33	0.22	0.46	1.45	0.93
L-35	19	0.5	1.4	4.4	8.6	16.1	16.1	14.4	24.8	13.4	0.2E	0.19	0.09	0.43	2.20	1.06
L-43	18	11.6	5.4	8.7	17.6	32.3	17.1	5.5	1.2	0.6	tr	0.54	0.36	1.05	1.71	1.30
L-44	12	4.4	10.7	20.6	39.1	18.7	5.6	0.8	tr	tr	tr	0.81	0.60	1.30	1.47	1.19
L-46	48	0.8	2.5	7.1	20.6	34.9	31.5	2.3	0.2	0.1	tr	0.44	0.32	0.67	1.45	1.10
L-52	27	0.1	0.6	2.6	6.8	53.8	34.6	2.6	0.2	tr	tr	0.39	0.32	0.46	1.20	0.97
L-53	22	1.1	0.2	0.7	2.2	4.7	12.1	18.7	49.8	10.4	0.1	0.11	0.084	0.19	1.50	1.44
L-56	8	0.9	1.6	2.4	7.1	12.3	11.9	13.5	18.5	30.4	1.4	0.13	0.066	0.35	2.34	1.48

TABLE I—(*continued*)

Grades		Gravel		Coarse Sand		Medium Sand		Fine Sand		Silt	Mud	Quartile Parameters				
Sub-Grades+		(a)	(b)	(c)	(d)	(e)	(f)	(g)	(h)	(i)		MD	Q₁	Q₃	So	Sk
Sample Nos.	Depth Ft.	%	%	%	%	%	%	%	%	%	%	mm.	mm.	mm.		
NORTHERN ELEUTHERA																
L-59	55	8.8	10.6	22.8	38.6	18.3	0.7	0.1	0.1	tr	tr	0.90	0.65	1.46	1.50	1.17
L-60	13	2.1	0.3	0.6	4.2	9.8	10.8	12.3	20.2	37.9	1.8	0.094	0.06*	0.24	2.0*	1.6*
L-62	1	0.1	0.5	7.3	31.3	49.3	10.9	0.3	0.1	0.2	tr	0.53	0.41	0.72	1.32	1.05
L-63	2	1.2	1.4	2.9	10.9	17.5	14.1	12.1	16.4	23.0	0.5E	0.19	0.079	0.46	2.41	1.00
L-67	2	7.5	2.7	2.0	2.7	4.4	12.8	36.4	28.6	2.8	0.1E	0.17	0.11	0.26	1.51	1.08
L-68	—	—	tr	0.1	0.7	5.0	20.0	46.5	26.5	1.2	tr	0.16	0.12	0.22	1.35	1.03

+ For explanation of sub-grade symbols see Figure 2.
E Mud content estimated by comparison with measured samples.
* Quartile values estimated by extrapolating the "cumulative curve."

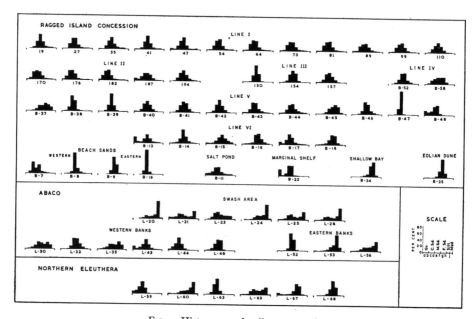

FIG. 4. Histograms of sediment samples.

The material passing the finest mesh sieve includes both the silt and mud grades. Quantitatively, it is of minor importance in all the sand samples; therefore it is not considered necessary to separate the two grades in each case. Microscopic examination shows that the silt is predominant, and measurements by repeated decantation after disaggregation in ammoniacal water, with a settling time appropriate to a nominal particle diameter of 5 microns, emphasized the quantitative insignificance of the mud fraction. For each of three samples (Nos. 35, 99, and 182), chosen as being representative of the shallow-water sediment over the Banks, the mud content is 0.01%. Due to its exceptional position sample B-37 shows a slightly higher value of 0.14%. Sample B-11, taken from a dried up salt pond on Nurse Cay, has a mud content of 0.21%, which is the highest of all the samples collected in the Ragged Island Concession.

These figures of the mud content are considerably lower than those given by Vaughan (1918) and Thorp (1936) for the sands from the east coast of Andros. Thorp's 14 samples had an average of 2.5% silt and 1.7% mud. His analyses were also done by sieving and a sedimentation method, adopting the same limit of 5 microns between silt and mud. His division between sand and silt was drawn at 50 microns, instead of 80 microns as here used, but although this lowers the silt-to-mud ratio, it has no bearing on the mud content. The difference between the two sets of results must therefore reflect a real difference in the mechanical composition of the sediments.

The particles that compose the traces of mud in the sands are all single crystals of calcium carbonate, but the bulk of them have no vestige of crystallographic shape. A few prisms with straight extinction occur, but there is no suggestion of the myriads of tiny needles of aragonite that make up drewite muds. The particles appear to be the "rock flour" produced by the abrasion of the sand grains against each other, and there is no evidence to suggest that they are produced by direct precipitation from sea water.

The particles with no crystallographic shape are probably derived from the matrix of cryptocrystalline aragonite, of which the bulk of the sand grains are composed. The rarer prisms are more likely to come from the break-down of organic tests.

X-ray photographs of the mud indicate the presence of both aragonite and calcite in comparable proportions, but it is not possible to separate the two microscopically. The abundance of broken foraminifera in the silt grade suggests that they contribute largely to the calcitic part of the mud.

SKELETAL MATERIAL

The calcareous sands covering wide areas of the Banks have a total skeletal content varying between about 10% and 40%. Slightly lower figures were found for some submarine banks of oölitic sand, and higher values were recorded from areas of abundant reefs. Table III shows the proportions of the chief components for each, and the average, of 7 typical bottom samples. In sharp contrast to this type of sediment is the predominantly skeletal facies found on the extreme edge of the Banks: it is commonly limited to the marginal shelf between the arcs of cays and the deep ocean. The analyses of 5 samples from this zone are also shown in Table III and their average is compared with that of 14 samples studied by Thorp (1936) from a similar environment off the eastern shores of Andros Island.

The skeletal content of a typical bottom sample is greatest in the extreme grade sizes that lie outside the two points of maximum curvature ("knees") of its mechanical cumulative curve. This is best illustrated by a well sorted sand,

TABLE III. AVERAGE PERCENTAGE CONTENTS OF BOTTOM SAMPLES
(Percentages rounded off to nearest whole number)

		On Banks	On Edges of Banks	
	Contents	A	B	C
		%	%	%
	Cay-rock	tr	2	
Non-skeletal	Lumps	55	5	9
	Grains	26	8	
	Aggregates	1	—	
	Faecal pellets	6	$\frac{1}{2}$	—
Skeletal	Mollusca	6	18	15
	Foraminifera	4	13	23
	Calcareous algae	1	39	27
	Madreporaria	tr	12	14
	Other forms	1	4	8
	Total skel. grains	12	86	87
Estimated total skeletal content		23	87	92

A. Average composition of sediment on Banks in Ragged Island area and between New Providence and Exuma Cays. Average of bottom samples Nos. 35, 99, 150, 182, 194, B-52, and B-58.
B. Average composition of sediment on edges of Banks ("marginal shelf") in Ragged Island area. Average of samples Nos. B-9, B-19, B-22, B-47, and B-48.
C. Average composition of sediment on edges of Banks ("marginal shelf") on east coast of Andros Island. Calculated from Thorp's analysis of 14 samples from Cocoanut Point and Golding Cay (Thorp, 1936, pp. 49, 58, and 110–113).

with a low total skeletal content. The small amount of gravel-grade may well be composed of one or two shells and segments of *Halimeda;* well rounded sub-spherical grains make up the abundant medium grades. Below the lower "knee," these become subordinate and unworn skeletal fragments are relatively abundant, particularly spicular forms such as Alcyonaria, segments of *Amphiroa* and *Penicillus*, echinoid spines, ostracods, serpulae, and so on. The change is a direct result of the sorting processes which control the mechanical composition of the sediment. It reflects a change in the mode of transport which occurs around this critical grain size, commonly within the fine-sand grade.

In the case of samples that are entirely skeletal, such changes are limited to variations in the proportions of the organisms. The grades finer than the lower "knee" are less rounded, and spicular forms are relatively abundant.

The skeletal remains of the various organisms found in the sands occur in all stages of mechanical disintegration, from fresh unworn tests to rounded detrital grains which may be clean, corroded, infilled, or encrusted.

MOLLUSCA

Lamellibranchs are commoner than gastropods, but both are widespread. Fragments of the large conch *Strombus gigas* are found in the sediment over the Banks, but the typical molluscan fragments are of thinner-shelled and smaller types. *Cerithidea costata, Olivella nivea*, and *Tellina (Strigella) candeana* are common and widespread. The latter is a thin delicate form which protects itself from all but the strongest scouring currents by burrowing deeply into the sand. Other delicate lamellibranchs, such as *Divaricella* cf. *quadrisulcata* with its open V-shaped ornament, are commonest in the shallow waters around the cays.

Colored shells are not commonly seen over the interior parts of the Banks, whereas the assemblage along the ocean margins includes many that are striped and highly ornamented, particularly among the smaller types. Various thick-shelled forms are abundant in this zone, notably species of *Lucina* which are sometimes thrown up by rough seas to form local accumulations on the shore, and various species of *Arca* including the common brown *Barbatia cancellaria*. These are but two of the commonest genera of a very varied assortment.

The gravel-grade contains whole tests. Immature forms and protoconchs occur in the sand-grades, together with characteristically translucent rounded grains derived from the larger thicker mollusks; worn fragments of massive corals are the only other organisms with which they may be confused, and unless the prismatic structure is revealed on crushing, a thin section may be needed to differentiate the two. The recognition of molluscan material in the finest sand and silt grades is difficult.

Empty tests of the land mollusk *Cerion* are abundant on parts of the cays and are in places found embedded in the sand dunes and dune-rock. They are a favorite protection for the small hermit crabs, by whose wanderings the shells may be incorporated in beach deposits.

FORAMINIFERA

The following brief account is based on the work of M. A. Illing (1950, 1952). In the course of her studies of the material from the Ragged Island Concession, she has recorded more than 100 different foraminiferal species, of which the majority occur only rarely. The forms that are quantitatively important in the shallow-water sediment of the Banks are given here, the classification adopted being that of Glaessner (1945, p. 89).

In the gravel and coarse-sand grades, Peneroplidae are predominant, and a few Miliolidae and Lituolidea occur. Common forms are *Archaias aduncus* (in several different habits), *A. compressus*, *Peneroplis proteus*, and *Valvulina oviedoiana*.

Less common forms include other Peneroplidae, *Clavulina tricarinata*, *Planorbulina*, *Amphistegina lessonii*, *Articulina mexicana*, *Vertebralina mucronata*, and several species of *Triloculina*.

Smaller Peneroplidae are outnumbered by Rotaliidea and Miliolidae in the medium- and fine-sand grades. Besides several of the genera aforementioned, the following are common: *Discorbis*, *Cibicides*, *Elphidium*, *Nonion*, *Asterigerina carinata*, indeterminate young Rotaliidea, and many species of *Triloculina*. Less common are *Cymbaloporetta*, *Quinqueloculina*, *Bolivina*, and many other genera.

The silt contains a very high percentage of foraminiferal fragments and tiny forms. Immature Miliolidae and Rotaliidea are predominant, and genera such as *Cornuspira*, *Spirillina*, *Buliminella*, and *Bolivina* are conspicuous.

The foregoing assemblage includes forms whose natural habitat ranges from quiet to turbulent conditions, but the mixing of the dead tests by marine currents masks such local environmental control. The skeletons become mere sand grains and are distributed and sorted as such in accordance with the usual hydrodynamic principles. The forms that occur in any sample are thus controlled far more by its mechanical grading than by the local indigenous foraminiferal assemblage.

On a broader scale, however, certain changes can be recognized. A few conspicuous species are confined to the margins of the Banks, where they are washed by the open ocean waters. They are predominantly colored forms, notably *Rotalia rosea* and the encrusting *Homotrema rubra*. On parts of the east coast of Abaco, these two are so common on the marginal shelf as to color the beaches pink. The prevalence of pigmented skeletons close to the ocean margins extends to other phyla, notably Mollusca and Alcyonaria, and presumably reflects some difference in the chemistry of the water as compared with that over the shallow Banks. Rare specimens of *R. rosea* found within the Banks have normally lost most of their color. Some uncolored species such as *Amphistegina lessonii* are similarly restricted to the marginal shelf; conversely, certain forms like *Articulina mexicana* are absent from this zone, but are common away from the ocean margin.

Due to the stronger oceanic waves and swell, the foraminifera and other organisms are more corraded on the edges of the Banks. Their worn tests become filled with calcareous matrix, and thin-section studies show the gradual oblitera-

tion of internal skeletal structure by recrystallization, until the grains become almost inseparable from those of non-skeletal origin.

Very few pelagic foraminifera belonging to the Globigerinidae were detected
in samples close to the open ocean. Their rarity is surprising.

CALCAREOUS ALGAE

The marine algae of Florida have been studied by W. M. Taylor (1928).
He lists eighteen calcified forms, belonging to the Codiaceae, Dascycladaceae,
and Corallinaceae, as well as many of the non-calcified Cyanophyceae.

Codiaceae.—By far the most important member of this group is *Halimeda*,
whose trilobed segments occur in all the samples studied. The spherical sporangia
are rarer. The external fine cellular pattern is easily recognized, but the irregularly arranged internal parts of the plant break up more readily, and may be confused
with other types. It is impossible to refer isolated segments to specific forms,
though it is known that several different species occur.

A few tiny perforated spicules were identified as pinnules of *Penicillus*. They
are patchily distributed among the bottom samples, avoiding the more current-
swept areas, and nowhere are they of any quantitative importance.

Dascycladaceae.—Several genera belonging to this group of plants are known
to grow in the Bahamas, but their remains have not previously been reported in
the calcareous sands. Worn featureless fragments, with a slightly milky luster
and a minutely granular texture (commonly due to boring algae), occur in many
samples and in some are abundant. Thin sections reveal spherical cells set in a
very dense groundmass of extremely finely crystalline aragonite. Comparison
with figured sections suggests that they are possibly *Acicularia*, but this identification is very tentative: some may be confused with the internal parts of *Halimeda*
segments after infill by calcareous mud. A few samples contain the very characteristic fragments of sporangial rings of *Neomeris*.

Corallinaceae.—The slender rod-like segments of plants such as *Amphiroa* or
Jania are minor contributors to the fine-sand grade of many of the bottom samples. They rarely preserve the details of their conceptacles on which the distinction between these two genera is based, and hence in most cases one can not do
more than refer to them as of "Amphiroa type." Small tufts of *A. fragilissima*
occur in one or two samples.

Quantitatively far more important, are the detrital grains of unsegmented
Corallinaceae—*Lithothamnion* and allied genera. No attempt was made to separate these various cerioid forms, and the term Lithothamnion is here used to include *Lithothamnion, Lithophyllum, Goniophyllum, et cetera*. When their characteristic smooth outer surface has been abraded off, they have a dense granularity
similar to the fragments of Dascycladaceae. Others may resemble smooth non-
skeletal grains. Thin sections were therefore used in their estimations.

Distribution.—Table III shows that one of the major differences between the
sands of the Banks proper, and those on their margins, lies in the content of cal-

51

careous algae. The change is both quantitative and qualitative. Over the Banks, the mere one or two per cent is almost entirely composed of segments of *Halimeda*. Traces of Amphiroa spicules are generally present and other Corallinaceae and Dascycladaceae occur locally around reefs, but they are of very minor importance over the area as a whole. In contrast, along the margins of the Banks calcareous algae become the largest single contributor to the sediment. Both *Halimeda* and Lithothamnion are abundant, here one and there the other being dominant. Dascycladaceae are still more variable: they are seldom the dominant form, but have been recorded as forming more than a tenth of a total sample.

Secondary infill.—All types of calcareous algae are liable to secondary infill by dense matrix, similar to that of their initial structure (Pl. 3.6). In the case of Lithothamnion which has a calcitic skeleton, such infilled grains lack the slightly brownish color associated with thin sections of cryptocrystalline aragonite. With Codiaceae and Dascycladaceae this criterion is absent, and, with the loss of the cellular pattern, it finally becomes impossible to separate them from non-skeletal grains. In analyzing the sections, the slightest trace of a cellular structure was taken to indicate a skeletal origin.

MADREPORARIAN CORAL

Despite numerous references to "coral sand" on the Admiralty Charts, madreporarian corals are chiefly confined to the marginal zone, and occur only in isolated and local reefs over the Banks as a whole. Table III shows that they were rarely seen in the Ragged Island Concession except in the immediate vicinity of the cays and on the marginal shelf where they form an average of 12% of the samples analyzed. The lower figure for the two beach sands as compared with the three bottom samples farther east supports Thorp's contention that coral grains break up rapidly, and are not found far from their place of growth.

There is no barrier reef on the outer edge of the marginal shelf along the Ragged Island arc of cays. Corals are distributed sporadically in patchy reefs on the shelf, usually close to the cays, and along some of the inter-cay channels. A particularly luxuriant growth occurs on the south shore of Hog Cay (Pl. 6.6) near locality B-35, and reefs are indicated in Figure 10 along the rocky western shore of Nurse Cay. With the aid of a glass-bottomed bucket or a diving mask, one can watch the countless organisms that inhabit these fringing reefs. It is a truly fascinating sight, with the white and golden corals and the delicate purple, gorgonian sea fans gently waving in the water, as the multi-colored fish dart in and out between them.

The most abundant corals are similar to those of Florida and Eastern Andros (Vaughan, 1915). They include various species of *Porites*, *Maeandrina clivosa* (Ellis and Solander), *M. cerrebrum* (Ellis and Solander), and *M. labyrinthiformis* (Linnaeus). *Orbicella* and *Acropora* are common, and many fragments of *Siderastrea* were found washed up on the eastern beaches. A fresh specimen of *Manicina areolata* (Linnaeus) was dredged up close to the reef southwest of Nurse Cay.

Recognition of madreporarian fragments in the bottom samples is difficult unless they are large enough to retain some of their septal structure. Worn and rounded translucent grains resemble those of molluscan origin. They may be recognized by their lack of prismatic structure on crushing, or by their trabecular structure which is seen in thin section. The latter technique was usually employed where abundant coral grains were suggested.

MINOR CONTRIBUTORS

Serpulae.—Hollow calcareous worm tubes of various shapes and sizes occur in minor quantities in most samples, either loose or cemented to other sand grains. The majority are formed entirely of a chalky white matrix of cryptocrystalline aragonite, but some are built of fine sand grains cemented together by this matrix. They are commonest where the sandy sea bottom is stabilized by sea grasses, and are rarest in submarine dunes of shifting sands. Local accumulations of their remains were found in shallow bays within the cays.

Small fragments of serpulae are difficult to identify, and certain bryozoan structures, if worn, are liable to be confused with them. They may also resemble worn pinnules of some calcareous algae.

Alcyonarian spicules.—Spindle-shaped spicules of gorgonians, or "flexible corals," appear in minor quantities in most samples. They include several different shapes and sizes, all having a rough papillate surface. Colorless spicules resembling those of *Briareum* are commonest, but pink forms are widespread. Near the ocean, purple forms are dominant, and there are a few of amber color.

Cary (1918) found that the Gorgonacea are an important agent in reef construction in the Tortugas (Florida), and that their remains are locally a major contributor to the skeletal sand. Such concentrations were not encountered in the present studies, and the spicules form only 1% or 2% of the fine-sand grade of most bottom samples; they may reach 5%–10% for the same grade on the marginal shelf and around the cays, where gorgonian sea fans and other forms are known to be common. However such a high figure in the finer grades is controlled by the mechanical composition of the sediment, and ordinarily reflects a paucity of other material of this size.

Sponge spicules.—Both calcareous and siliceous sponge spicules are widely distributed in the finer grades of sediment. Siliceous forms include well preserved monaxonid, triaxonid, and tetraxonid types; rare heteractinellid and lithistid spicules are also present. They all occur in the silt grade, and are of very minor quantitative importance. Together with diatoms, they form most of the 0.1%–0.3% SiO_2 content of the sands (Table IV).

Calcareous spicules of the same types are only slightly more abundant, though many are large enough to fall within the medium-sand fraction. In the silt grade, tiny spherical multi-rayed Calcarea are invariably present, and are ordinarily common. They measure about 50 microns in diameter, and are easily seen if the silt is mounted in oil. Due to their calcitic nature, they may be recognized

TABLE IV. CHEMICAL ANALYSES OF BOTTOM SAMPLES AND OÖLITES

(Percentages rounded off to nearest whole number)

	Ragged Island Area						Andros and Florida[0]		
Sample No.	35	99	150	182	B-48	35(j)			
Chemical Analyses†									
	%	%	%	%	%	%	Oölite Boca Grande Key, Florida %	Oölite Sharp Point Andros I. %	B.S. 71 Great Bahama Bank‡ %
SiO_2	0.22	0.14	0.30	0.12	0.10	1.00*			
Al_2O_3	0.10	0.30	0.04	0.10	0.14				
Fe_2O_3	0.012	0.011	0.010	0.011	0.011				
CaO	53.65	53.82	54.22	52.64	52.62	50.20			
MgO	0.61	0.80	0.62	1.68	1.92	1.88			
CO_2	42.20	42.55	42.13	41.86	43.07	41.12			
H_2O & organic	2.68	2.13	2.63	3.28	1.77				
Total	99.47	99.75	99.95	99.69	99.63	94.20			
Reduced Analyses, excluding H_2O and organic matter									
SiO_2	0.23	0.14	0.30	0.12	0.10	1.06*	0.03	0.07	0.13*
$(Al, Fe)_2O_3$	0.12	0.32	0.05	0.11	0.15		0.42	0.13	
$CaCO_3$	98.34	97.84	98.33	96.17	95.66	94.78	99.05	99.56	99.49
$MgCO_3$	1.31	1.70	1.32	3.60	4.09	4.16	tr	tr	0.38
$Ca_3P_2O_8$							tr	tr	
$CaSO_4$							0.50	0.24	
Total	100.00	100.00	100.00	100.00	100.00	100.00	100.00	100.00	100.00

† Analyst: Riley, Harbord, and Law, London. All samples washed clean of soluble salts with ammoniacal water, and dried at 105° C.
 [0] From Vaughan (1918), pp. 269, 270.
 ‡ Locality northwest of Andros Island, and 20 nautical miles east of Cat Cay.
 * Total insolubles.

in thin sections of sand grains in which they have been incorporated (Pls. 3.3 and 4.9). They are featured in the *Challenger Reports* (see "Deep Sea Deposits," Pl. XIV, Fig. 3b) but appear to have been overlooked or misinterpreted by many workers on recent calcareous sediments.

Millepora.—Large branching fragments of the hydroid *Millepora* occur in some reef assemblages on the marginal shelf. Unless their cellular pattern is preserved, they may be confused with Corallinaceae which are commonly abundant in such environments.

Echinoidea.—Some echinoids grow in colonies on the sand in quiet areas. Other commoner forms make and inhabit crevices in the rocky sea bottom and especially in the coral and algal reefs. Whole tests of *Echinometra lucunter* (Linné) and several Spatangina are commonly washed up on the eastern beaches of the cays. Fragments of their spines and plates of their tests are easily recognized.

Bryozoa.—Traces of both Cyclostomata and Cheilostomata were seen in many samples. The former may resemble Serpulae, but can be recognized by their assemblage of many sub-parallel tubes, some of which can be seen to divide. Cheilostomatous forms commonly encrust molluscan shells. Their unattached buoyant skeletons are common in the sediment left along the beach cusps at high-tide level on the protected cay beaches.

Crustacea.—Fragments of the appendages and carapaces of various crabs, crayfish, and other decapods are widespread but quantitatively unimportant.

Ostracods are common in the fine-sand and silt grades. No attempt has been made to identify the several different forms present.

Pelagic organisms.—Examination of the finest silt grades revealed one or two doubtful coccoliths, but these tiny forms do not appear to be common in the plankton over the Banks.

Of a rather scarce and poorly silicified diatom assemblage (Mann, 1936), *Navicula* and *Mastogloia* are by far the commonest genera. One or two Radiolaria and rare anchor-shaped tunicate spicules are seen in the decalcified silt residues.

Non-Skeletal Calcareous Sand Grains

The main problem in the study of these sediments is the origin of the sand grains formed of a compact groundmass of cryptocrystalline aragonite. As individual grains and combined into lumps, they form three-quarters of the total sediment. Their former interpretation as being derived from the Pleistocene basement and the consolidated limestone of the cays has already been mentioned. Such secondary fragments are, in fact, remarkably scarce in the bottom sediment: even in samples collected near the cays, lumps with the granular calcitic cement characteristic of cay-rock are but a minor constituent. Derived ovoids released by the break-up of the cement undoubtedly occur in the sand: they are very similar to those now forming on the sea bed, but all the evidence suggests that quantitatively they are of small importance.

For the purposes of tabulation, the non-skeletal sand grains have been separated into faecal pellets, aggregates, grains, and lumps.

FAECAL PELLETS

Photographs and thin sections of what he termed "ellipsoidal grains" from the Bahama sands were published by Vaughan (1924, Pls. I and II, Figs. 1, 2, and 3). Although he never called them faecal pellets, he linked their origin with similarly shaped glauconite grains.

The study of faecal pellets in the muds of the Clyde by H. B. Moore (1933) has led many workers to stress their importance and abundance in both recent and ancient sediments. Eardley (1938, p. 1401) described the calcareous, rod-like, faecal pellets found in such profusion in the recent sediments of the Great Salt Lake, Utah, and proved that they come from the brine shrimp *Artemia gracilis* which is very common in these super-saline waters. Thorp (1936, pp. 61–63) gives a good account of the pellets in the calcareous sands and muds of Andros Island. Though most of his estimates are between 1% and 10%, he gives figures of 20% and 30% for the pellet content of two samples from South Bight. He describes the Andros pellets as "somewhat loose and the soft friable nature is shown when they are easily broken down with a needle point" (*op. cit.*, p. 61). Pellets of this description occur in the Ragged Island Concession samples, but the vast majority are firm, well cemented ellipsoids as shown in Plate 1.1 (stage 4). Their longest diameter ordinarily lies between 0.5 and 0.7 mm., though smaller forms occur. The ratio between their maximum and minimum diameters averages 2.0. They occur in almost all samples of sediment, but in greatly varying amounts. There is no suggestion of a concentration in shore deposits as Thorp found in South Bight. They are commonest where the sandy sea bottom is partly stabilized, and are scarcer where it is being piled into submarine dunes by strong tidal currents. Sample B-52 with 19% is the most pellet-rich sediment analyzed, but such figures are largely controlled by the abundance of the medium-sand grade into which the pellet dimensions put them. The tabulated figures take no account of pellets included in larger lumps.

Cementation.—When first excreted faecal pellets are soft and friable aggregates of silt particles, bound together by a certain amount of organic mucus. They are rarely seen in this condition in the bottom sediments as a whole, but some samples near the cays were found to contain all transitions from those that disintegrate at a pin prick, to completely cemented forms, as illustrated in Plate 1.1. The fragility of the first group is shown in the photograph by their rather broken appearance due to some disintegration that occurred during the drying and subsequent laboratory treatment of the samples. The scarceness of friable pellets indicates that their cementation is rapid. It is probably aided by bacteriological precipitation of aragonite within the pellet.

Gray pellets.—In some samples (notably Line IV and the eastern part of Line II) many of the pellets vary from light gray to almost black. The color is due to tiny flakes of a black opaque substance disseminated through the internal parts of the pellet (Pl. 3.2). In some cases they are concentrated in a vague concentric layer. When the calcareous matrix is dissolved in acid, the black material is left as tiny specks enmeshed in mucus. The specks resemble the ground-up remains of chitinous fragments, and are similar to the much rarer flakes seen in non-colored pellets. The white and the gray pellets are alike in size and shape, and the difference may be merely due to a change in diet rather than in parent organism.

Friable uncemented pellets commonly break into two, and hard forms that have developed from such fragments may be recognized by their bee-hive shape. Some smaller and more spherical grains, which show by their gray color that they are of faecal origin, have developed by subsequent rounding of such broken pellets. In the absence of their distinctive color, they could not be separated from non-faecal ovoids. Indeed the common white pellets resemble other well rounded grains of matrix[3] in everything but shape, emphasizing the difficulty in distinguishing between faecal and non-faecal grains.

Thin sections of several pellets shown in Plate 3.1 are very similar to those published by Vaughan. They contain small opaque patches, white by reflected light, formed of ultra-fine-grained calcareous mud, probably with some admixed organic material. A few blebs that are black in both reflected and transmitted light are thought to be tiny chitinous flakes. Small skeletal remains such as foraminifera and sponge spicules (particularly the calcareous heteractinellid forms) can be identified in them. Siliceous remains occur here and there, and when ex-

[3] The term "matrix" as used in this paper refers to a compact homogeneous groundmass of cryptocrystalline aragonite.

tracted by decalcification in acid, they may appear corroded as if partly resorbed. Some pellets have a vague clearer border zone free of inclusions; apart
from this they have no regular internal structure.

Parent organism.—No direct evidence is available to determine what animal
or animals are responsible for their excretion. Thorp tentatively suggests certain
gastropods or maldanid worms. Echinoids have also been put forward, but their
limited distribution is against such an origin.

Various mollusks, including both lamellibranchs and gastropods, appear to be
the most likely agents, but further observations are needed in this field.

FRIABLE AGGREGATES

Many of the bottom samples contain irregularly shaped, loosely bound aggregates of silt particles, such as those shown in Plate 3.5. They have a chalky white
texture, and are easily disrupted with a pin point. They contain all manner of tiny
skeletal and non-skeletal particles: calcareous and some siliceous spicules and
fragments of foraminifera can be recognized, and here and there a pink fragment
of a gorgonian spicule stands out by reason of its color.

The aggregates occur on their own, or attached to larger sand grains. They
commonly form around some sticky organic material, such as a thread of algal
mucilage, or a young foraminifera. Other aggregates contain no obvious organic
tissue to bind them, and seem to be held by a very slight cementation. It appears
that under suitable conditions, these tiny calcareous particles need very little
persuasion to make them adhere together.

The aggregates are typically seen in sandy bottom sediments containing
lumps of grapestone (see later) with chalky white cement in the crevices between
component grains. They commonly occur where the bottom has a light sea
grass population, and no doubt the associated smaller thread-like algae play a
large part in their formation. Where the sand grains show no evidence of accretion, friable aggregates are absent.

Wherever they occur, these aggregates of silt particles can be found in all
stages of cementation, from their initial very friable condition, through more
firmly bound forms in which the component particles can still be recognized, to
well cemented and rounded grains whose composite nature is completely obscured. The process is illustrated in six stages in Plate 1.2. The same features
are shown in Plate 1.3 and 1.4, in which the larger component particles, though
less typical, are rather easier to photograph.

The whole sequence is an exact parallel of the changes that occur in the soft
faecal pellets after excretion, and it must likewise be ascribed to the precipitation
of aragonite cement within the aggregate by bio-chemical and purely physico-
chemical processes. Further evidence on this point is given later in dealing with
the sediments at the western end of Gun Point Channel.

While becoming less friable, the aggregates are rolled around on the sea
floor and lose their very irregular shape. The final product is a well rounded firm

grain, composed of a matrix of cryptocrystalline aragonite with little evidence of its original composite nature.

In some samples collected within a few miles of the edge of the Banks, these aggregates form as much as a quarter of the fine-sand grade and range into the medium-sand fraction. Considering the samples as a whole, they are a very minor contributor. Their importance lies not so much in their quantity, as in the evidence they show of the mode of formation of well cemented grains of matrix.

GRAINS OF ARAGONITE MATRIX

The abundance of sand grains formed of cryptocrystalline aragonite has already been repeatedly mentioned. They are the fundamental unit in the formation of the vast spreads of calcareous sand that cover the sea bottom on the Banks.

Description.—The grains are dominantly rounded and smooth surfaced (Pl. 1.5). Their mean diameter ordinarily lies between 0.1 and 0.6 mm. though both smaller and larger grains occur. The most common size is about 0.4 mm. The average degree of rounding and sphericity varies with environmental conditions, being greatest where tidal currents are strong enough to pile the sand into submarine dunes. Where currents are very slight, the grains, while remaining rounded, adopt less regular shapes. The most spherical forms have been termed "ovoids," more for convenience and to avoid repetitive description than to emphasize any singular quality. They occur in varying amounts in all environments. The degree of rounding of the grains begins to fall off where their diameter is less than 0.1 mm., though abrasion is still effective in the coarser part of the silt fraction. Though this appears to agree with the experience of workers on the rounding of quartz sand grains by water action (Twenhofel, 1945) the rounding with calcareous sands is far more perfect and rapid. There is also the important difference that the rounding of calcareous sand grains can be both by abrasion and accretion, and it seems possible for the two processes to operate at the same time.

The external texture of the grains is dependent on local conditions. If examined in strong illumination, they commonly have a rather greasy luster difficult to reproduce in a photograph. In some, a matt texture is seen, as in Plate 2.9. They have a very faint grayish tint, as opposed to the creamy color of those with oölitic surfaces.

Thin sections show that they are composed of a dense, structureless groundmass, with a particle size of the order of 1 to 4 microns, making optical mineralogical identification impossible. X-ray photographs indicate that it is formed of aragonite with up to about 5% of calcite as an impurity, no other minerals being recognizable. Staining tests give an aragonite reaction, but are unreliable because of the fine state of division.

The specific gravity of very well rounded sub-samples of the medium-sand grade varies between 2.83 and 2.85, confirming the aragonitic as opposed to calcitic nature. The quoted values for crystals of calcite and aragonite are 2.71 and 2.94, respectively. The slight drop from the latter theoretical value can readily be attributed to incomplete cementation within some of the lumps in the sample.

Typical sections are shown in Plates 3.4 and 3.5. They have a light brown color when sufficiently thin. Darker patches, which are white by reflected light, commonly occur. They represent ultra-fine material, which may contain traces of organic impurities, possibly bacterial. These irregular inclusions tend to disappear by resorption and recrystallization in the grains which have undergone prolonged rolling. Such recrystallization does not change the mineralogical nature of the grains which remains aragonitic: it merely tends to produce a uniform texture throughout the matrix. Blebs that are dark in both transmitted and reflected light are rare. Some grains can be seen to contain tiny skeletal remains, as was noted in the case of faecal pellets. Those formed of aragonite tend to be resorbed into the matrix and obliterated. Calcite structures are more resistant. In particular, clear tiny heteractinellid Calcarea spicules are commonly seen, as in Plate 3.3, perfectly preserved and surrounded by the matrix. They resemble tiny centers of radial recrystallization, and photographs of similar structures interpreted in this way have been published by Richards and Hill (1942, Pl. I) from the Great Barrier Reef, Australia. Without examination of their material, it is impossible to decide whether the two occurrences are analogous. Rare siliceous remains, principally diatoms and sponge spicules, can be extracted by decalcification. Due to the almost complete recrystallization of the material forming the grains, it is impossible to differentiate the component silt particles and the cement; yet as the silt is dominantly skeletal (Table V), it follows that the grains (and lumps) must contain a large proportion of calcium carbonate which is ultimately of skeletal origin.

ORIGIN

It is clear that these grains of matrix are of diverse origin. Five main sources are recognized.

1. Grains formed by the cementation of friable aggregates of silt particles
2. Faecal pellets
3. Those formed from skeletal grains
4. Derived grains
5. Composite grains

1 and 2. The first two of these processes have already been described. Both depend on the aggregation of silt and mud particles in the surface layers of the sediment, and both readily account for the skeletal remains that can be identified in the cemented grains.

3. Grains composed of a precisely similar matrix may also be produced by the infill and recrystallization of detrital organic skeletons. Thin sections show the gradual obliteration of the original structure. Calcareous algae are particularly susceptible to this process, and the larger foraminifera, particularly the Peneroplidae, may also be affected but rarely lose all vestiges of their cellular pattern.

4. Some grains come from the mechanical disintegration of cay-rock, and detrital fragments of the latter are easily identified by their calcite cement. Their rarity, even close to the cays, is due to two factors. Firstly, they tend to break up and release their component rounded sand grains; secondly, and more important, they are swamped by sand of the three preceding types.

The derived grains are externally indistinguishable from their recent equivalents. The internal structure may also be identical, but as cay-rock is commonly composed of strongly oölitic sands, many derived grains show evidence in thin section of a concentric structure, which is not confined to a mere thin outer skin.

5. In the present study, the term "lumps" is used for those composite sand grains which have superficial re-entrants between their component calcareous "grains," whereas the latter have no such re-entrants. Thus, besides the four principal types, "grains" may be released by the break-up of "lumps" into which they have been cemented; alternatively, by the infill of their superficial re-entrants, such "lumps" may be turned into "grains" under the classification here used, with no external evidence of their composite origin.

Abundance.—The majority of the free grains fall within the fine-sand grade; hence, their abundance in a sample of sediment is largely controlled by its mechanical composition. It is impossible to estimate accurately the relative proportions of grains of the four sources, as they all tend toward a similar end product. The dominantly skeletal sands on the edges of the Banks contain minor quantities of grains of matrix classified as non-skeletal, which probably include all five groups, particularly types 3 and 4. Over the Banks proper, it is considered that type 1—cemented aggregates—is the dominant source, and faecal pellets and composite grains (types 2 and 5) are also important: the relative scarcity of calcareous algae precludes an abundance of type 3, and all the evidence suggests that derived grains (type 4) are comparatively rare. This conclusion is directly opposed to Vaughan's view that the bulk of the calcareous sands are derived from the oölitic Pleistocene basement, a conclusion which has recently been challenged by Newell (1951, p. 14) who is of the opinion that the surface sediments are dominantly oölitic. In the light of the present studies, this too seems unlikely; although no samples are available from the northwestern part of the Great Bahama Bank where Vaughan's specimens were collected, there is every reason to believe that they are similar to those of the Ragged Island Concession, and that they are formed principally by the processes of aggregation of silt particles here discussed.

LUMPS

Grains lying in contact with each other on the sea bed tend to become cemented together and form a composite sand grain, or "lump," under the definition here adopted. The various habits assumed by these lumps are typical of certain environmental conditions, and the following specific forms are recognized; grapestone, botryoidal lumps, encrusted lumps, and irregular well cemented forms that show no outstanding feature to label them. These are estimated separately in the tabulations of sample contents (Tables V and VI), but as there is no sharp division between one habit and another, precise numerical relations are of little significance.

The component rounded grains commonly protrude from the lumps, giving an appearance resembling that of a bunch of grapes. The term "grapestone" is therefore proposed for this type of lump, though it must be admitted that the analogy is weak where the lumps are small and have only a few component grains. The habit of the larger grapestone lumps is illustrated in Plate 2.1. Mechanical disintegration due to current action prevents excessive growth, and the common size, shown in Plate 2.2, is about 1 mm. diameter. The processes of grapestone formation are not selective, and small sand grains of all types, skeletal and non-skeletal, whole and fragmentary, become incorporated.

The cement that joins the grains is finely divided aragonite of varying texture. It forms first around the points of contact, and is friable and chalky white, in contrast to the greasy or matt-textured grains. It can easily be scraped off with a pin point, and is composed of aggregated particles of mud dimensions. From its occurrence, it is clear that it is being precipitated from sea-water, yet the particles show no recognizable crystalline shape, and are similar to the material that forms the matrix of the grains themselves.

Externally, the chalky white cement is restricted to the crevices between the protruding grains. As cementation proceeds, the cement beneath this surface layer becomes firmer and matt-textured. Further grains or small lumps may be joined on by the same sequence of stages. However, the forces of mechanical disintegration prevent unlimited growth, and finally all traces of the chalky white texture are lost. Prolonged rolling and abrasion of such completely cemented lumps remove the protruding grains, and destroy the grapestone habit. In the end they may be reduced to composite grains, whose multiple origin can only be seen in thin section.

The deposition of cement is dominantly a surface phenomenon, forming a relatively impermeable outer crust on the lumps (Pl. 3.7). Fluid movement into the internal parts may thus be sufficiently inhibited for the aragonite to be precipitated in well formed prismatic crystals, large enough for microscopical identification. They grow radially outward on the grain surfaces into the cavities, giving a drusy appearance on a very fine scale. The crystals vary in size according to their rate of growth. The larger ones shown in Plate 3.8 and 3.9 are up to 50 microns in length and 10–15 microns across, and have square terminations. In size and shape, they are not unlike the aragonite crystallites forming the deposit of "fur" in a household kettle. Others form layers of parallel tiny blade-like crystals, about 10 microns long. Further growths of similar acicular aragonite fill the remaining interstices, and at this stage, the lumps will fracture across rather than around their component grains, revealing the clear fringe of cement surrounding each of them. Commonly this concentric structure is emphasized by an opaque layer of ultra-fine-grained matrix, white by reflected light, coating the grains beneath the acicular zone (Pls. 3.9 and 4.1). All transitions can be found from this relatively coarsely crystalline condition to the common form where the deposition of cement was more rapid and formed as unorientated cryptocrystalline aragonite. By prolonged cementation the component grain boundaries become obscured (Pl. 3.10) until finally the whole lump is composed of the normal uniform matrix, with no signs of its original composite structure.

Botryoidal lumps.—Where grapestone is subjected to oölitic accretion (discussed in detail later) each protruding grain is coated with a shiny layer, and the sutures between them are partly filled in. The habit is thereby altered to an approximately botryoidal form, illustrated in Plate 1.9. Apart from the pearly texture, it commonly resembles worn grapestone: botryoidal lumps are therefore only tabulated separately in the case of oölitic samples. The "very irregular, somewhat nodose, white grains" described and figured by Eardley (1938, p. 1389, and Pl. 15) from the oölitic sands of the Great Salt Lake, Utah, appear to be similar to the botryoidal lumps of the Bahaman sediments.

ENCRUSTED LUMPS AND RELATED HABITS

In certain environments grapestone is replaced by a type of lump in which surface cementation largely obscures the component grains, and imparts a slightly grayish powdery texture with no trace of chalky white cement. The lumps are irregularly shaped, suggesting accretion rather than corrasion (Pl. 2.4). In the tabulations of lithological sediment contents, this rather featureless habit is included in the "total lumps" without being separately listed.

The tendency for the interior parts of the lumps to remain porous beneath a firm outer crust is more marked than it is in grapestone, and the same processes leading to the slow formation of delicate drusy aragonite needles may take place in the intergranular cavities. Thin sections show that the surface layers harbor countless tiny boring algae. Their thread-like filaments vary in diameter between 5 and 10 microns and are up to about 300 microns long. By destroying the outlines of the component grains near the surfaces of the lumps, they aid the formation of the well cemented crust composed of the usual irresolvable aragonite matrix, and are responsible for the powdery external texture (Pl. 4.1). Somewhat similar algal threads are found boring into skeletal grains (Vaughan, 1919b, Pl. 34) where they destroy the structure of the surface layers, and impart a similar powdery etched texture. They differ from the shorter unicellular algae commonly seen boring into oölitic grains.

Algally encrusted lumps.—Occurring with the foregoing type of lump are others which reveal their algal encrustation by a faint green or brown discoloration (Pl. 2.3). Clear cyanophyceous threads may be seen growing on them, and entrapping small grains and silt particles. Isolation of the algae by decalcification was not very satisfactory, but they appear to ramify throughout the lumps from the rather friable surface to the more densely cemented material beneath. Very thin sections of isolated fragmental grains reveal the remains of algal threads which may be similar. They are about 5 microns in diameter, and are closely packed together, as in Plate 4.2 (the latter is very similar to illustrations of algal threads from Pennsylvanian and Permian limestones of Kansas by Harlan Johnson,—1946, p. 1107, Pl. 9, Fig. 3, 4—who regards them as the remains of several common genera of very simple form). It seems likely that they play an important part in the biochemical extraction of the calcium carbonate which surrounds them. As the distinction between this type and the broader filamentous boring form is not clear, there is no sharp division between "algally encrusted lumps" and the featureless type here described. They have been separated in the tabulations of sediment contents, in order to avoid classifying lumps as being formed by algal action in the absence of external visible evidence.

Attempts to grow cultures of the algae in the laboratory were not successful, and it has not been possible to proceed with their identification. Probably several different genera and species are present.

Other binding agents.—The cementation of grains trapped in the root meshes of *Halimeda* and various marine grasses leads to the formation of some of the largest lumps of grapestone habit.

Grains may be bound together into lumps by various other organic agencies of doubtful affinities. Some are covered by a thin transparent mucilaginous sheath, which may be of algal origin. Certain grapestone lumps reveal a cellular

---

»»→

EXPLANATION OF PLATE 1

1.1   Stages in cementation of faecal pellets, from friable pellets (stage 1) to well cemented pellets (stage 4). ×10.

1.2 and 1.3   Stages in conversion of friable aggregates of silt particles (*1.2*) and fine sand grains (*1.3*) into well cemented grains of cryptocrystalline aragonite matrix. ×10.
Stage 1.   Friable aggregates.
Stages 2, 3. Firmer aggregates.
Stages 4, 5. Grains of matrix.
Stage 6.   Ovoids.

1.4 Sand grains selected from sediment forming sand-tongue of Gun Point Channel to show stages in conversion of friable grapestone lumps into composite grains by cementation and oölitic accretion. ×10.
Stage 1. Friable grapestone, typical of sample B-44.
Stage 2. Well cemented grapestone, typical of sample B-42.
Stage 3. Botryoidal lumps, typical of sample B-41.
Stage 4. Composite oölitic grains, typical of sample B-39.

1.5   Well rounded grains of aragonite matrix from eolian sand dune: sample B-35(g). ×10.

1.6   Shiny oölitically coated sand grains, typical of western beaches of cays of Ragged Island area: sample B-8(e). ×10.

1.7   Four fragments of a single botryoidal lump, showing internal porous and drusy nature due to radial growth of aragonite needles beneath impervious shiny oölitic surface: from sample B-7(a). ×10.

1.8   Shiny botryoidal lumps, including rectangular oölitically coated segment of *Halimeda* (lower center): from sample B-7(c). ×10.

1.9   Oölitically coated grapestone and botryoidal lumps: from sample B-39(d). ×10.

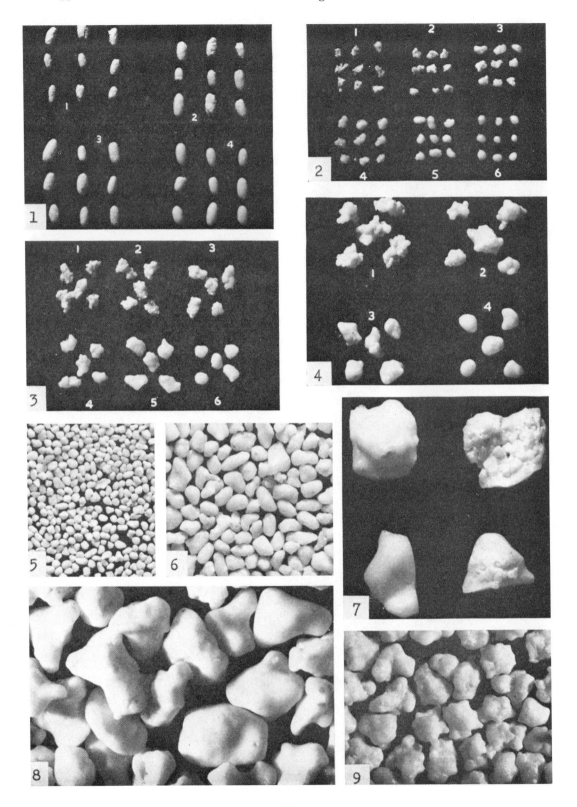

partly calcified, unidentified organism between the grains, serving to bind them together. It may be an irregular foraminifera, or it too may be algal.

OÖLITIC SAND GRAINS

Where the sediment is swept along by strong currents of water supersaturated with calcium carbonate, accretion takes the form of superficial oölitic coats of orientated aragonite, seldom more than 30–40 microns thick. The grains become very well rounded, and in extreme conditions resemble cave-pearls in their creamy pearly luster (Pl. 1.6). Black (1933a, p. 462) described the same phenomenon from the western part of the Great Bahama Bank:

The grains are spherical or ellipsoidal and possess a very characteristic structure, with a large nucleus of extremely fine-grained limestone, enclosed within a few oolitic layers of aragonite usually two or three.

Terminology.—The terms "oölite," "oölith," and "oölitic" have been used in widely differing senses by different authors. If interpreted literally according to their derivation from the Greek "oön" and "lithos," meaning "eggstone" or "roe stone," they would refer to any sand or sandstone composed of smooth spheroidal grains. Such, however, is not the normal usage of the term, which connotes some degree of regular internal structure. Some authors would accept either a radial or a concentric pattern, or both. Others use the term spherulite for the radial arrangement, and define oölites as rocks composed of "small rock particles of elliptical or spherical shapes with concentrically laminated structure" (Twenhofel, 1932, p. 757). To this definition must be added the division of each "oölith" (Rastall, 1933, p. 484) into a nucleus and an outer zone of concentric layers. It is impracticable to lay down limiting ratios between the thicknesses of the nucleus and the concentric laminae, for in a typical oölite (in the sense of a rock type) all gradations exist from non-oölitic ("non-oölithic") grains to those made up almost entirely of oölitic layers. Nevertheless, a useful distinction may

---

←⋘

EXPLANATION OF PLATE 2

2.1   Large grapestone lumps: from sample B-42(b). ×10.

2.2   Grapestone lumps with chalky white cement (barely discernible in photograph) in re-entrants between component grains: from sample B-58(d). ×10.

2.3   Algally encrusted lumps: from sample 176(b). ×10.

2.4   Well cemented, irregularly shaped lumps, including some corroded grapestone: from sample 99(d). ×10.

2.5   Sample 99(g), showing many well rounded grains of matrix. ×10.

2.6   Sample 35(g), showing friable aggregates, grains of matrix and a few skeletal fragments. ×10.

2.7   Sample B-19 (unsieved): well sorted skeletal sand grains, typical of eastern beaches of cays of Ragged Island area. Foraminifera, sponge spicules and a few fragments of cay-rock are conspicuous among worn fragments of calcareous algae. ×10.

2.8   Friable lumps of grapestone with chalky white cement (barely discernible) between component grains, most of which are finer than usual: from sample B-45(e). ×10.

2.9   Matt textured, well rounded grains of matrix and faecal pellets: from sample B-46(e). ×10.

2.10   Granular textured skeletal grains, chiefly calcareous algae: sample B-47(e). ×10.

be made between a "superficial oölite," in which most of the sand grains are "superficial oöliths" with only thin external oölitic layers, and a true "oölite" which is dominantly composed of "oöliths" with well developed concentric structure. Most of the modern Bahaman oölitic sands are therefore composed of superficial oöliths, in contrast to the true oölites of the consolidated Pleistocene cayrock.

The processes of oölitic accretion affect both grains and lumps. In grapestone, the protruding component grains are coated with shiny oölitic layers; the sutures between them first become constricted, and then filled in by similar oölitic material, giving the lumps a pseudo-botryoidal appearance on a small scale. This term has therefore been adopted for such conditions. It is illustrated in Plate 1.9 and in extreme form in Plate 1.8, which includes an oölitically covered segment of *Halimeda*, still recognizable by its shape.

Thus, creamy shiny ovoids and botryoidal lumps are characteristic of oölitic sediments.

STRUCTURE

The oölitic layers have crystallographic properties consistent with their being formed of minute orientated prisms of aragonite, too fine to be separated microscopically. Optically, the slow vibration direction is normal to the surface of the sheath, that is, radial with respect to the grain on which the sheath is formed; the fast direction is tangential on the same basis, and the sign of elongation is thus negative. These observations suggest that the component aragonite needles or prisms lie with their major crystallographic axes (which correspond

EXPLANATION OF PLATE 3

3.1 Faecal pellets with long axes in plane of section: from sample B-46. Ordinary light, ×40.

3.2 Section of gray faecal pellet, showing scattered tiny black particles beneath outer clear border: from sample 194. Reflected light, ×40.

3.3 Well rounded grain of aragonite matrix, with paler border, and containing heteractinellid Calcarea spicule: from sample 99(g). Ordinary light, ×100.

3.4 Well rounded grains (ovoids) of cryptocrystalline aragonite matrix. Most have clearer borders in which faint relics of superficial oölitic structure may be seen: from sample 99(g). Ordinary light, ×40.

3.5 Thin section prepared from representative part of sample B-35(g) shown in Plate 1.5. Several well rounded grains of matrix are superficial oöliths, and a few are composite. Crossed nicols, ×40.

3.6 *Halimeda* segments. Smaller fragment, which has been attacked by boring algae on right-hand side, shows considerable infill by aragonite matrix, leading to partial obliteration of cellular structure: from sample B-22. Ordinary light, ×40.

3.7 Grapestone lump (with chalky textured cement) showing porous interior and better cemented surface layer: from sample B-45. Ordinary light, ×40.

3.8 Part of interior of lump, similar to 3.9, showing aragonite prisms growing into pore space between component grains of matrix: from sample 99. Crossed nicols, ×250.

3.9 Interior of lump with densely cemented surface (not shown). Component grains are surrounded by dark layer of ultra-finely crystalline aragonite matrix, and are cemented together by clear prismatic aragonite crystals: from sample 99. Crossed nicols, ×40.

3.10 Grapestone lump cemented by clear acicular aragonite crystals in interior (white areas), and by dense cryptocrystalline aragonite in outer part leading to obliteration of component grain boundaries: from sample 150. Crossed nicols, ×40.

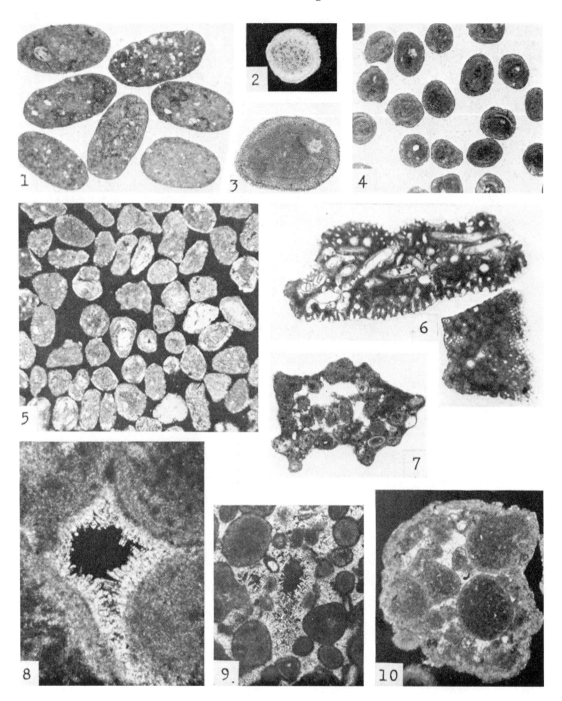

with the fastest vibration direction) randomly disposed in the surface of the grains. The same conclusion was reached as long ago as 1879 by Sorby, who likened their formation to the mechanical accumulation of the layers of a large rolled snowball (Sorby, 1879, p. 74).

The two refractive indices of the oölitic sheaths are 1.605, and a rather variable value averaging 1.640 for the slower vibration. The measurements were made on laminae surrounding grains seen in equatorial section. The former value, corresponding with the tangential vibration direction, was confirmed by observations on loose fragments of the sheaths, separated by crushing oölitically coated grains. It is precisely equal to the mean of 1.525 and 1.686, which are the respective values of α and γ for aragonite (Larsen, 1921). This is in conformity with the structural interpretation previously given. The larger refractive index is, however, considerably lower than would be expected on this hypothesis, which would imply a value equal to, or between β and γ of aragonite (1.684 and 1.686, respectively). It must be remembered, however, that we are dealing with an aggregate of crystals which is probably porous, particularly in the radial direction; this is thought to account for the low and variable radial refractive index. A certain amount of imperfect orientation of the crystallites may be a further contributing factor, though it can not by itself account for the wide discrepancy.

Lacroix (1898) described the structure of the well known oöliths from Carlsbad and elsewhere, and postulated a new variety of aragonite "ktypeite" to account for their anomalous optical properties. However, they are probably mineralogically like the Bahaman oöliths. Lacroix found a double refraction of 0.020, compared with the present figure of 0.035. He determined the specific gravity

---

⟫⟫→

EXPLANATION OF PLATE 4

4.1 Algally bored lump, showing denser surface cementation with holes left by algal filaments. Interior cemented by prismatic aragonite: from sample 99. Ordinary light, ✕40.

4.2 Part of lump of matrix containing ramifying tubes, about 5 microns in diameter, left by algal filaments: from sample 182. Ordinary light, ✕100.

4.3 Recent superficial oöliths from western end of sand-tongue of Gun Point Channel. Note lenses of matrix between oölitic laminae on central grain: sample B-39(f). Ordinary light, ✕40.

4.4 Recent oölith. In places, lowermost laminae of orientated aragonite crystallites merge inward into matrix with no sharp dividing line: sample B-39. Ordinary light, ✕100.

4.5 Beach rock, showing thin layer of blades of microcrystalline aragonite cementing two sand grains, upper of which is fragment of cay-rock; besides its typical clear calcite cement, it shows aragonite in three habits—cryptocrystalline matrix of component grains, oölitic surface laminae, and faint acicular layer growing at right angles on surface of lump. From specimen B-29, west shore of Hog Cay. Crossed nicols, ✕250.

4.6 Botryoidal lump, showing oölitic layers of clear aragonite filling superficial re-entrants. Component grains in lump are cemented by dark dense matrix (white by reflected light). From beach-rock specimen B-29, west shore of Hog Cay. Crossed nicols, ✕40.

4.7 Oölith in cay-rock. Outermost oölitic layers have been altered to unorientated matrix, similar to material of nucleus. From specimen B-3, Gt. Ragged Island. Crossed nicols, ✕100.

4.8 Cay-rock showing typical calcite cement between component oöliths and non-oölitic well rounded grains of matrix. One oölith shows alternate shells of oölitic aragonite and clear calcite formed by selective replacement. From specimen B-31, northwest Raccoon Cay. Crossed nicols, ✕40.

4.9 Poorly consolidated oölitic cay-rock, showing calcite cement in typical granular habit. Central oölith contains heteractinellid Calcarea spicule. From specimen B-36, Hog Cay. Ordinary light, ✕100.

4.10 Strongly oölitic cay-rock. From Hunt's Cave, New Providence Island. Ordinary light, ✕40.

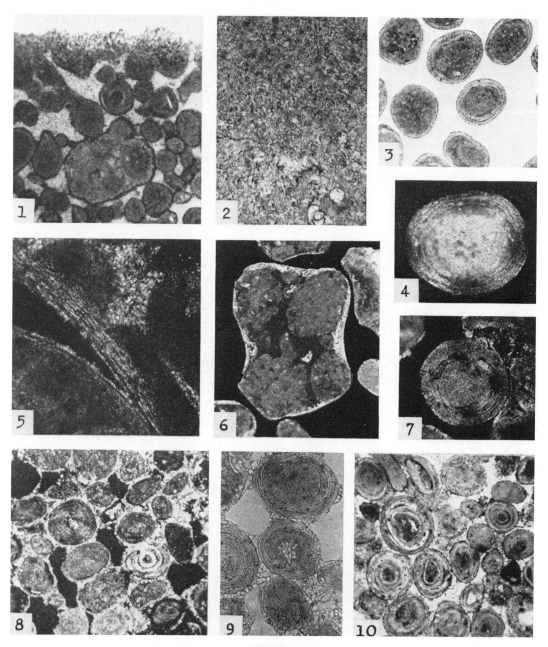

PLATE 4

of the oöliths as being between 2.58 and 2.70. This could be due to the porous nature of the sheath aggregates, in support of the foregoing view concerning the low radial refractive index.

It is interesting to note that sections cut at right angles to the oölitic layers sometimes show a very faint pleochroism, from light honey-brown corresponding with the slow direction, to a slightly richer similar color for the fast vibration. Oölitic sheaths, viewed normal to their surface, appear isotropic apart from a very slight mottling on a minute scale. They sometimes give a very ill-defined positive pseudo-uniaxial figure, as would be expected from the aggregate structure described.

Attempts to investigate the pattern of the oölitic layers by X-rays failed to indicate any preferred orientation of the crystallites. A flake of an oölitic sheath was used in these tests, but it seems likely that its structure was masked by reflections from adhering unorientated particles which could not be excluded from the X-ray beam. By showing that no other mineral is present, the photographs do confirm the aragonitic nature of the oölitic laminae.

Alteration.—The nuclei of the superficial oöliths are normally sharply differentiated from the outer concentric layers, but in some the laminae contain elongate lenslike inclusions, lacking any trace of orientated structure, and similar to the matrix of the nuclei and other non-oölitic ovoids (Pl. 4.3). In some oöliths, layers of this matrix, enclosed between orientated laminae, completely surround the grains. Ordinarily they are sharply defined, but they may interfinger laterally with oölitic aragonite. In the same way it is commonly found that the innermost oölitic layer merges imperceptibly into structureless matrix as in Plate 4.4, so that it is impossible to draw a sharp line between nucleus and laminae. Some superficial oöliths which have developed around skeletal grains, show a narrow zone of matrix between the skeletal structure and the sheaths.

In comparing the Bahama oölites with the Sprudelstein of Carlsbad and elsewhere, Sorby considered that their turbidity and enclosed unorientated granules were due to material enmeshed during growth in muddy water, by his snowball action previously referred to (Sorby, 1879, p. 74). However, his observation that the outer layers are "usually more crystalline than the inner," scarcely supports such a hypothesis, and it seems more likely that the turbidity is due to the alteration of oölitic laminae into unorientated matrix. The conversion may be complete and detected only by a difference in texture, such as a pale or dark border on the grain; it may be localized as already described: or it may be more or less uniformly incomplete, leaving a faint concentric structure in the matrix, which reveals itself between crossed nicols as a broad ill-defined black cross.

The factors which destroy the orientation of the aragonite are not known. It is commonly a surface phenomenon, and accounts for the loss of their shiny texture when oölitic grains are transported to environments where oölitic accretion is not taking place. It seems likely that the porous structure inferred from the optical properties (probably in the form of hydrous layers included between the crystallites) may be breaking down. Such an explanation was offered by Johnston, Merwin, and Williamson (1916, pp. 486–90) to account for the several varieties of skeletal and non-skeletal calcium carbonate such as "conchite," "ktypeite," and "vaterite," which have anomalous refractive indices, *et cetera*. The change may be related to some kind of bacterial action, though this is based on conjecture rather than any concrete evidence. Boring algae are common, and their dark blebs (white by reflected light) concentrate in, or just below, the innermost oölitic layers. They are of the short stubby type, about 15 microns in diameter, found boring into other lumps and grains. There is nothing to suggest that they are concerned with the loss of orientation of the aragonite. All that can be concluded is that certain unknown conditions facilitate the recrystallization of the oölitic laminae into unorientated cryptocrystalline aragonite matrix, which, being denser, is the more stable form. The conversion is not universal, and the oölitic orientation is found perfectly preserved in many of the oölites of the Pleistocene cay-rock.

ORIGIN

There is little general agreement with respect to the ways in which oolites develop, probably due to the fact that their origins may be various and may result from several different combinations of conditions. (Twenhofel, 1932, p. 764.)

Part of the confusion arises out of the widely differing use of the term "oölite." It is not proposed to review here the vast literature on the origin of oölites, but merely to consider those factors which have a bearing on the formation of the Bahaman oölites.

With the possible exception of Alexander Agassiz, all geologists who have examined the oölites of Florida and the Bahamas, have thought them to be marine in origin. A. Agassiz, writing on the Florida Elevated Reef (1896), ascribed the accumulation of the oölitic rocks to eolian action. Griswold, in an appendix to Agassiz's paper, described the internal concentric structure of the oöliths as supporting his view that the hummocky dunes were water laid. Agassiz comments (*op. cit.*, p. 55) that the oölitic structure "applies in every way to aeolian rock blown from a coral reef sand tract." He has consequently been interpreted by Vaughan as implying an eolian origin for the oöliths; but Agassiz is not clear on this point, being more interested in the mode of formation of the dune bedding of the rocks, than of their component sand grains.

Vaughan himself (1910, 1914a) considered that the calcium carbonate was first precipitated with the help of denitrifying bacteria in gelatinous form. It later changed into spherulitic grains, growing in size from 4 to 6 microns up to the dimensions of oöliths, by chemical agencies which caused precipitation of calcium carbonate in concentric shells outward from the interior nucleus. Sanford, Clapp, and Matson reached similar conclusions, and Vaughan maintained that:

all marine oolites, originally composed of calcium carbonate of whatever geologic age, may confidently be attributed to this process. (Vaughan, 1914a, p. 54.)

However, it appears that at this time he considered all spherical grains as oöliths, regardless of internal structure, for he refers to:

Quantities of soft, non-indurated as well as indurated oolite grains, the former easily crushed by a touch with the point of a needle. (Vaughan, 1918, p. 272.)

Soft oöliths are also mentioned by Newell (1951, p. 14) in discussing the sand bars near Cat Cay West of Andros. Vaughan later reconsidered his opinion and concluded that:

The cores of the oolite grains are similar to the grains in the muds, and it may yet be shown that the muds represent a stage in the formation of the oolitic limestone. (1924, p. 327.)

The classic experiments of Linck (1903) on the mineralogy and habit of calcium carbonate precipitated from sea water, led him to believe that all oölites are of inorganic physico-chemical origin. This was directly opposed to Rothpletz's view based on studies of the oölites of the Great Salt Lake, which he considered to be formed by the activity of cyanophyceous algae identified as *Gloeocapsa* and *Gloeotheca* (Rothpletz, 1892). Similar conclusions were reached by Wethered for the Jurassic (and some Carboniferous) oölites in Britain. Noting the common occurrence of threads of *Girvanella* and other algae in the oöliths, he concluded that they were the cause of the concentric structure. He drew attention to the rather vaguely defined radial pattern of dark and light zones intersecting the more definite concentric structure, which is a common feature in oölites of all geological ages, as well as in those now forming in the Great Salt Lake. Whereas most workers ascribe the light and dark zones to recrystallization and localization of impurities, Wethered considered they are fundamentally different, and that:

the clear [portion] represents the walls or skeletal structure of the organisms forming the granules; and the granular [dark portion] has either been secreted by these organisms, as in the case of the calcareous algae, or has filled in spaces left by the decomposition of the living matter. (Wethered, 1895, p. 199 and Pl. VII.)

A similar origin is supported by Dangeard, who claims that all Mesozoic, Tertiary, and Recent oölites can be shown to contain algal fibers (verbal communication). After slow decalcification, the algal remains may be emphasized by staining with methylene blue. In this way:

on aperçoit à l'intérieur des écorces oolithiques de nombreux filaments ramifiés colorés en bleu. . . . Certains filaments courent parallèlement aux zones concentriques, d'autres s'enfoncent obliquement ou perpendiculairement. Les filaments mesurent 12, 15, 20 μ de diamètre. (Dangeard, 1936, p. 240 and Pl. XVI.)

By working on fresh material brought up from boreholes, Dangeard ruled out the possibility of the algae being recent boring forms. But neither his observations, nor those of Wethered, or Rothpletz, can preclude the very rational interpretation that these filaments were left by algae which bored into the grains as they lay on the sea floor.

A brief mention must be made of the colloidal theory of Schade and Bucher. The latter maintains that:

most if not all, oolitic and spherulitic grains were formed by at least one constituent substance changing from the emulsoid state to that of a solid. (Bucher, 1918, p. 602.)

73

Bradley has developed this explanation to account for the ferruginous oölites of the Green River formation of Wyoming and Colorado. He claims that the oöliths, some of which are now embedded in lumps of cryptocrystalline calcite, grew *in situ* when their surrounding matrix was in a partly colloidal state. Yet he makes the significant observation that:

one of these beds consists almost wholly of a gently cross-bedded mixture of small oolite grains and fine grains of quartz and feldspar. . . . (Bradley, 1929, p. 221.)

This association of oölites and current-bedding is a common feature of rocks of all ages.

Eardley's work on the oölites of the Great Salt Lake, Utah, disposed of Rothpletz's theory that they grow in slimy masses of algae. He also disproved Mathew's suggestion of subaerial oölitic growth, and showed that the latter's supposed soot bands were white by reflected light and were merely layers of ultra-fine calcareous matrix (Eardley, 1938, pp. 1384–85). He found that the oölites are restricted to areas of strongly agitated waters near the shore line, where they are subjected to an oscillatory motion on the ripple-marked lake bottom. Eardley's theory for their formation is based on the observation that particles of calcium carbonate less than 2 microns in size are more soluble than larger grains. He suggested that the carbonate is first precipitated in the surface layers of the water as minute crystals. These sink to the bottom, where they redissolve: at the same moment an equivalent number of molecules come out of solution: and build on the larger grains, which thus grow oölitically as they are rolled backward and forward (*op. cit.*, pp. 1372–73).

New evidence.—In the case of the Bahaman deposits, it is clear that the differences between the recent superficial oöliths and the Pleistocene true oöliths are more a matter of degree than of kind, so that an origin established for one is applicable to the other.

Oölitically coated sands are forming only where the sediment is subjected to strong marine currents (normally tidal). They do not occur in the extreme marginal areas, as oölitic accretion can not commence until the fresh oceanic waters sweeping on to the shallow Banks have been sufficiently warmed and stirred up to become appreciably supersaturated with calcium carbonate. Superficial oöliths are thus typically developed on the western beaches of the Ragged Island Cays, where the shallow warm water washes back and forth over the beach sands, turning them into beautifully smooth pearly grains (Pl. 1.6), and also in some of the sand banks and sand-tongues formed by the tidal currents surging onto the Banks through the gaps in the marginal line of cays. Further details of these environmental conditions are given later.

The examination of any of these superficial oölites reveals the significant fact that the oölitic nature is best developed in those grades which are most abundant: it tails off very noticeably beyond either knee of the cumulative curve into the extreme coarse or fine grades of the sediment. The rounding of the grains varies in very similar fashion, and, as both this and the sorting are controlled by the method of transportation, it is evident that oölitic accretion depends fundamentally on the movement of the sand grains under the impetus of marine currents. The rounding of the grains can be due to both corrasion and accretion, and both processes may continue at the same time in these current swept environments.

As already emphasized, there is no evidence to suggest that algae, whose filaments are commonly found in oöliths (particularly in beach sands), have anything to do with their formation. Rather do they seem to play a part in breaking up the oölitic structure by boring into it.

These findings therefore support what may be called the "classical theory" of the origin of oölites, which has been well expressed by Hatch, Rastall, and Black (1938, p. 177):

In Florida and the Bahamas oolites are growing where water saturated with calcium carbonate is drifted backwards and forwards over a shallow sea floor in the track of tidal currents . . . which have sufficient transporting power to winnow away all traces of finer sediment. The grains which remain are kept in motion, and, acting as nulcei, become encased in concentric layers of precipitate.

PRIMARY NATURE OF NON-SKELETAL SAND GRAINS

The surface texture of the sand grains is sensitive to changes in bottom conditions and in those factors which control the precipitation of calcium carbonate from the sea water. The relative proportions of grains and lumps in a sample of sediment are largely controlled by its mechanical composition. In a fine sand, grains predominate: in a coarse sand, lumps are the more abundant. The tendency toward grain and lump formation is opposed by the disintegrating effects of marine currents, and both the mechanical composition and the superficial texture of the sediment are an expression of the balance between the two opposed factors, accretion and corrasion. The latter becomes dominant in the coarser grades, preventing the formation of unduly large lumps. The insignificance of the silt and mud grades has normally been attributed to the winnowing action of currents. While this is undoubtedly true, there is some evidence to show that it is helped by accretionary processes, which use up such fine material in the initial stages of the formation of calcareous sand grains.

The lumps were previously interpreted by Sorby, Vaughan, Thorp, and others as detrital fragments of the limestone that forms the cays: a similar derived origin was proposed for the rounded grains. Objections to this view have been raised by Newell (1951, p. 13) based on his observation that marine organisms destroy the oölitic structure of the surface layers of cay-rock. The full evidence against the derived hypothesis has been given. It may be summarized as follows.

1. All stages may be seen in the formation of grains from aggregated silt particles.

2. The irregular shape and friable nature of some of the grapestone lumps make it clear that they are locally formed. The growth of the lumps and the associated changes in the texture of their cement in different environments can be seen.

3. Cay-rock is composed of a calcareous sand cemented by clear, relatively

coarse, granular calcite, and fragments of cay-rock showing these features, though rare, do occur in the sediment. Grapestone and other primary lumps, on the other hand, are cemented by cryptocrystalline aragonite.

4. The distribution of the various types of lumps, and, in particular, the virtual absence of grapestone from the marginal platform, is not consistent with a derived origin.

5. Encrusted lumps are primary because their internal structure is related to their external impervious layer. The same applies in less degree to other lump habits.

6. Thin sections may reveal a lump enclosed within a larger lump.

7. Cay-rock contains a high percentage of oöliths, whose concentric structure is not confined to a few superficial layers as in most recent oölitic sands. The oölitic laminae may be replaced by calcite, and the same feature occurs patchily in other grains in cay-rock: it is not seen in primary sand grains, oölitic or otherwise. Hence, the rarity of signs of recrystallization of aragonite into calcite in the sand grains of the bottom sediments is contrary to a derived origin.

The conclusion is inevitable that such secondary material is swamped by the rapid accumulation of primary skeletal and non-skeletal calcareous sand grains.

[*Editors' Note:* Material has been omitted at this point.]

TABLE V. LITHOLOGICAL ANALYSES SHOWING PERCENTAGE COMPOSITION OF EACH GRADE OF BOTTOM SAMPLES B-22 AND NOS. 35 AND 99, TYPICAL, RESPECTIVELY, OF MARGINAL SHELF AND EASTERN AND WESTERN HALVES OF LINE I

(Percentages rounded off to nearest whole number)

Sample No. — Grade content	Marginal Shelf — B-22						Line I (East) — 35						Line I (West) — 99					
	Gr 23.6	CS 41.7	MS 33.1	FS 1.0	Slt 0.5	Total	Gr 1.9	CS 34.5	MS 51.4	FS 9.4	Slt 2.6	Total	Gr 4.2	CS 42.3	MS 45.7	FS 6.9	Slt 0.8	Total
Lumps — Cay-rock		tr	tr			tr	tr	tr				tr						—
Grapestone			tr			tr	43	91	35			50	10+	15+	60+			34+
Botryoidal						—	tr					tr	30+	4				3+
Algally encrusted	3	2±				2±												
Total lumps	5	13	11			10#	43	91	35			50	64	94	72			75#
Grains			5	30	C	2		1	51	58	A	33		I	8	77	A	9
Friable aggregates						—			3	25		4			I	I		½
Faecal pellets			2	tr		⅛			7	tr		4			6	I		3
Skeletal grains — Mollusca	50	17	20±	5	VC	26±	53	5	3	2	C	5	34	4	4	5	C	5
Foraminifera	3	24	20	25		18	1	2	1	12	C	3	I	tr	6	13	VC	4
Calcareous algae	15	16	20	25	C	17H			tr	2	R	⅓H	tr	I	2	2	R	2H
Madreporaria	15	25±	17±	tr		20±	1	tr	tr	tr*		tr	I					tr
Serpulae	2	1				1	2	1	tr	I*	C*	1	tr	I	I	I*		I
Other forms	10M	4	4	15	VC	6		1	tr	I*		1	tr	tr	tr		C*	tr
Total skeletal grains	95	87	82	70	VA	88	57	8	4	17	VC	9	36	6	13	21	A	I2
Estimated total skeletal content						92						14						28

+ The relative proportions of the lump habits are only approximate, as they merge into each other.
# The lumps that are not separately listed are irregularly shaped and their component grains do not protrude. In B-22 many contain filamentous algae.
′ The aggregates are less friable.
H *Halimeda* is the dominant form with some large fragments of Lithothamnion in B-22. Amphiroa spicules are commonest in fine grades of Nos. 35 and 99.
* Chiefly spicular organisms.
M Millepora and a few bryozoa fragments.

Symbols used in the estimation of the silt grade

<5%	R: rare
6–20%	C: common
21–50%	VC: very common
51–90%	A: abundant
>91%	VA: very abundant

TABLE VI. LITHOLOGICAL ANALYSIS OF SEDIMENT SAMPLES,
(Percentages rounded off

		Sediment on Banks						Beach Sands				Marginal Shelf
Sample No.	Line I		Line III	Line II		Line IV		Western		Eastern		Marginal Shelf
	35	99	150	182	194	B-52	B-58	B-7	B-8	B-9	B-19	B-22
Cay-rock	tr	—	1	—	—	—	—	4±	5±	1	8	tr
Lumps — Grapestone	50	34+	48	tr	42	23+	52	8	7	7	4	tr
Botryoidal	—	—	8°	—	—	—	—	51°	18°	—	—	—
Algally encrusted	tr	3+	—	14+	tr	1	1	—	—	—	—	2±
Total lumps	50	75#	59φ	70#	50#	28φ	55φ	59	25	7	9	10#
Skeletal grains — Grains	33	9	34	10	33	35	27±	24°	52°	28″	5″	2
Friable aggregates	4	½	—	tr	—	1	½	4	8			½
Faecal pellets	4	3	2	½	7	19	6±	2°	4°	1	tr	
Mollusca	5	5	2	10	7	7	5	3	1	11	18	26±
Foraminifera	3	4	1	6	2	7	4	1	2	9′	28′	18
Calcareous algae	½H	2H	1H	3H	1H	1H	½H	2H	1H	37±L	27L	17H
Madreporaria	—	tr	tr	tr	—	—	—	—	—	3	4	20±
Serpulae	tr	1	tr	1	tr	1	2	} ½	1	4	3	1
Other forms	1	tr	tr	½	½	tr	½		1	4	3	6
Total skeletal grains	9	12	4	20	10	16	11	6	5	64	80	88
Estimated total skeletal content	14	28	10	44	17	23	25	13	10	65	82	92

+   Relative proportions of lump habits are only approximate, as they merge into each other.
\#   Lumps that are not separately listed are irregularly shaped and their component grains do not protrude.
φ   Lumps that are not separately listed are well cemented grapestone, subsequently worn and smoothed.
!   Cement includes both aragonite and calcite.
°   Shiny oölitic surfaces.
′   Partly derived from cay-rock.

SHOWING PERCENTAGE CONTENT OF VARIOUS COMPONENTS

to nearest whole number)

	Nurse Cay Channel Line VI						Gun Point Channel Line V									Eolian Dune
	B-13	B-14	B-15	B-16	B-17	B-18	B-34	B-37	B-39	B-41	B-43	B-45	B-46	B-47	B-48	B-35
	2±	½	½	tr	2	tr	tr	—	—	—	—	—	tr	tr	2	1
	—	—	—	½	—	32	11	36	15+	54+	80	21	9	—	—	5!
	—	—	—	—	—	—	—	—	26°+	9°+	—	—	—	—	—	tr
	—	—	½	—	2	—	—	—	—	—	—	—	—	tr	tr	—
	—	2	1	1	8	34	11	36	41	64	80	22φ	11φ	tr	tr	5
	1	4	10	5	3	34	53	39	56°	33°	12	64	57	1	2	79
	—	—	tr	—	½	2	—	2	—	—	1	1	2	—	—	—
	tr	½	tr	tr	tr	tr	tr	2	tr°	tr	1	2	11	1	tr	3
	21	10	11±	9	20	9	5	6	2	2	3	4	6	11	23	5
	17	12	8	12	18	7	6	9	1	1	3	3	5	6	4	2
	49H	63H	56H	66H	42H	11H	17H	2H	trH	trH	1H	3H / 2H	5 / 4H	67D	45L	2H
	7	3±	10±	3	5	½	—	—	—	—	—	1	4	12	20	tr
	½	1	½	tr	1	tr	} 8	3	tr	tr	tr	1*	1*	3	4	tr
	2	3	3	3	2	2										2
	97	92	89	93	88	30	36	20	3	3	7	11	20	99	96	11
	97	93	89	93	89	40	40	27	4	6	16	13	23	98	96	12

' Includes pink forms *H. rubra* and *R. rosea.*
* Chiefly spicular organisms.
H *Halimeda* is dominant form. Amphiroa spicules are common in fine-sand and silt grades.
L Corallinaceae (Lithothamnion) dominant, with lesser Dascycladaceae and Codiaceae.
D Dascycladaceae and Codiaceae (*Halimeda*) dominant, with lesser Corallinaceae (Lithothamnion and Amphiroa).

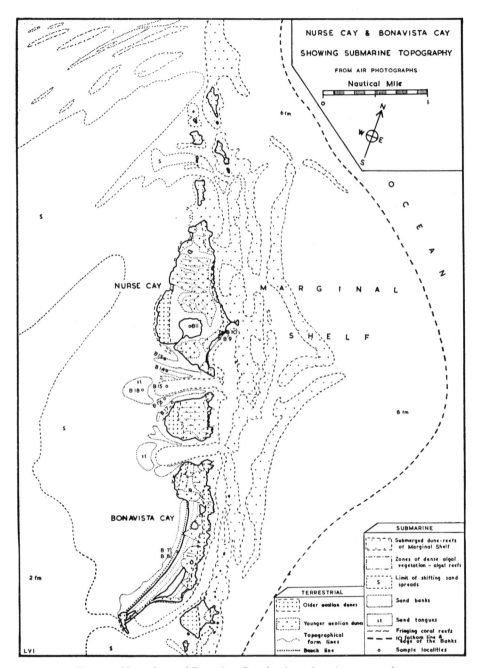

FIG. 10.—Nurse Cay and Bonavista Cay showing submarine topography.

6.6 Fossil dune resting on 5-foot raised marine-cut platform. Dune has been consolidated into cay-rock and its convex eolian bedding has been picked out by weathering and erosion. Light patches in water are due to luxuriant submerged colony of corals. Looking west to southernmost tip of Hog Cay, opposite Gun Point.

7.1 Air view of part of Exuma arc of cays. Flying height about 1,000 feet. Line of cays measures approximately 6 miles from side to side of photograph.

With exception of black cays, whole area is under water. Darker zones mark main inter-cay channels, where sea bottom is stabilized by marine vegetation. They are 10–25 feet deep. Marginal shelf beyond arc of cays is somewhat deeper (25–60 feet). Faint change in color in distance beyond cays marks outer edge of Banks.

7.2 Circular gorgonian-algal colonies ("micro-atolls") on Great Bahama Bank, north of southeastern lobe of "Tongue of the Ocean" (locality "R" in Fig. 1). Flying height about 1,000 feet. Rings measure 30–100 yards across. Whole area is under water.

[*Editors' Note:* Figures 1 through 5 of Plate 6 and Figures 3 through 6 of Plate 7 have been omitted.]

Regional Pattern of Sedimentary Changes

The sediment samples that have been described cover but a small fraction of the 40,000 square miles of the Banks, yet with the help of additional material from New Providence Island, from Abaco and Northern Eleuthera collected by C. S. Lee, and from the coasts of Andros kindly lent by M. Black, it is possible to establish a general sequence of facies changes in the surface sediments from the margins of the Banks toward their center.

Along the ocean edge the sea bottom is covered with patchy deposits of sand made up almost entirely of skeletal fragments, the most important of which are calcareous algae. This so-called "marginal shelf" is commonly bounded on its inner side by a broken line of cays and submerged reefs, within which the sediment is dominantly non-skeletal. The sharp drop in skeletal content may be less abrupt where there are no cays, but in such areas air photographs reveal a well defined limit to the spreads of shifting sand, a few miles from the ocean edge. Probably this marks the beginning of the non-skeletal deposits, though skeletal accumulations may be incorporated in their oceanward extremities.

Within the line of cays, a broad belt of the sea bottom is covered with calcareous sand containing abundant grapestone. In the Ragged Island area the belt is 8–12 miles wide, but these limits may not apply elsewhere. Several changes in the nature of the deposits occur within this zone. In the outer shallower regions, the sand grains show traces of the chalky white, friable, superficial cement that has been found to indicate active accretion. Loosely bound aggregates of silt particles occur in the same area, and all stages in their conversion to normal grains of matrix can be followed. Superficial oölites and botryoidal lumps occur in the track of the more powerful tidal currents sweeping through the channels between the cays. Farther from the ocean the chalky white cement is normally absent, and the grapestone has a uniform matt texture.

Finally, the environment producing the grapestone habit is lost, and is followed by one in which surface accretion is intimately related to the action of minute algae. Most of these appear to be of the sediment-boring kind, but by their activity they play a part in consolidating the surface layers of the lumps. In these interior parts of the Banks, some undoubted algal encrustation occurs in which the growth of the non-skeletal lumps is essentially due to algal action, both by sediment binding, and by biochemical extraction of aragonite from sea water.

Observations from the air show that the tidal streams become less important over the central parts of the Banks, until they finally die out and the submarine pattern of sand banks and marine vegetation loses its linear streaking. In this interior sector, the plant and animal life is commonly concentrated into separate, approximately circular clusters, randomly distributed over the sea bottom, each having a roughly concentric arrangement. Such "microatolls" are particularly abundant north of the lobe of the Tongue of the Ocean in the vicinity of 24° 20′ N. and 76° 55′ W., where the photograph of Plate 7.2 was taken (point "R" in Fig. 1).

The samples of Line II from the Ragged Island Concession reach only 20 miles into the Banks from their eastern margin. Farther west, the navigation charts record shelly sands, and it is probable that the general nature of the deposits remains similar to those described, apart from a possible increase in their skeletal content. Black has described a corresponding change in the deposits east of the Florida Straits, where traced toward Andros. There the sequence goes a stage further, and passes from a shelly sand to a calcium carbonate ooze (Black, 1930, p. 110). This is the zone of drewite mud, composed of minute aragonite needles which accumulate in the protected shelf lagoon. It is thought by Bavendamm and Field to be formed by bacterial precipitation in the fetid mangrove swamps on the western shores of Andros (Field, 1931, p. 776): Black, however, attributes the precipitation to the concentration of the sea water by evaporation (1933a).

Calcareous marls also occur in the swamps west of Abaco (*cf.* Cloud and Barnes, 1948, p. 37): mangroves and other vegetation are again abundant, and the bottom deposits of the anastomosing lagoons reek of hydrogen sulphide due to the action of sulphate-reducing bacteria. The sediment contains abundant organic matter. It is a highly calcareous gray-brown or buff-colored marly silt, in which aragonite needles occur akin to those of drewite. The color is due to ferric hydroxides which are readily absorbed by the finely divided aragonite. Many of the abundant foraminiferal tests are selectively stained so as to emphasize their cellular structure (*cf.* Dangeard, 1936, p. 238). The rich red earths that occur locally on the islands have been formed by further concentration of these insoluble ferruginous residues.

Figure 11, showing the locations of Lee's line of samples across the eastern

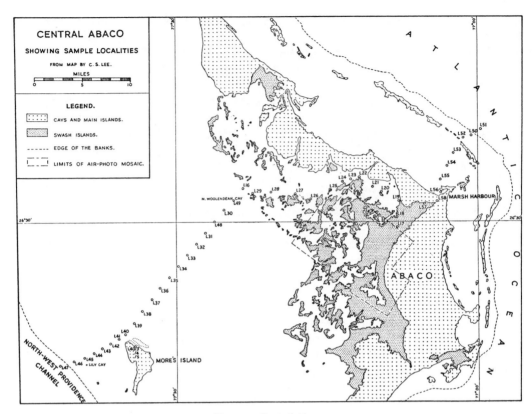

FIG. 11.—Central Abaco.

limb of the Little Bahama Bank, indicates that this huge swash area of brackish swamps is in direct connection with the open sea on the west and with streams from Abaco Island on the east. An analysis of the stream water at station L-17 on the edge of the swamp (Table VII) shows it to be a very diluted sea water with some extra dissolved calcium carbonate, iron, and silica, picked up during its percolation through the marsh sediments and limestones of Abaco Island. Its calcium carbonate content is less in total quantity than that in the same volume of any of the more saline waters, and therefore, as shown by Eardley (1938, p. 1327), there is no chance of its being deposited by the mixing of the fresh water and sea water in the brackish swamp. A part of the inner swamp zone is illustrated in Plate 8 showing the curious kidney-shaped lagoons of shallow water between the scalloped swash islands lying just above sea-level. It is thought that each pool is centered on a group of submarine fresh-water springs emerging from solution holes in the submerged cay-rock basement (Lee, 1951, p. 653 and Fig. 8). A similar pattern occurs on a smaller scale in the shallow Lake Killarney on New Providence Island. The depth of the Abaco swamp rarely exceeds 3 feet, and, at the time of sampling, the most rapid change in salinity occurred between stations L-22 and L-25.

In the deeper water between the swash area and More's Island (localities L-16 to L-40), the bottom sediment is silty calcareous sand in some places (*e.g.*, L-32) lightly iron-stained. The mud content is nowhere more than 1 or 2% (Table I), but it contains traces of tiny aragonite needles similar to those that compose drewite. More's Island marks a submerged line of reefs. West of it (localities L-42 to L-46) the skeletal content rises to about three-quarters of the total sediment, thus conforming with findings elsewhere on the margins of the Banks. In the same area the silt fraction becomes negligible from the winnowing action of the more powerful currents. Histograms of a selection of Lee's Abaco samples are included in Figure 4.

The calcareous marly silts from Abaco are an essentially different type of sediment from those from the more open parts of the Banks described in this paper. The mud content of the samples of the swash area rarely exceeds 5%. Thus, although drewite needles occur in them, they are not strictly comparable with the pure white marls on the west coast of Andros, described by Field (1928) and others.

[*Editors' Note:* Material has been omitted at this point.]

TABLE VII. CHEMICAL ANALYSIS OF DISSOLVED SALTS IN WATER SAMPLES FROM ABACO AND NORTHERN ELEUTHERA

Water Sample No.*	L-17W	L-24W₁	L-24W₂	L-28W₁	L-35W₁	L-35W₂	L-51W₁	L-51W₂	L-53W₁	L-53W₂	L-62W₂	L-65W	Challenger Mean†
Cl	51.05	54.86	}55.62	55.43	{54.94	}55.57	{55.51	}55.93	55.60	55.35	55.43	{54.83	55.29
Br	0.17	0.19			0.19		0.20					0.22	0.19
SO_4	6.38	7.99	7.85	8.03	8.14	7.89	7.83	7.49	8.20	8.03	7.94	8.24	7.69
CO_3	4.63	0.25	0.26	0.17	0.17	0.17	0.17	0.20	0.18	0.17	0.18	0.25	0.21
Na	28.56	30.42	}30.94	31.42	{30.40	}31.36	31.25	31.23	30.99	31.44	31.61	{29.99	30.59
K	1.55	1.29			1.11							1.13	1.11
Ca	3.69	1.20	1.27	1.13	1.18	1.17	1.18	1.22	1.22	1.22	1.13	1.26	1.20
Mg	3.22	3.74	4.05	3.82	3.86	3.83	3.86	3.91	3.81	3.79	3.70	4.05	3.72
SiO_2	0.74	0.06	0.01	tr	0.01	0.01	0.01	0.02	tr	tr	0.01	0.03	
Fe	0.01	tr	tr	tr	tr	tr	tr	—	tr	tr	tr	tr	
P_2O_5	tr	tr	tr	tr	tr	tr	tr	—	tr	—	tr	tr	
Total	100.00	100.00	100.00	100.00	100.00	100.00	100.00	100.00	100.00	100.00	100.00	100.00	100.00
Total solids (parts per thou.)	1.49	21.66	20.50	35.36	35.99	35.99	35.99	35.76	35.97	36.13	34.64	26.94	33.01 to 37.37

Analyst: R. F. Rackham, of Counties Public Health Laboratories, London.
* Water samples with suffix W1 were collected on sea bottom.
Water samples with suffix W2 were collected at surface.
For sample localities see Figures 11 and 13.
† Mean of 77 analyses of ocean water from many localities, collected by "Challenger" expedition. W. Dittmar, analyst. *Challenger Rept., Physics and Chemistry*, Vol. 1 (1884), p. 203.

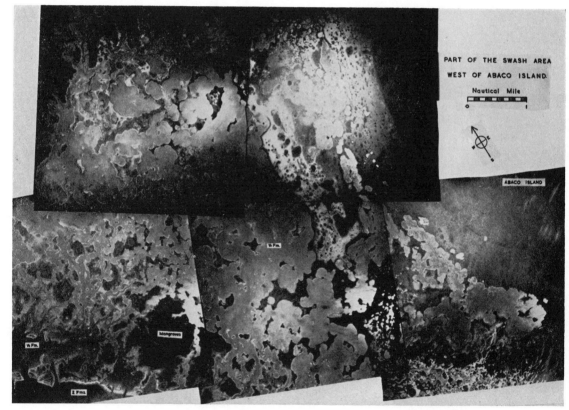

PLATE 8

Part of swash area west of Abaco Island (see Fig. 11 for position of mosaic).
Each lobe of shallow kidney-shaped pools of swamp in southeast part of mosaic is thought to be centered on group of fresh-water springs coming from forested mainland of Abaco. Farther west, swamp becomes brackish between scalloped islands of swash sediment. These rarely rise more than a few feet above sea-level, and are stabilized by grass, mangroves, and other vegetation.

CALCAREOUS SAND, AND BUILDING OF BAHAMA BANKS

The whole conception of non-oölitic non-skeletal sand grains of calcium carbonate being authigenic, as described in this paper, throws a new light on the origin of other calcarenites, both in the Bahamas and elsewhere, which, in the past, have been regarded as detrital clastic deposits derived from the mechanical breakdown of pre-existing limestones.

The recent Bahaman unconsolidated sand is thin, and the Pleistocene limestone comes to the surface of the sea bed in some areas. It was itself formed by the subaerial cementation of similar calcareous sands, and beneath its compact surface layer, it is rubbly and commonly cavernous. The meager information available suggests that it is only a few hundred feet thick (Field and Hess, 1933). The exploratory well drilled for oil at Stafford Creek, North Andros Island, passed at approximately 400 feet into sparkling white saccharoidal and dense dolomitic limestone, some of which shows relics of an initial calcarenitic composition. Rhombs of dolomite line cavities left by solution of fossils, and the matrix has been recrystallized on a finer scale. A selection of samples, made available through the kindness of E. J. White of the Anglo Iranian Oil Company, showed that the dolomites predominate down to 2,200 feet. From there down to 2,640 feet, the drill cuttings again indicate an almost unaltered rubbly calcarenite, very similar to the Pleistocene cay-rock at the surface. Partial dolomitization has occurred in the beds below, but there is again every indication that the original sediment was a calcareous sand. The samples examined were all from the top 4,000 feet, but repeated reference to "oölitoid bodies" in a log of the cuttings suggests that similar calcarenites extend much deeper.

It is thus likely that the processes leading to the formation and accumulation

87

of the recent non-skeletal calcareous sands described in this paper have been a major factor in the formation of the Bahama Banks. The Andros well passed through a continuous sequence of more than 14,500 feet of Tertiary and Cretaceous limestones and dolomites. This great thickness of calcareous sediments shows that the Banks have been slowly subsiding. The non-skeletal, well sorted, calcareous sands are an essentially shallow-water deposit; hence, their formation and accumulation has kept pace with the subsidence, and has helped to build up the huge submerged plateau as seen to-day.

REFERENCES

AGASSIZ, A., 1894, "A Reconnaissance of the Bahamas and of the Elevated Reefs of Cuba in the Steam Yacht 'Wild Duck,' January to April, 1893," *Bull. Mus. Comp. Zool.*, Vol. xxvi, No. 1, pp. 1–203.

——, 1896, "The Elevated Reef of Florida," *ibid.*, Vol. xxviii, No. 2, pp. 1–62.

BAVENDAMM, W., 1932, "Die mikrobiologische Kalkfällung in der tropischen See," *Archiv. Mikrobiologie*, Vol. iii, pp. 205–76.

BLACK, M., 1930, "Great Bahama Bank—A Modern Shelf Lagoon" (abstract), *Bull. Geol. Soc. America*, Vol. xli, pp. 109–10.

——, 1933a, "The Precipitation of Calcium Carbonate on the Great Bahama Bank," *Geol. Mag.*, Vol. lxx, pp. 455–66.

——, 1933b, "The Algal Sedimentation of Andros Island, Bahamas," *Philos. Trans.*, Ser. B, Vol. ccxxii, pp. 165–92.

BØGGILD, O. B., 1930, "The Shell Structure of the Molluscs," *D. Kgl. Danske. Vid. Selsk. Skrifter*, Naturv. og Mathem. Afd., 9 Raekke, Vol. ii, pp. 233–326.

BRADLEY, W. H., 1929, "Algae Reefs and Oolites of the Green River Formation," *U. S. Geol. Survey, Prof. Paper 154-G*, pp. 203–23.

BRAMLETTE, M. N., 1926, "Some Marine Bottom Samples from Pago Pago Harbor, Samoa," *Pap. Tortugas Lab.*, Vol. xxiii, Pub. 344, pp. 1–36.

BUCHER, W. H., 1918, "On Oolites and Spherulites," *Jour. Geol.*, Vol. xxvi, pp. 593–609.

CARY, L. R., 1918, "The Gorgonaceae as a Factor in the Formation of Coral Reefs," *Pap. Tortugas Lab.*, Vol. ix, Pub. 213, pp. 341–62.

CHAPMAN, F., AND MAWSON, D., 1906, "On the Importance of Halimeda as a Reef-Forming Organism," *Quar. Jour. Geol. Soc. London*, Vol. lxii, pp. 702–10.

CLARKE, F. W., 1916, "The Data of Geochemistry," 3d ed., *U. S. Geol. Survey Bull. 616*.

——, AND WHEELER, W. C., 1922, "The Inorganic Constituents of Marine Invertebrates," *ibid.*, *Prof. Paper 124*, 2d ed.

CLOUD, P. E., AND BARNES, V. E., 1948, "Paleoecology of the Early Ordovician Sea in Central Texas," *Natl. Res. Council, Div. Geol. Geog., Rept. of Committee on Marine Ecology as Related to Paleontology* (No. 8), pp. 29–83.

CUMINGS, E. R., 1932, "Reefs or Bioherms?" *Bull. Geol. Soc. America*, Vol. 43, pp. 331–52.

CUSHMAN, J. A., 1922, "Shallow-Water Foraminifera of the Tortugas Region," *Pap. Tortugas Lab.*, Vol. xvii, Pub. 311, pp. 1–85.

DALL, W. H., 1892, "Neocene of North America," *U. S. Geol. Survey Bull. 84*, p. 101.

DANGEARD, L., 1936, "Étude des calcaires par coloration et décalcification. Application à l'étude des calcaires oolithiques," *Bull. Soc. Geol. France*, Ser. 5, Vol. vi, pp. 237–45.

DOLE, R. B., 1914, "Some Chemical Characteristics of Sea-Water at Tortugas and around Biscayne Bay, Florida," *Pap. Tortugas Lab.*, Vol. v, Pub. 182, pp. 69–78.

DOLE, R. B., AND CHAMBERS, A. A., 1918, "Salinity of Ocean-Water at Fowey Rocks, Florida," *ibid.*, Vol. ix, Pub. 213, pp. 299–315.

DREW, G. H., 1914, "On the Precipitation of Calcium Carbonate in the Sea by Marine Bacteria, and on the Action of Dentrifying Bacteria in Tropical and Temperate Seas," *ibid.*, Vol. v, Pub. 182, pp. 7–45.

EARDLEY, A. J., 1938, "Sediments of Great Salt Lake Utah," *Bull. Amer. Assoc. Petrol. Geol.*, Vol. 22, pp. 1305–1411.

——, 1951, *Structural Geology of North America*. Harper and Bros.

EDGEWORTH DAVID, T. W., AND SWEET, G., 1904, "The Geology of Funafuti. The Atoll of Funafuti," Sect. V, *Roy. Soc. London*.

EMERY, K. O., TRACEY, J. I., LADD, H. S., AND HOFFMEISTER, J. E., 1946, "Operation Cross Roads, Bikini Atoll," *Natl. Res. Council, Div. Geol. Geog., Rept. of Committee on Marine Ecology as Related to Paleontology* (No. 6).

EVANS, O. F., 1944, "Some Structural Differences between Wind-Laid and Water-Laid Deposits on the West Shore of Lake Michigan," *Jour. Sed. Petrology*, Vol. 14, pp. 94–96.

EWING, M., WOOLLARD, C. P., VINE, A. C., AND WORZEL, J. L., 1946, "Recent Results in Submarine Geophysics," *Bull. Geol. Soc. America*, Vol. 57, pp. 909–34.

FAIRBRIDGE, R. W., AND TEICHERT, C., 1948, "The Rampart System at Low Isles, 1928–1945," *Repts. of the Great Barrier Reef Committee*, Vol. vi, Pt. 1.

FIELD, R. M., 1928, "The Great Bahama Bank. Studies in Marine Carbonate Sediments,"*Amer. Jour. Sci.* (5), Vol. 16, pp. 239–46.

———, 1931, "Geology of the Bahamas," *Bull. Geol. Soc. America*, Vol. 42, pp. 759–84.

———, AND HESS, H. H., 1933, "A Bore Hole in the Bahamas," *Amer. Geophys. Union Trans. Ann. Mtg.*, pp. 234–45.

FOWLER, J. W., AND SHIRLEY, J., 1947, "A Method of Making Thin Sections from Friable Materials and Its Use in the Examination of Shales from the Coal Measures," *Geol. Mag.*, Vol. lxxxiv, pp. 354–59.

GEE, H., 1932, "Inorganic Marine Limestone," *Jour. Sed. Petrology*, Vol. 2, pp. 162–66.

GINSBURG, R. N., 1953, "Beachrock in South Florida," *ibid.*, Vol. 23, pp. 85–92.

GLAESSNER, M. F., 1945, *Principles of Micropaleontology*. Melbourne.

GOLDMAN, M.I., 1918, "Composition of Two Murray Island Bottom Samples According to Source of Material," *Pap. Tortugas Lab.*, Vol. ix, Pub. 213, pp. 249–62.

———, 1926, "Proportions of Detrital Organic Calcareous Constituents and Their Chemical Alteration in a Reef Sand from the Bahamas," *ibid.*, Vol. xxiii, Pub. 344, pp. 37–66.

GRISWOLD, L. S., 1896, "Notes on the Geology of Southern Florida," *Bull. Mus. Comp. Zool.*, Vol. xxviii, No. 2, pp. 52–59.

HARVEY, H. W., 1945, *Recent Advances in the Chemistry and Biology of Sea Water*. Cambs. Univ. Press.

HATCH, F. H., AND RASTALL, R. H., 1923, *The Petrology of Sedimentary Rocks*, 2d ed. London.

———, AND ———, 1938, *The Petrology of the Sedimentary Rocks*, 3d ed., revised by BLACK, M. London.

HINDE, G. J., 1904, "Report on the Materials from the Borings at the Funafuti Atoll. The Atoll of Funafuti," Sec. xi, *Roy. Soc. London*.

HOWE, M. A., 1918, "Calcareous Algae from Murray Island, Australia, and Cocos-Keeling Islands," *Pap. Tortugas Lab.*, Vol. ix, Pub. 213 pp. 291–96.

ILLING, M. A., 1950, "The Mechanical Distribution of Recent Foraminifera in Bahama Banks Sediments," *Ann. Mag. Nat. Hist.*, 12th Ser., iii, pp. 757–61.

———, 1952, "Distribution of Certain Foraminifera within the Littoral Zone on the Bahama Banks," *ibid.*, v, pp. 275–85.

IRVINE, R., AND WOODHEAD, G. S., 1889, "On the Secretion of Lime by Animals," *Proc. Roy. Soc. Edinburgh*, Vol. xv, pp. 308–20.

———, AND ———, 1890, *ibid.*, Vol. xvi, pp. 324–54.

JOHNSON, J. H., 1943, "Geologic Importance of Calcareous Algae with Annotated Bibliography," *Quar. Colorado Sch. Mines*, Vol. 38, pp. 1–102.

———, 1946, "Lime-Secreting Algae from the Pennsylvanian and Permian of Kansas," *Bull. Geol. Soc. America*, Vol. 57, pp. 1087–1120.

JOHNSTON, J., 1915, "The Solubility Constant of Calcium and Magnesium Carbonate," *Jour. Amer. Chem. Soc.*, Vol. xxxvii.

———, AND WILLIAMSON, E. D., 1916, "The Role of Inorganic Agencies in the Deposition of Calcium Carbonate," *Jour. Geol.*, Vol. xxiv, pp. 729–50.

JOHNSTON, MERWIN, AND WILLIAMSON, 1916, "The Several Forms of Calcium Carbonate," *Amer. Jour. Sci.*, Vol. xli, pp. 473–512.

JUDD, J. W., 1904, "General Report on the Materials Sent from Funafuti. The Atoll of Funafuti," Sec. x, *Roy. Soc. London*.

KINDLE, E. M., 1923, "Nomenclature and Genetic Relations of Certain Calcareous Rocks," *Pan-Amer. Geologist*, Vol. xxxix, pp. 365–72.

KRUMBEIN, W. C., AND PETTIJOHN, F. J., 1938, *Manual of Sedimentary Petrography*. New York.

LACROIX, A., 1898, "Sur la ctypéite, etc.," *Compt. Rend.*, Vol. cxxvi, p. 601.

LADD, H. S., TRACEY, J. I., LILL, G., WELLS, J. W., AND COLE, W. S., 1947, "Drilling on Bikini Atoll, Marshall Islands" (abstract), *Bull. Geol. Soc. America*, Vol. lviii, pp. 1201–02.

LARSEN, E. S., 1921, "The Microscopic Determination of the Non-Opaque Minerals," *U. S. Geol. Survey Bull. 679.*

LEE, C. S., 1951, "Geophysical Surveys on the Bahama Banks," *Jour. Inst. Petrol.*, Vol. 37, No. 334, pp. 633–57.

LEITMEYER, H., AND FEIGL, F., 1934, "Eine einfache Reaktion zur Unterschiedung von Calcit und Aragonit," *Zeit. Krist. Min. und Pet. Mitt.*, Abt. 8, B. 45, pp. 447–56.

LINCK, G., 1903, "Die Bildung der Oolithe und Rogensteine," *Neues Jahrb. Min.*, B.B. xvi, pp. 495–513.

Lipman, C. B., 1924, "A Critical and Experimental Study of Drew's Bacterial Hypothesis on CaCO₃ Precipitation in the Sea," *Pap. Tortugas Lab.*, Vol. xix, Pub. 340, pp. 179–91.
————, 1929, "Further Studies on Marine Bacteria with Special Reference to the Drew Hypothesis on CaCO₃ Precipitation in the Sea," *ibid.*, Vol. xxvi, Pub. 391, pp. 231–48.
Mann, A., 1936, "Diatoms in Bottom Deposits from the Bahamas and the Florida Keys," *ibid.*, Vol. xxix, Pub. 452, pp. 121–28.
Matson, G. C., 1910 (See Vaughan, T. W., 1910).
Matthai, G., 1928, *Catalogue of the Madreporarian Corals in the British Museum (Natural History)*, Vol. vii, "Recent Maeandroid Astraeidae."
Mayer, A. G., 1918, "Ecology of the Murray Island Coral Reef," *Pap. Tortugas Lab.*, Vol. ix, Pub. 213, pp. 1–48.
Mayor, A. G., 1924a, "Structure and Ecology of Samoan Reefs," *ibid.*, Vol. xix, Pub. 340, pp. 1–25.
————, 1924b, "Causes Which Produce Stable Conditions in the Depths of the Floors of Pacific Fringing Reef-Flats," *ibid.*, Vol. xix, Pub. 340, pp. 27–36.
————, 1924c, "Inability of Stream Water to Dissolve Submarine Limestones," *ibid.*, Vol. xix, Pub. 340, pp. 37–49.
McClendon, J. F., 1918, "On Changes in the Sea and Their Relation to Organisms," *ibid.*, Vol. xii, Pub. 252, pp. 213–59.
Moberg, E. G., Greenberg, D. M., Revelle, R. R., and Allen, E., 1934, "The Buffer Mechanism of Sea-Water," *Bull. Scripps Inst. Oceanog.*, Tech. Ser., iii, pp. 231–320.
Moore, H. B., 1933, "Faecal Pellets from Marine Deposits," *Discovery Reports*, vii, pp. 17–26.
Murray, J., and Irvine, R., 1891, "On Coral Reefs and Other Carbonate of Lime Formations in Modern Seas," *Proc. Roy. Soc. Edinburgh*, Vol. xvii, pp. 79–109.
Murray, J., and Renard, A. F., 1891, *The Voyage of H.M.S. Challenger. Deep-Sea Deposits*. London.
Nelson, R. J., 1853, "On the Geology of the Bahamas, and on Coral-Formations Generally," *Quar. Jour. Geol. Soc. London*, Vol. ix, pp. 200–15.
Nevin, C., 1946, "Competency of Moving Water to Transport Debris," *Bull. Geol. Soc. America*, Vol. 57, pp. 651–74.
Newell, N. D., 1951, "Organic Reefs and Submarine Dunes of Oölite Sand around Tongue of the Ocean, Bahamas," *Bull. Geol. Soc. America*, Vol. lxii, p. 1466.
————, Rigby, J. K., Whiteman, A. J., and Bradley, J. S., 1951, "Shoal Water Geology and Environments, Eastern Andros Island, Bahamas," *Bull. Amer. Mus. Nat. Hist.*, Vol. 97, Art. 1.
Pettijohn, F. J., 1949, *Sedimentary Rocks*. New York.
Rastall, R. H., 1933, *The Geological Magazine*, p. 484.
Revelle, R., 1934, "Physico-Chemical Factors Affecting the Solubility of Calcium Carbonate in Sea Water," *Jour. Sed. Petrology*, Vol. 4, pp. 103–10.
————, and Fleming, R. H., 1934, "The Solubility Product Constant of Calcium Carbonate in Sea-Water," *Proc. 5th Pacific Sci. Cong.*, iii, pp. 2089–92.
Rich, J. L., 1948, "Submarine Sedimentary Features on Bahama Banks and Their Bearing on Distribution Patterns of Lenticular Oil Sands," *Bull. Amer. Assoc. Petrol. Geol.*, Vol. 32, pp. 767–79.
Richards, H. C., and Hill, D., 1942, "Great Barrier Reef Bores, 1926 and 1937," *Reports of Great Barrier Reef Committee*, v., pp. 1–122.
Rothpletz, A., 1892, "On the Formation of Oölite," *Amer. Geol.*, Vol. x, pp. 279–82, translated by Cragin, F. W.
Sanford, S., 1910, "The Topography and Geology of Southern Florida," *2d Ann. Rept. Florida Geol. Survey*, pp. 177–231.
Sayles, R. W., 1931, "Bermuda during the Ice Age," *Proc. Amer. Acad. Arts and Sci.*, Vol. 66, pp. 381–468.
Schalk, M., 1946, "Submarine Topography off Eleuthera Island, Bahamas" (abstract), *Bull. Geol. Soc. America*, Vol. lvii, p. 1228.
Schuchert, C., 1935, *Historical Geology of the Antillean-Caribbean Region*. New York.
Sellards, E. H., 1908, "Administrative Report," *Florida Geol. Survey Ann. Rept. 1*, pp. 7–16.
Shepard, F. P., 1948, *Submarine Geology*. New York.
Smith, C. L., 1940, "The Great Bahama Bank," *Jour. Marine Research*, Vol. 3, pp. 147–89.
Smith, N. R., 1926, "Report on a Bacteriological Examination of 'Chalky Mud' and Sea-Water from the Bahama Banks," *Pap. Tortugas Lab.*, Vol. xxiii, Pub. 344, pp. 69–72.
Sorby, H. C., 1879, "The Structure and Origin of Limestones," *Proc. Geol. Soc. London*, Vol. xxv, pp. 56–95.
Stark, J. T., and Dapples, E. C., 1941, "Near Shore Coral Lagoon Sediments from Raiatea, Society Islands," *Jour. Sed. Petrology*, Vol. 11, pp. 21–27.
Stetson, H. C., and Upson, J. E., 1937, "Bottom Deposits of the Ross Sea," *ibid.*, Vol. 7, pp. 55–66.
Suess, E., 1904–1909, *The Face of the Earth*. English translation by Sollas.

TAYLOR, W. M., 1928, "The Marine Algae of Florida," *Pap. Tortugas Lab.*, Vol. xxv, Pub. 379, pp. 1–219.

THORP, E. M., 1936, "Calcareous Shallow-Water Marine Deposits of Florida and the Bahamas," *ibid.*, Vol. xxix, Pub. 452, pp. 37–120.

TRACEY, J. I., LADD, H. S., AND HOFFMEISTER, J. E., 1946, "Reefs and Islands of Bikini, Marshall Islands" (abstract), *Bull. Geol. Soc. America*, Vol. lvii, p. 1238.

TRASK, P. D., ET AL., 1939, *Recent Marine Sediments*, Amer. Assoc. Petrol. Geol., Tulsa, Oklahoma.

———, 1932, *Origin and Environment of Source Sediments of Petroleum*. Gulf Publishing Company, Houston, Texas.

TWENHOFEL, W. H., 1932, *Treatise on Sedimentation*, 2d ed. Baltimore.

———, 1945, "The Rounding of Sand Grains," *Jour. Sed. Petrology*, Vol. 15, pp. 59–71.

UMBGROVE, J. F. H., 1947, "Coral Reefs of the East Indies," *Bull. Geol. Soc. America*, Vol. 58, pp. 729–78.

VAUGHAN, T. W., 1910, "A Contribution to the Geologic History of the Floridian Plateau," *Pap. Tortugas Lab.*, Vol. iv, Pub. 133, pp. 99–185.

———, 1914a, "Preliminary Remarks on the Geology of the Bahamas, with Special Reference to the Origin of the Bahaman and Floridian Oölites," *ibid.*, v, Pub. 182, pp. 47–54.

———, 1914b, "Building of the Marquesas and Tortugas Atolls and a Sketch of the Geologic History of the Florida Reef Tract," *ibid.*, pp. 55–67.

———, 1915, "On Recent Madreporaria of Florida, the Bahamas, and the West Indies, and on Collections from Murray Island, Australia," *ibid.*, Year Book 14, p. 220.

———, 1918, "Some Shoal-Water Bottom Samples from Murray Island, Australia, and Comparisons of Them with Samples from Florida and the Bahamas," *ibid.*, Vol. ix, Pub. 213, pp. 235–88.

———, 1919a, "Fossil Corals from Central America, Cuba, and Porto Rico, with an Account of the American Tertiary, Pleistocene, and Recent Coral Reefs," *U. S. Natl. Mus. Bull. 103*, pp. 189–524.

———, 1919b, "Corals and the Formation of Coral Reefs," *Smithsonian Rept. for 1917*, pp. 189–276.

———, 1924, "Oceanography in Its Relation to Other Earth Sciences," *Jour. Washington Acad. Sci.*, Vol. 14, No. 14, pp. 307–33.

———, AND SHAW, E. W., 1915, "Geologic Investigations of the Florida Coral Reef Tract," *Pap. Tortugas Lab.*, Year Book 14, pp. 232–38.

WATTENBERG, H., 1933, "Kalziumkarbonate und Kohlensäuregehalt des Meerswassers," *Wissensch. Ergeben. 'Deutschen Atlantischen Exped.' auf des forsch. und vermessungeschiff 'Meteor,' 1925–27*, iii.

WELLS, R. C., 1918, "The Solubility of Calcite in Sea-Water in Contact with the Atmosphere, and Its Variation with Temperature," *Pap. Tortugas Lab.*, Vol. ix, Pub. 213, pp. 316–18.

WENTWORTH, C. K., 1922, "A Scale of Grade and Class Terms for Clastic Sediments," *Jour. Geol.*, Vol. xxx, pp. 377–92.

West Indies Pilot, Vol. III, 1946. London.

WETHERED, E., 1890, "On the Occurrence of the Genus Girvanella in Oolitic Rocks, and Remarks on Oolitic Structure," *Quar. Jour. Geol. Soc. London*, Vol. xlvi, pp. 270–281.

———, 1895, "The Formation of Oolite," *ibid.*, Vol. li, pp. 196–206.

3

Reprinted from pages 472–485 and 492–497 of *Jour. Geology* **71**:472–497 (1963),
by permission of The University of Chicago Press

RECENT CALCIUM CARBONATE FACIES OF THE GREAT BAHAMA BANK. 2. SEDIMENTARY FACIES[1]

EDWARD G. PURDY

Rice University, Houston, Texas

ABSTRACT

Statistical analysis of the constituent composition and grain-size data accumulated for 218 Bahamian sediment samples resulted in the delineation of the following five facies: (1) coralgal facies—characterized by a relative abundance of skeletal grains, particularly coral and calcareous algae fragments; (2) oölitic facies—characterized by an abundance of oölitically coated grains; (3) grapestone facies—typified by an abundance of grapestone and cryptocrystalline grains; (4) pellet-mud facies—characterized by an abundance of fecal pellets and particles smaller than $\frac{1}{8}$ mm.; and (5) mud facies—typified by an abundance of particles smaller than $\frac{1}{8}$ mm. and a relative paucity of indurated fecal pellets.

The coralgal facies owes its distinctiveness to a relative lack of non-skeletal calcium carbonate precipitation and to the large areas of rock bottom that characterize this depositional environment. In contrast, differences among three of the other four facies are considered to be primarily the product of differential current strength, with current velocities progressively decreasing in intensity from the oölitic to the pellet-mud facies. The contrast between pellet-mud and mud deposits appears to be due largely to a decreased rate of formation and/or induration of fecal pellets in the mud environment.

Ideally one might expect Bahamian facies to be distributed in a series of five concentric bands parallel to the bank's margin. The karst surface of the submerged Pleistocene basement rock, however, creates local current conditions that cause the development of a somewhat patchlike facies pattern.

INTRODUCTION

Part 1 of this paper (Purdy, 1963) described the petrography of the constituent particles occurring on the Andros platform and depicted the relationships between these constituent particles as a reaction group hierarchy. Part 2 delineates sedimentary facies and discusses the factors contributing to the origin and distribution of the quantitatively important grain types.

In order to take full advantage of the constituent composition and grain size data compiled for each of the 218 samples considered in this study, it was decided that Bahamian facies should be delimited not on the basis of one or two constituents as is frequently the case with the determination of stratigraphic facies but rather on the basis of the distribution of *all* the quantitatively important constituent particles. The desirability of delineating facies in this manner has been indicated by Imbrie and Purdy (1962), and the statistical calculations resulting in the facies diagram of figure 1 are presented in the same article.[2] It is necessary

only to indicate here that these facies boundaries are based not only on statistical computations but also on interpretations of aerial photographs, bathymetric data, and field observations. It also should be noted that eighteen of the 218 samples were classified subjectively as to facies on the basis of their constituent composition, a procedure necessitated by the limitations of the available electronic computer program. These eighteen samples consisted of those in which there were no data available on the weight percentage of material smaller than $\frac{1}{8}$ mm. and those in which the analytical error was considered to be relatively large.

Tables 1–6 present the mean constituent composition and grain size of each of the facies.

CORALGAL FACIES

The coralgal facies is restricted in occurrence to the outer platform of the bank, but

[1] Manuscript received December 30, 1960; revised January 17, 1962.

[2] The reader will note that the facies map illustrated in fig. 1 differs slightly from that presented by Imbrie and Purdy (1962, fig. 10). These differences are a product of additional field information obtained during the summer of 1961 and incorporated into fig. 1.

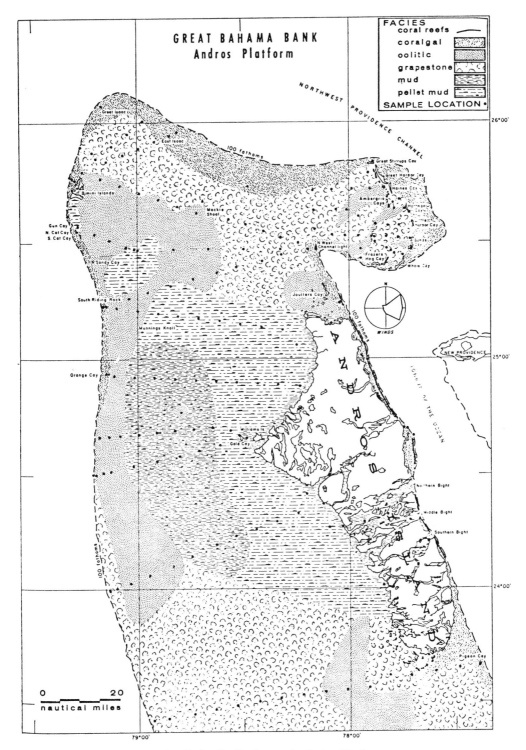

FIG. 1.—Facies distribution on the Andros Platform

93

not all outer platform areas are mantled by coralgal sediment (fig. 1). The mean constituent composition of the facies is characterized by an abundance of skeletal grains (table 1). Corals and calcareous algae are particularly abundant, hence the name "coralgal facies." Many of the coral fragments have their pores infilled by silt- and clay-sized carbonate particles similar to that which comprises fecal pellets and mud ag-

TABLE 1

MEAN CONSTITUENT COMPOSITION AND
GRAIN SIZE OF CORALGAL FACIES*

	Mean	Observed Range	Standard Deviation
Coralline algae	1.8	0- 8.7	2.2
Halimeda	9.8	2.6-38.2	9.3
Peneroplidae	3.3	0.2-16.9	3.8
Other Foraminifera	3.8	0-11.9	2.5
Corals	5.9	0-27.5	7.1
Mollusks	7.1	0.1-31.0	5.5
Miscellaneous	2.1	0.4- 6.2	1.6
Unknown	7.4	2.8-13.6	3.0
Total skeletal	41.1	6.9-82.6	17.9
Fecal pellets	4.7	0-17.8	14.3
Mud aggregates	6.1	0-23.7	5.3
Grapestone	5.4	0-28.0	5.9
Organic aggregates	2.7	0-11.7	3.5
Oölite	6.4	0-23.3	7.2
Cryptocrystalline grains	20.9	3.4-55.3	12.8
Cay rock	1.9	0-23.1	5.4
Calcilutite	0	0- 1.0	0.2
Weight percentage less than ⅛ mm	10.8†	0-57.0	13.8

* No. of samples from coralgal facies = 33.

† This mean was computed from two less than the total no. of samples from the coralgal facies due to sample loss during preparation.

gregates. This material was observed in all stages of transition to cryptocrystalline carbonate; however, the coral skeleton, itself, apparently does not recrystallize to cryptocrystalline carbonate. Alteration of coral fragments is accomplished chiefly by boring algae that cause gradual obliteration of coral microstructure and transform the initially transparent fragments to ones having a light grayish or brownish tint (Ginsburg, 1956, p. 2425). The calcitic skeletons of coralline algae also are infilled occasionally. The majority of the aragonitic *Halimeda* fragments showed signs of recrystallization

to cryptocrystalline carbonate, and many *Halimeda* grains were markedly recrystallized. The internal tubes of these fragments were frequently filled by brownish-colored silt- and clay-sized carbonate particles and more rarely by acicular aragonite crystals growing normal to the tube walls. The Peneroplidae, in particular, are largely recrystallized to cryptocrystalline carbonate. Among the characteristic Foraminifera occurring in the facies are the red-colored *Rotalia rosea* and the red-encrusting species of *Homotrema*, *Acervulina*, and *Planorbulina*. Infrequently the red-colored pelagic foraminifer *Globigerinoides rubra* was encountered, but pelagic Foraminifera are extremely rare in the coralgal facies and practically nonexistent in other Bahamian facies. Among the miscellaneous constituents alcyonarian spicules and fragments of *Millepora* and echinoderms are particularly common.

The most abundant non-skeletal constituents in the facies are cryptocrystalline grains. These consist largely of recrystallized non-skeletal constituents, chiefly grapestone and fecal pellets, and more rarely mud aggregates and oölite. The smaller oölitically coated grains have a relatively thick oölitic coating and have been transported from the adjacent oölitic facies. The more abundant, larger, thinly coated grains (superficial oöliths of Illing) may have formed essentially in place as discussed in a later section. Some of the fecal pellets and grapestone in the sediment are sufficiently "fresh" to warrant the interpretation of recent formation, but most of these same constituents are severely recrystallized to cryptocrystalline carbonate. Most of the mud aggregates present are relatively free from recrystallization effects and appear to have formed recently. The organic aggregates in the facies are of the type considered to be agglutinated worm tubes. The proximity of the depositional environment to islands explains the relatively high cay rock content.

Depositional environment.—In contrast to other Bahamian environments the coralgal depositional environment is characterized by extensive areas of rock bottom. This rock substrate is colonized by a distinctive community of organisms (see Newell *et al.*, 1959) among which the more conspicuous skeletal contributing elements are coralline algae, *Millepora*, alcyonarians (sea fans and sea whips), and corals. With the exception of a few species of corals, such as *Manicina areolata*, which occur on sediment substrates, all the above-mentioned organisms are restricted in occurrence to rock bottoms. Thus it is evident that the relatively high content of coralline algae and coral fragments in the coralgal facies is due primarily to the presence of large areas of rock bottom. However, the smaller areas of rock bottom that sporadically occur within the shelf lagoon *do not* support the same organism assemblage as the outer platform rock substrates. Conspicuously absent or reduced in abundance in these shelf lagoon rock areas are coralline algae, corals, and *Millepora*. Gorgonians (sea fans) are also absent, but plexaurids (sea whips) appear to be abundant. Evidently then, the presence of a rock substrate is not the only factor controlling the distribution of coralline algae and corals in the Bahamas.

A second limiting factor is turbulence. The dependence of reef development on water agitation is too well known to require discussion. In Pacific reefs encrusting coralline algae (*Lithothamnion* and *Poralithon*) buttress the reef front in the surf zone where coral development is greatest, other things being equal. In the Bahamas the preference of coralline algae and corals for agitated water is indicated by the negative correlation (Purdy, 1963, table 2) between the abundance of particles smaller than $\frac{1}{8}$ mm. and both coralline algae ($-.119$) and coral ($-.141$) fragments. Since the coralline algae category includes both crustose corallines and articulate corallines, the negative correlation suggests that both types of corallines prefer current-agitated conditions. Field observations indicate that the crustose corallines prefer a greater degree of water turbulence than the articulate variety. The coral identification category includes fragments derived from scleractinians inhabiting sediment bottoms (e.g., *Manicina areolata*) as well as rock bottoms. Therefore the negative value of the correlation between corals and the abundance of particles smaller than $\frac{1}{8}$ mm. should be taken as a minimum value for coral fragments derived from rock-inhabiting corals. As previously noted, current velocities generally decrease toward the center of the bank. Consequently, it is not surprising to find a dearth of corals and coralline algae on the small local areas of rock bottom that occur within the shelf lagoon. Another hydrographic factor that might explain the prevalence of corals and coralline algae in the coralgal facies as compared to other Bahamian facies is the previously mentioned increase in salinity bankward. The paucity of plankton in the waters of the shelf lagoon (Fish, *in* Field, 1931) also might effectively limit the distribution of corals, and the relatively cool temperatures of shelf lagoon waters during winter might constitute an additional limiting factor.

The close approximation between the distribution of these rock-bottom-inhabiting organisms and the distribution of the skeletal fragments derived from these organisms suggests that the transportation of skeletal debris may be less extensive than has commonly been assumed. There are occasional instances of long-distance transportation of skeletal material. For example, the shell of the intertidal gastropod, *Batillaria minima*, was found in a sample locality in the shelf lagoon located approximately 10 miles from the nearest intertidal area. These isolated examples of extensive transportation, however, are rare, and the results available from the present study and from Ginsburg's work (1956) in south Florida certainly are encouraging for paleoecological studies.

The abundance of non-skeletal grain types in the shelf lagoon sediments of the Bahamas indicates that the bank waters overlying this area favor non-skeletal carbonate precipitation. Conversely, the paucity of non-recrystallized examples of these same grain types in the coralgal facies makes it clear that non-skeletal carbonate precipi-

95

tation is at a minimum in this depositional environment. These attributes apparently result from several interrelated factors. Cooler oceanic waters are heated by the sun as they move onto the shoal bank. Consequently the solubility of carbon dioxide in the bankward-moving waters progressively decreases, and finally a point is reached where the warmed oceanic waters are saturated with carbon dioxide. Further heating then causes carbon dioxide to be driven off.

TABLE 2

MEAN CONSTITUENT COMPOSITION AND GRAIN SIZE OF OÖLITIC FACIES*

	Mean	Observed Range	Standard Deviation
Coralline algae.........	0.1	0– 1.1	0.2
Halimeda.............	1.8	0–13.1	2.7
Peneroplidae.........	0.5	0– 8.3	1.2
Other Foraminifera.....	0.5	0– 3.4	0.7
Corals...............	0.1	0– 2.0	0.3
Mollusks.............	1.4	0– 4.8	1.4
Miscellaneous.........	0.3	0– 1.6	0.4
Unknown............	1.6	0– 6.9	1.3
Total skeletal.........	6.2	0.1–25.9	6.3
Fecal pellets..........	7.2	0–28.5	9.1
Mud aggregates........	2.8	0–15.1	3.2
Grapestone...........	4.5	0–18.4	5.3
Organic aggregates.....	0.2	0– 1.9	0.4
Oölite...............	66.6	8.4–98.0	25.0
Cryptocrystalline grains	7.4	0.2–34.5	6.8
Cay rock.............	0.1	0– 3.3	0.5
Calcilutite...........	0.0	0.0– 0.4	0.0
Weight Percentage less than $\frac{1}{8}$ mm..........	5.0	0.8–25.0	5.0

* No. of samples from oölitic facies = 71. The oölite and mixed oölite samples from the South Cat Cay–Browns Cay area are included in this facies

It is well known that, as the carbon dioxide content of a given water mass decreases, the solubility of calcium carbonate also decreases, so that progressive loss of carbon dioxide from the cooler water masses moving onto the bank tends to promote conditions of calcium carbonate supersaturation and ultimately of calcium carbonate precipitation. In addition the carbon dioxide content of oceanic waters flooding the bank is reduced by photosynthesizing plants. The increase in salinity from the outer platform to the shelf lagoon further reduces the solu-

bility of calcium carbonate in bankward-moving waters because of the accompanying increase in concentration of calcium ions and decrease in solubility of carbon dioxide. Because the waters of the shelf lagoon are warmer and more saline than those of the outer platform, non-skeletal precipitation should be relatively great in the shelf lagoon as compared to the outer platform, *as in fact it is*. Those non-skeletal constituents present in the coralgal facies are, for the most part partially or completely recrystallized and appear to have been inherited from a previous sedimentary cycle, the constituents of which are currently being diluted in abundance by skeletal debris.

It will be noted in figure 1 that the outer platform is mantled in places by grapestone sediment rather than coralgal sediment. It may be that in these areas skeletal dilution of the pre-existing constituents has been minimized owing to the lack of extensive areas of rock bottom. Absence or limited extent of rock bottoms in these areas is suggested by the general lack of islands or cays. On the other hand it is possible that the sampling density is such that the existence of a relatively narrow strip of coralgal sediment on the outer platform has been overlooked.

OÖLITIC FACIES

The oölitic facies is typified by an abundance of oölite and a paucity of particles smaller than $\frac{1}{8}$ mm. The skeletal content of these deposits is less than that of any other bank facies (table 2). In the South Cat Cay–Browns Cay area shown in figure 2, the oölitic facies was subdivided into an oölite and mixed oölite facies. These two facies are distinguished from each other by the following features: (1) the oölite facies has a higher content of oölite and lower percentage of particles smaller than $\frac{1}{8}$ mm. than the mixed oölite facies and (2) the skeletal content of the mixed oölite facies is approximately 9 per cent higher than that of the oölite facies (table 3). The importance of this distinction is that the distribution of the oölite facies is practically coincident with that of the

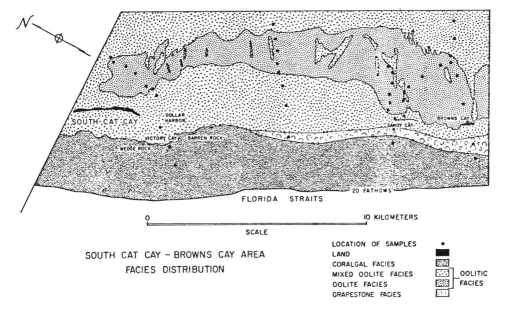

FIG. 2.—Facies distribution in the South Cat Cay-Browns Cay area

TABLE 3

MEAN CONSTITUENT COMPOSITION AND GRAIN SIZE OF OÖLITE AND MIXED OÖLITE
FACIES FROM THE SOUTH CAT CAY-BROWNS CAY AREA*

	OÖLITE FACIES			MIXED OÖLITE FACIES		
	Mean	Observed Range	Standard Deviation	Mean	Observed Range	Standard Deviation
Coralline algae..................	0	0 – 0.3	0.1	0.3	0 – 1.1	0.4
Halimeda.......................	0.4	0 – 3.7	0.8	2.5	0.7– 6.5	1.8
Peneroplidae....................	0.1	0 – 0.5	0.1	1.5	0 – 8.3	2.3
Other Foraminifera..............	0.1	0 – 0.5	0.1	0.9	0.2– 2.0	0.6
Corals..........................	0	0 – 0.4	0.1	0.3	0 – 2.0	0.6
Mollusks........................	0.4	0 – 2.8	0.2	2.4	0.6– 7.3	1.8
Miscellaneous...................	0.1	0 – 0.8	0.2	0.3	0.1– 1.0	0.3
Unknown.......................	0.7	0.1– 1.5	0.4	2.4	1.2– 3.4	0.7
Total skeletal...................	1.7	0.1– 6.8	1.5	10.5	3.8–25.9	6.1
Fecal pellets....................	0.1	0 – 1.9	0.2	1.4	0 – 6.9	2.0
Mud aggregates.................	0.2	0 – 1.7	0.4	2.9	0.4– 8.9	2.4
Grapestone.....................	1.0	0 – 6.9	1.5	8.1	1.6–18.4	6.0
Organic aggregates..............	0	0 – 0	0	0.2	0 – 0.8	0.3
Oölite..........................	92.6	80.2–98.0	1.9	60.1	42.3–73.7	12.4
Cryptocrystalline grains..........	2.6	0.5– 6.9	1.7	9.1	3.5–18.8	4.9
Cay rock.......................	0	0 – 0.4	0.1	0.7	0 – 3.3	1.3
Calcilutite......................	0	0 – 0	0	0	0 – 0	0
Weight percentage less than ⅛ mm...	1.9	0.8– 5.5	1.2	7.1	1.7–25.0	7.6

* No. of samples from oölite facies = 25; no. of samples from mixed oölite facies = 11.

97

shoals where optimum conditions for oölite formation are attained (Purdy, 1961). Because of the small size of the oölite facies in areas outside of the South Cat Cay–Browns Cay region, it was decided to combine the oölite and mixed oölite samples on the bank under the general heading of oölitic facies. This facies is restricted largely to that part of the shelf lagoon margined by the outer platform.

The environmental conditions attending the formation of oölite have been described by several investigators (Illing, 1954; Newell *et al.*, 1960; and Purdy, 1961), and the origin of Bahamian oölite has been considered most recently by Newell *et al.* (1960). Optimum conditions for oölite formation in the Bahamas apparently occur only where sea water supersaturated with calcium carbonate is subjected to relatively great current agitation (Newell *et al.*, 1960). The abundance of non-skeletal grain types at and bankward of the barrier rim indicates that sea water in this region is supersaturated with calcium carbonate. Consequently the distribution of oölite in the Bahamas must be related to sites of relatively great current agitation. The dependency of oölite accretion on current agitation and the general decrease in tidal current velocities bankward of the barrier rim would lead one to expect a rapid decrease in ooid abundance bankward of these marginal oölite deposits. Figure 1 shows that these expectations generally are fulfilled, for the oölitic facies is largely limited in occurrence to the area immediately bankward of the barrier rim of cays and shoals. Several factors, however, tend to complicate the distribution pattern. The first of these concerns the presence of relict oölite deposits. It is evident that, in some localities within the oölitic facies, grains are no longer being coated. For example, many of the ooids in sample 310 have their outer laminae recrystallized and most of the cryptocrystalline grains in the same sample consist of recrystallized ooids (e.g., Purdy 1963, pl. 5, *A* and *B*). Clearly oölitic accretion is no longer taking place in this locality. A similar relict oölite deposit occurs at

sample locality 395. Here the extent of ooid recrystallization is not nearly as great as that of locality 310, but the position of this deposit amid lime muds and the non-agitated condition of the overlying, relatively deep (14 feet) water mass made its age suspect. Consequently the sample was subjected to radiocarbon analysis, which revealed that the deposit was 1560 ± 100 years old. This age is approximately twice as great as the 740 ± 100 years date reported by Newell *et al.* (1960) for the Browns Cay ooids. It could be argued that these contrasting dates are due to differences in nuclei ages rather than in laminae ages. There is, however, no evidence to support such an interpretation, and therefore it seems more reasonable to conclude that the oölite at locality 395 represents a relict deposit. The occurrence of these old deposits causes the oölitic facies to extend farther bankward than anticipated in some areas.

A more important factor in explaining the facies' bankward extent is the occurrence of shoal, unsheltered areas within the shelf lagoon. In contrast to the 12 and 18 feet of water that covers most of the shelf lagoon, these shoals are usually submerged less than seven feet. Their shallow nature causes a local increase in current velocity, an effect similar to that resulting when a submerged barrier is placed across the width of a stream. The increased current agitation, in turn, causes the precipitation of oölitic laminae. Sample locality 314 (Purdy, 1963, fig. 2) exemplifies the interior shelf lagoon formation of oölite. This deposit is covered by approximately 3–6 feet of water at mean low-water spring tides and is given the formal designation of Mackie Shoal on U.S. Navy Hydrographic Office (1867) Chart 26a. The Mackie Shoal depositional environment duplicates that described for the Browns Cay oölite ridge. There are many other small shelf lagoon areas which both are covered by 6 feet or less of water and are unprotected from current action by Andros Island (e.g., see Purdy, 1963, fig. 2). Oölite also probably is forming in these areas.

In other parts of the bankward extension of the oölitic facies the ooids are neither as severely recrystallized as in the two relict oölite deposits nor as abundant as on Mackie Shoal. The sediments contain less than 65 per cent oölite and are usually covered by a thin brownish-colored organic film. Samples collected from such localities frequently emit a hydrogen sulfide odor. These characteristics evidence a lack of bottom agitation and suggest that the diminished abundance of ooids is perhaps a reflection of a reduced rate of oölitic accretion or a product of recent sedimentary dilution of a relatively old oölite deposit. A third possibility is that the reduced oölite content has resulted from transportation of ooids from environments of oölitic accretion into environments of non-oölitic accretion. It seems probable that all three factors are important, although the relative importance of each cannot as yet be determined.

GRAPESTONE FACIES

An abundance of grapestone and cryptocrystalline grains characterize this facies. The mean cryptocrystalline grain content is approximately 6 per cent greater than that of the coralgal facies. Organic aggregates of the type shown in plate 2, *D*, of Purdy (1963) are particularly common. The total skeletal content is relatively low, though not as low as in the oölitic facies and the principal skeletal contributors are mollusks (table 4). Limestone flakes are found only within this facies. Though absent from most parts of the grapestone facies, the flakes are particularly common near sample localities 514 and 515. On some parts of the Andros platform the grapestone facies occurs bankward of the coralgal facies; in other areas it is bordered on its seaward margin by the oölitic facies. The occasional extension of the grapestone facies to the marginal escarpment has been mentioned previously.

Depositional environment.—Bankward of the arcuate group of islands on the northwestern lobe of the Andros platform (Great Stirrups Cay Whale Cay) are localities in which the grapestone facies is subaerially exposed at low tide (e.g., sample locality 323). At the other extreme a maximum depth of 42 feet was recorded for the same facies at sample locality 528. In the shoaler areas the bottom usually exhibits a sparse floral covering of the marine angiosperms *Thalassia testudinum* and *Cymodocea manatorum. Thalassia* tends to be more common in occurrence than *Cymodocea.* In the deeper parts of the depositional environ-

TABLE 4

MEAN CONSTITUENT COMPOSITION AND GRAIN SIZE OF GRAPESTONE FACIES*

	Mean	Observed Range	Standard Deviation
Coralline algae	0.1	0– 2.7	0.4
Halimeda	2.7	0–13.0	3.0
Peneroplidae	0.9	0– 5.3	1.1
Other Foraminifera	1.7	0.2– 4.7	1.1
Corals	0.1	0– 2.1	0.3
Mollusks	4.1	0.2–20.4	3.6
Miscellaneous	0.7	0– 7.4	1.2
Unknown	2.5	0.6– 6.6	1.2
Total skeletal	12.7	1.7–37.6	7.5
Fecal pellets	4.9	0.3–17.2	4.1
Mud aggregates	2.3	0– 8.2	1.5
Grapestone	32.0	10.3–64.4	11.1
Organic aggregates	1.7	0– 7.0	1.7
Oölite	14.9	0–44.7	11.4
Cryptocrystalline grains	26.8	11.2–46.8	8.7
Cay rock	0.1	0– 3.7	0.5
Calcilutite	0	0– 0	0
Weight percentage less than $\frac{1}{8}$ mm	4.5	1.3–14.5	2.3

* No. of samples from Grapestone facies = 64.

ment patches of dense growths of *Thalassia* were observed. Some of these form small topographic banks or mounds approximately 1 foot above the surrounding sand bottom. These probably originated in a manner similar to that described for the grass covered banks and mounds of Florida Bay (Ginsburg, 1953; Ginsburg and Lowenstam, 1958). Mounds of burrowing organisms also were noted frequently in these deeper areas as were occasional small sand ripples. Particularly frequent in occurrence was a tan organic film that covered the bottom. The organic firm is similar to that which covers local areas within the oölitic facies and is identical to that described by Kornicker (1959, p. 224-225) from the Bimini region. Included within the film are such things as diatoms, algal filaments, unidentified plant fragments, and sponge spicules (Kornicker, 1959, fig. 32). Kornicker (1959, p. 224) has demonstrated that this material also occurs suspended in the water, thus indicating that the bottom film or pelogloea results from the deposition of suspended organic detritus. In general the organic detritus film appears to be thicker in the shoaler parts of the grapestone facies. Most of the above-mentioned characteristics indicate bottom stability and therefore a depositional environment which is characterized by a lack of appreciable current action. The abundance of grapestone, in itself, evidences relatively little bottom agitation, for grains can only be cemented together where there is a lack of differential grain movement. The dependency of grapestone formation on bottom stability also was noted by Illing (1954, p. 85).

Origin of grapestone.—In many of the sample localities bankward of the Great Stirrups Cay–Whale Cay group of islands the grapestone grains were observed to have dark-brown to black organic material between the several sand-sized constituents constituting each aggregate. Upon partially decalcifying the grapestone grains, it was found that this dark organic matter consists of abundant algal cells that apparently all belong to the same species. The identity of the alga was not determined, but thin-section observations indicate that it is not one of the boring algae, for the tubules left by boring algae within grapestone are con-

centrated neither in the cement nor in the component grains. Slow decalcification of thin-sectioned grapestone grains with dilute hydrochloric acid reveals a relative concentration of organic matter within the cement of the aggregate (pl. 1, *B* and *C*). The origin of much of this organic matter is difficult to determine. Some of it consists of algal cells identical to those comprising the dark organic matter previously mentioned, and some of it is represented by filaments of boring algae, but most of the organic material is of a nondescript nature and apparently represents an accumulation of organic detritus.

The localization of carbonate precipitation between the component grains of grapestone indicates that the intergrain chemical environment is more favorable for precipitation than elsewhere on grain surfaces. The observed concentration of organic matter between the component grains of grapestone and within the grapestone cement itself suggests that the organic material may somehow be a factor in localizing this precipitation. To express it another way, grapestone cement may be a product of organically induced precipitation. Non-recrystallized grapestone cement commonly contains numerous inclusions of skeletal fragments (e.g., pl. 1, *A*), which appear to have adhered to some initially sticky substance. This attribute of the cement suggests that organic detritus originally occupied the space between the component grains and that it is chiefly the presence of organic detritus rather than algal colonies that causes the precipitation of the grapestone cement. Perhaps it is the bacterial decomposition of

---

PLATE 1

A, Magnified view of part of grapestone grain shown in Purdy (1963, pl. 2, *C*). Note the similarity between the carbonate particles comprising grapestone cement and those constituting the fecal pellet and mud aggregate illustrated by Purdy (1963, pls. 1, *A* and *B*, respectively). Crossed nicols, ×100.

B, Decalcified, thin-sectioned grapestone grain. White areas represent voids left by dissolving component carbonate constituents. Dark areas consist of an insoluble residue of organic matter. Concentration of organic material between component grains and within areas formerly occupied by grapestone cement is noteworthy. Plane polarized light, ×100.

C, Decalcified, thin-sectioned grapestone grain. Concentration of organic matter between component sand-sized grains is again in evidence. Plane polarized light, ×35.

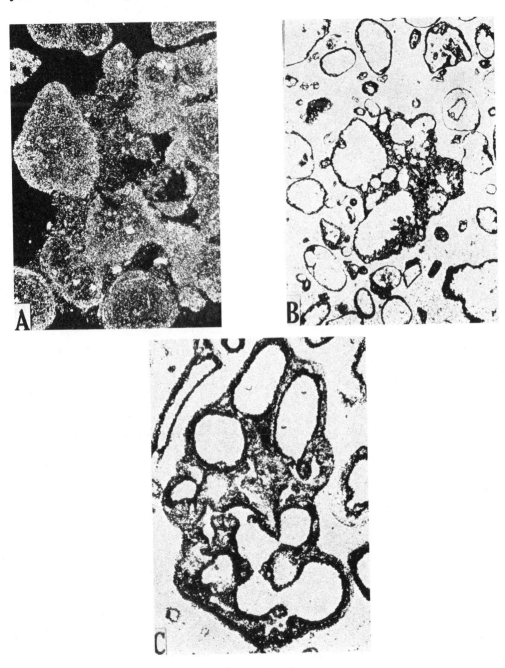

Grapestone grains

organic detritus that causes the formation of the cement. For example, the decomposition of organix detritus by ammonifying and nitrate-reducing bacteria would produce ammonia which would react with the calcium bicarbonate in the immediately surrounding water to cause precipitation of calcium carbonate. Regardless of the precise reactions, it seems probable that the decomposition of organic detritus between grains causes the precipitation of a carbonate cement. It should be emphasized that this precipitation of cement is dependent upon two factors: (1) the supersaturated nature of bank waters with respect to calcium carbonate, and (2) the lowering of calcium carbonate solubility in the immediate microenvironment through certain unspecified decompositional reactions. The calcium carbonate supersaturated bank waters causes precipitation to occur when the organic detritus is decomposed, and the prevalence of organic detritus between grains determines where the carbonate will be precipitated. As a working hypothesis it may be supposed that in some areas where current action is slight, organic detritus settles out of the water and accumulates between the grains on the sediment bottom. If the current is not too weak or if there is an occasional increase in current velocity, organic detritus will be prevented from accumulating on the current-swept grain surfaces and will only accumulate in the interstices between grains at the depositional interface. For similar reasons silt- and clay-sized carbonate particles would be expected to occur in these same interstices where they would adhere to, and become incorporated with, the accumulating organic detritus. Decomposition of the organic detritus would result in the precipitation of a binding carbonate cement in which inclusions of silt- and clay-sized particles would be evident (pl. 1, A). Thus a grapestone grain would originate. Occasional bottom agitation by infrequently occurring stronger currents or by storm waves would prevent unlimited growth of the aggregate. Short periods of bottom agitation followed by longer periods of

bottom stability and cement precipitation would produce the typical grapestone structure shown in plate 2, D, of Purdy (1963). In areas that are characterized by still greater bottom stability, organic detritus might be expected to accumulate as a thin film covering the depositional interface. In these instances carbonate precipitation would occur on the surfaces of grains as well as in the interstices between grains, resulting in a diminishment of the original differential grain relief at the depositional interface of the forming grapestone grains (e.g., see Purdy, 1963, pl. 4, A). As the precipitation of carbonate cement is confined largely to the sediment surface, it would follow that areas of extreme bottom stability would be characterized by large flattened grapestone grains. The limestone flake shown in plate 5, C, of Purdy, (1963) represents such a flattened aggregate. Thus as bottom stability increases, grapestone grains become both larger and flatter, and conversely.

Although many of the grapestone grains exhibit a cement identical to that shown in plate 1, A, the majority of the grapestone aggregates are cemented by cryptocrystalline carbonate identical in appearance to that of recrystallized ooids. The possibility of direct precipitation of this cryptocrystalline carbonate cannot be entirely negated; however, the frequently observed transition from cement containing an aggregation of silt- and clay-sized carbonate particles to cryptocrystalline carbonate makes it seem more likely that the cryptocrystalline cement is a recrystallization product. The abundance of partially or severely recrystallized grapestone grains suggests that grapestone formation was more widespread in the past than it is today. Certainly if there are relict oölite deposits, there are probably also relict grapestone deposits.

Grapestone has a relatively high negative correlation ($-.391$) with the abundance of particles smaller than $\frac{1}{8}$ mm. (Purdy, 1963, table 2). This indicates that as the abundance of grapestone increases, the abundance of particles smaller than $\frac{1}{8}$ mm. decreases. Seemingly this relationship contradicts what has been said concerning the dependency of grapestone formation on bottom stability. This inverse relationship can be interpreted as a product of three factors. First, relict grapestone deposits may be attended by greater current agitation now than when the grapestone was forming at lower stands of sea level. Consequently the initially present fine-grained material may have been subsequently winnowed from the relict deposits. Second, many of the small particles deposited in grapestone-forming areas become incorporated within the aggregates, as indicated by the presence of small carbonate particles in grapestone cement. This, of course, reduces the apparent abundance of the smaller sedimentary constituents. Third, current strength may be sufficiently great to prevent the deposition of smaller carbonate particles in some grapestone-forming areas without causing appreciable bottom agitation. All three factors would tend to cause a decrease in the quantity of fine-grained carbonate particles associated with grapestone. Thus the relatively high negative correlation between the abundance of grapestone and particles smaller than $\frac{1}{8}$ mm. does not vitiate the general conclusion that grapestone formation is dependent upon bottom stability.

Cryptocrystalline grains.—It has been repeatedly mentioned that cryptocrystalline grains result from the recrystallization of various Bahamian constituents. The similarity between cryptocrystalline grains derived from different grain types makes it difficult to determine the relative importance of sources of cryptocrystalline constituents. Some qualitative observations were made, however, on the basis of the shape of the cryptocrystalline grains and on the frequency of occurrence of transitional recrystallization stages in which the source material could be identified. These observations indicate that the great majority of the cryptocrystalline grains have resulted from the recrystallization of non-skeletal constituents. Among these constituents grapestone is the source of most of the cryptocrystalline grains. Fecal pellets, ooids, and

mud aggregates, in that order, constitute source material for smaller quantities of cryptocrystalline carbonate. The organic aggregates associated with grapestone (Purdy, 1963, pl. 2, *D*) are also susceptible to recrystallization, and indeed when such recrystallization has proceeded to an advanced stage, it becomes extremely difficult to identify the binding cement as being skeletal or non-skeletal in origin.[3] Some cryptocrystalline grains may be derived from the disintegration of cay-rock fragments, but the rarity of identifiable cay-rock fragments even close to the cays indicates that such derived grains must be of minor quantitative importance. Illing (1954, p. 29) has arrived at a similar conclusion. Among the skeletal constituents *Halimeda* and, to a lesser extent, peneroplids constitute the chief skeletal sources of cryptocrystalline grains.

Within the grapestone facies the dominant source of cryptocrystalline grains is grapestone. In the oölitic facies fecal pellets and, more rarely, ooids comprise the main source of cryptocrystalline grains. Infilled *Halimeda* and peneroplid grains become quantitatively important source material in the coralgal facies; however, their importance even here is overshadowed by the more abundant recrystallized grapestone constituents. Recrystallized mud aggregates and fecal pellets also comprise some of the cryptocrystalline grains in the coralgal facies. Curiously enough both the pellet mud and the mud facies are characterized by a relative paucity of cryptocrystalline grains. Indeed, the extent of recrystallization is at a minimum in these two facies. This general lack of recrystallization could be explained if the pelllet-mud and mud environments

[3] It will be recalled that a given aggregate was identified as grapestone only when less than half the binding carbonate was recognized as being skeletal in origin. Consequently there may be a tendency in this study to underestimate grossly the volumetric importance of the type of aggregate that appears to be bound by an irregular foraminifer or alga. Future investigations may demonstrate that the skeletal-bound aggregates are quantitatively as important or more important in Bahamian sediments than the type of composite grain herein termed grapestone.

were inimical to recrystallization processes or if the constituent particles of the pellet-mud and mud facies were, on the average, younger than those of the other Bahamian facies. In order to determine which alternative was correct, two samples were selected for radiocarbon age determinations. Sample 528 was dated from the grapestone facies and a composite sample (samples 382 and 384) was dated from the pellet-mud facies. In both samples only the constituents larger than $\frac{1}{8}$ mm. were analyzed for their radiocarbon content. The total skeletal content of the composite sample and sample 528 was approximately the same. Thus differences in age between the two samples cannot be attributed to differences in total skeletal content. The constituent composition of the non-skeletal fraction of the two samples contrasts markedly. The composite pellet-mud sample is characterized by an abundance of fecal pellets (60.5 per cent), absence of grapestone, and a low percentage of cryptocrystalline grains (1.7 per cent). In contrast, sample 528 has a low percentage of fecal pellets (5.8 per cent) and relatively high percentages of grapestone (24.4 per cent) and cryptocrystalline grains (30.8 per cent). The composite pellet mud sample yielded a radiocarbon date of 700 ± 150 years, while the grapestone sample dated 2,500 ± 150 years. These limited data suggest that the extent of grain recrystallization is determined largely by the age of the initial constituents rather than by environmental differences.

The cause of recrystallization is unknown. Lowenstam and Epstein (1957, p. 372) have noted that the cryptocrystalline nuclei of Bahamian ooids have the same O^{18}/O^{16} ratios as the surrounding oölitic laminations. This suggests that the recrystallization of the nuclei and the precipitation of the oölitic laminations occurred in similar environments (*ibid.*). However, the factors controlling recrystallization remain obscure.

PELLET-MUD FACIES

This facies is typified by an abundance of fecal pellets and particles smaller than $\frac{1}{8}$

mm. Particularly noteworthy is the occurrence of innumerable clay-sized aragonite needles. The skeletal content of the facies is lower than that of the coralgal and grapestone facies but higher than that found in the oölitic facies. Mollusks and peneroplids are the major skeletal contributors. Samples collected from the northern part of the pellet-mud facies west of Andros contain appreciable quantities of oölite that probably were transported into the area by

TABLE 5

MEAN CONSTITUENT COMPOSITION AND
GRAIN SIZE OF PELLET-MUD FACIES*

	Mean	Observed Range	Standard Deviation
Coralline algae.......	0	0– 0	0
Halimeda...........	1.3	0– 3.9	0.8
Peneroplidae........	2.7	0.4– 4.9	1.1
Other Foraminifera....	1.2	0.5– 2.0	0.5
Corals.............	0	0– 0	0
Mollusks............	2.7	1.3– 4.6	1.0
Miscellaneous.......	0.4	0– 1.5	0.4
Unknown...........	2.5	1.3– 6.6	1.1
Total skeletal........	10.8	4.3–17.3	2.8
Fecal pellets.........	32.6	13.4–49.8	8.6
Mud aggregates.......	5.1	1.1–10.3	2.5
Grapestone..........	0.3	0– 1.2	0.3
Organic aggregates....	0.2	0– 1.1	0.3
Oölite.............	6.3	0.1–37.8	10.7
Cryptocrystalline grains	1.7	0– 4.5	1.1
Cay rock...........	0.2	0– 1.1	0.3
Calcilutite..........	0	0– 0	0
Weight percentage less than $\frac{1}{8}$ mm.........	42.8	7.9–63.7	17.1

* No. of samples from pellet-mud facies = 23.

occasional storm waves from the oölitic facies to the northeast and west. The relatively high oölite content of these few samples accounts for the 6 per cent mean oölite content of the facies (table 5). The paucity of cryptocrystalline grains in pellet-mud deposits has been mentioned previously. With the exception of two relatively small areas in the vicinity of Gun Cay and Sandy Cay, the pellet-mud facies is limited in distribution to the shelf lagoon region west of Andros.

Depositional environment.—The depth of water covering the pellet-mud deposits varies between 1 and 4 fathoms, and all pellet-mud deposits are characterized by bottom stability as evidenced by the abundance of particles smaller than $\frac{1}{8}$ mm., the abundance of worm mounds, and the absence of ripple marks on the bottom. Locally the bottom is covered by dense growths of *Thalassia testudinum*, but in general the floral covering is sparse; this is particularly true of the area immediately west of Andros. White streaks of water frequently were observed in the pellet-mud area west of Andros. These streaks or whitings, as they are called, generally are elongated in an east-west direction, having a length of the order of magnitude of $\frac{1}{2}$ mile and a width which approximates 200 yards. The marked milky coloration of the whitings distinguishes them from the less turbid surrounding waters. The contrasting turbidness indicates that more suspended carbonate particles occur within the whitings than elsewhere. Whitings also occur in Florida Bay where "schools of mullet . . . produce unusually turbid water by expelling undigested sediment picked up from the bottom and by stirring the bottom in feeding" (Ginsburg, 1956, p. 2398). Cloud (*in* Cloud and Barnes, 1957, p. 176) reports the presence of sediment-ingesting fish in the area west of Andros. Perhaps bottom disturbances caused by these fish or by the frequent water spouts observed in this region result in the formation of local patches of white water that subsequently are elongated by tidal currents. Certainly there is little available evidence to suggest that the whitings are a product of the localized precipitation of aragonite needles.

Fecal pellets and mud aggregates.—It is evident that the grain size of constituents included within any type of fecal pellet is dependent both on the particle sizes available for pelleting and on the maximum particle size that the pellet-producing organism is capable of ingesting. The abundant fecal pellets that typify the pellet-mud facies are composed mainly of silt- and clay-sized particles. Consequently the formation of these pellets necessitates the presence of a considerable quantity of silt- and clay-sized grains. This abundance of fine-grained material, in turn, precludes the possibility of much bottom agitation, so that indirectly the formation of these fecal pellets is dependent upon relatively low bottom-current velocities.

It will be recalled that the pellets are initially friable and that subsequent to their formation they become indurated by the interstitial precipitation of a calcium carbonate cement. Any bottom agitation during the period when the pellets are still friable would, of course, cause pellet disaggregation and resulting loss of the typical ellipsoidal pellet shape. Such pellet disaggregation does, in fact, occur and results in the formation of mud aggregates. The sequence of events leading to the formation of mud aggregates seems to be as follows. Friable, recently formed fecal pellets are disaggregated by some type of bottom disturbance, perhaps caused by storm waves or by the movement of organisms, such as crabs, across the surface of the substrate. The disaggregation destroys the ellipsoidal shape of the mucus-bound aggregates and produces irregular-shaped aggregates, mud aggregates, identical in petrographic appearance to fecal pellets. The mud aggregates are, of course, initially smaller than the fecal pellets from which they have originated; however, subsequently they may reunite and be joined by similarly formed mud aggregates to form a grain that is much larger than the pellets. Individual constituents may adhere to the sticky mucus-bound aggregate, and fecal pellets themselves may become incorporated within the growing composite grain (Purdy, 1963, pl. 2, A). Some mud aggregates may be formed through the adhesion of fine-grained particles to organic detritus, but it seems likely that most of the mud aggregates have resulted from processes of disaggregation and aggregation of friable pellet material. Thus the preservation as well as the formation of pellets is dependent upon minimal bottom agitation.

The process or processes resulting in the induration of friable fecal pellets and mud aggregates is poorly understood. Individual skeletal grains occurring in the same samples as indurated pellets and mud aggregates lack any sort of carbonate encrustation. Consequently the cement that indurates the pellets and mud aggregates must be considered a product of locally induced precipitation. Calcium carbonate precipitation evidently takes place within pellets and mud aggregates in response to changes in the immediate microchemical environment. The only obvious reason for such changes is the decomposition of the organic mucus contained within these grains; however, the chemical reactions resulting in the interstitial precipitation of cement are unknown.

Not all types of fecal pellets become indurated. The loosely bound feces of holothurians rapidly disintegrate to nondescript organic slimes and never become cemented. Even the firmly bound feces of some organisms fail to become indurated. For example the tapered, rod-shaped fecal pellets of the marine gastropod *Batillaria minima* were observed frequently in sediment samples collected near the northern extremity of the Bimini Islands' lagoon, but very few of these were indurated (Kornicker and Purdy, 1957). In contrast a great number of indurated ellipsoidal pellets was found associated with their non-cemented equivalents in the pellet-mud deposits west of Andros. Perhaps this contrast between indurated and non-indurated pellets can be explained by differences in the composition of the organic mucus that initially binds the friable feces. The bacterial decomposition of one type of organic matter might lead to an increase in alkalinity in the immediate microchemical environment that would induce the precipitation of a carbonate cement, providing the sea water had previously been saturated or near-saturated with respect to calcium carbonate. The decomposition of another type of organic mucus might result in a decrease in the alkalinity of the microchemical environment, in which case a carbonate cement would not be precipitated. Alternatively it may be the character of the depositional environment that determines whether or not the decomposition of organic matter within fecal material results in the precipitation of a carbonate cement. Perhaps both factors account for the preponderance of indurated ellipsoidal pellets in the pellet-mud facies.

The formation of numerous ellipsoidal fecal pellets, then, is dependent upon negligible current action and the consequent accumulation of a large quantity of fine-grained particles. The induration of the pellets necessitates both sea water that is at least saturated or near-saturated with calcium carbonate and physicochemical changes within and around the pellets that cause the precipitation of a carbonate cement. These physicochemical changes are believed to result from the bacterial decomposition of the organic mucus within the pellets.

Lime mud.—The abundance of fine-grained carbonate particles in the sediment west of Andros has attracted the attention of many geologists who see in these deposits a recent counterpart of the many calcilutites of the geologic record. Although these sediments have been the subject of a number of investigations, there is considerable disagreement concerning the origin of the abundant small silt- and clay-sized constituents.

Four general theories of origin have been proposed for the lime-mud deposits: (1) a bacterial origin; (2) a derived origin; (3) a physicochemical origin; and (4) a skeletal disintegration origin.

[*Editors' Note:* Material has been omitted at this point.]

MUD FACIES[8]

The mud facies is distinguished from the pellet-mud facies by a higher content of particles smaller than $\frac{1}{8}$ mm. and a lower content of fecal pellets. The pellets comprise approximately 20 per cent less of the sediment than in the pellet-mud facies, while the reverse is true with respect to the quantity of particles smaller than $\frac{1}{8}$ mm. The abundance of erosional debris in the Bight samples from Andros Island accounts for the small amount of cay rock in the sediment (table 7). The mud facies is restricted in occurrence to the Andros Bights and to parts of the shelf lagoon west of Andros.

Depositional environment.—The water covering this facies frequently exhibits whitings and tends to be less than 3 fathoms deep. There is some suggestion that salinity fluctuations are greater here than in the pellet-mud facies, but other than this vague indication, there are no obvious environmental differences between the pellet-mud and mud deposits. The contrasting constituent composition and grain size of these two facies indicates that some environmental differences exist, but the nature of these is not definitely known.

There are three possible explanations for the relatively high proportion of particles smaller than $\frac{1}{8}$ mm. in the mud facies. First, it is possible that pellet-producing organisms are less abundant in the mud environment than in the pellet-mud environment. The lack of extensive aggregation of silt- and clay-sized constituents to form sand-sized grains would account for the abundance of grains smaller than $\frac{1}{8}$ mm. and also would account for the contrast in pellet content between the pellet-mud and mud deposits. Second, a possible lack of induration of fecal pellets might account for the smaller grain size of the mud facies. The use of hydrogen peroxide in disaggregating the samples for sieve analysis would effectively release the silt- and clay-sized components of the noncemented pellets from their organic matrix and thus would account for the difference in grain size between mud and pellet-mud

[8] This is the skeletal mud facies of Imbrie and Purdy (1962). The name of the facies was changed to emphasize the most obvious and important distinction between the two mud facies, namely, the abundance of indurated pellets.

samples. Third, the small grain size of the mud facies possibly might be the result of excessive deposition of fine-grained sediment in areas attended by minimal current agitation.

An approach to the solution of this problem can be made by totaling the amount of fecal pellets, mud aggregates, and particles smaller than $\frac{1}{8}$ mm. for each of the two facies. Since the pellets and mud aggregates are composed of fine-grained carbonate particles, the totals will reflect the amount of fine-grained material in each of the two facies prior to aggregation. A comparison of the two totals indicates that there is approximately 3 per cent more fine-grained material in the pellet-mud facies as compared to the mud facies. If it is assumed that this difference is significant in terms of indicating relative magnitude of current action, then it is the pellet-mud environment rather than the mud environment that is attended by minimal current velocities. To express it another way the totals indicate that the high proportion of particles smaller than $\frac{1}{8}$ mm. in the mud facies cannot be explained by assuming excessive deposition of fine-grained sediment in that environment. This suggests that the contrast in grain size between the pellet-mud and mud facies is due largely to a decreased rate of formation or induration, or both, of fecal pellets in the mud environment. Further speculation as to the nature and cause of this phenomenon must await the accumulation of additional data.

An interpretation of the genetic relationships among the quantitatively important Bahamian sediment constituents is shown in figure 3.

FACIES PATTERN

Peat deposits are found overlying Pleistocene limestone and underlying unconsolidated sediments both in the Bahamas and in southern Florida (Newell *et al.*, 1959, p. 192–193). Radiocarbon dates indicate that these deposits formed approximately 4,000 to 5,000 years ago when the Pleistocene post-Wisconsin rise in sea level first flooded these areas (*ibid.*). Prior to this inundation the low stand of the Wisconsin sea had caused the Great Bahama Bank to be subaerially exposed, as indicated by the numerous drowned sinkholes occurring on the bank (Newell and Rigby, 1957). The combination of a rising sea level flooding a karst surface gave rise to the somewhat quiltlike pattern of Bahamian facies.

As sea level rose, the uneven karst surface superimposed local increases and decreases in current velocity on the regional tidal regimen. The shoaler unsheltered areas were characterized by considerable turbulence and consequent oölite formation. The deeper or more sheltered localities were characterized by bottom stability and consequent grapestone or lime-mud formation. Initially shoal areas might subsequently become less shoal, provided the rate of oölite accumulation did not keep pace with the rise of sea level. In these instances the gradual increase in water depth would result in a progressive decrease in bottom agitation and a consequent deceleration in the rate of oölite formation. Such a sequence of events would eventually result in a relict oölite deposit. Further decrease in turbulence might cause the ooids to become cemented together to form grapestone. The gradual rise in sea level also would submerge successively higher parts of the subaerially exposed platform, and as a result the main sites of oölite formation might be expected to shift to the newly inundated shoals.

The sheltering effect of initially shoal areas might be expected to decrease as sea level rose to its present position, assuming, of course, that the rate of oölite accumulation did not maintain the shoalness. In these instances formerly quiet water environments would gradually become more current-agitated. An initial lime-mud deposit might have its smaller constituents winnowed away and the residual fecal pellets and mud aggregates might become cemented together to form grapestone; or the change toward increasing bottom agitation might be sufficiently rapid to cause oölitic accretion around pellet and mud aggregate nuclei. Such a change would account for the large

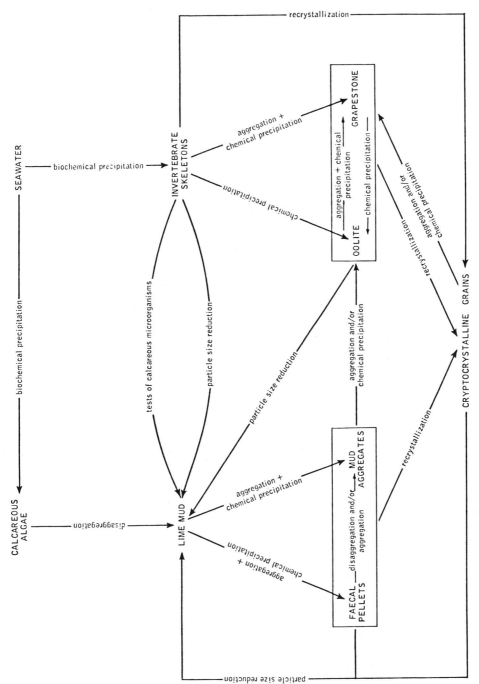

Fig. 3.—Genetic relationships among quantitatively important Bahamian sedimentary constituents

proportion of pellet nuclei in Bahamian ooids. The relict nature of some oölite deposits has been noted previously. Some parts of the grapestone facies also appear to be relicts. For example, the 2,500 ± 150 years date of grapestone sample 529 approximates the 2,330 ± 100 years date obtained by Newell et al. (1959, p. 195) for the top of the

mud core taken in an Andros mangrove swamp. Moreover both these dates compare favorably with the 2,530 ± 100 years age of the inner 20 per cent of the ooids collected in the Browns Cay area (Newell et al., 1960, p. 489). In contrast the 740 ± 100 years age of intact ooids from the same area and the 700 ± 150 years date of the composite

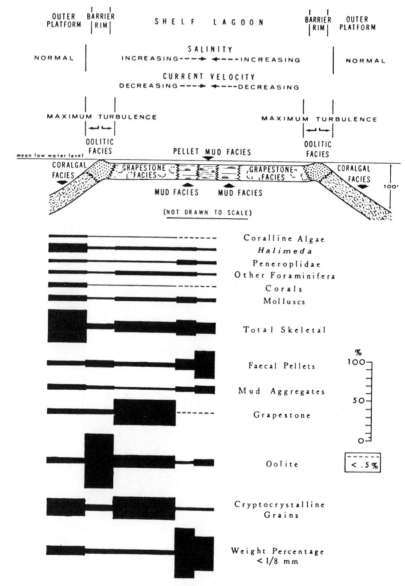

FIG. 4.—Variations in mean constituent particle composition and grain size of Bahamian facies on an idealized Bahamian platform.

pellet-mud sample (samples 382 and 384) demonstrate a more recent origin for these constituents. These data indicate the relict nature of grapestone sample 529. Similarly the severely recrystallized aspect of many of the non-skeletal constituents occurring on the outer platform in both the coralgal and grapestone facies suggest that these constituents also are products of a lower stand of sea level.

Although relict deposits certainly complicate the facies pattern, it is primarily the karst surface of the submerged Pleistocene limestone that causes the pattern to depart most from an idealized sequence of concentric facies bands. With the exception of the coralgal facies the development of one or the other of the remaining four facies is dependent at least in part upon relative current intensity. Consequently the local increases and decreases in current velocity induced by the karst surface cause the local development of facies. Thus only a crude approximation to a concentric pattern is evidenced by the distribution of Bahamian facies.

CONCLUSION

The smooth limestone surface of an idealized Bahamian platform such as that shown in figure 4 would be expected to have a series of five facies concentrically arranged with respect to the periphery of the bank. The outer platform is characterized by large areas of rock bottom colonized chiefly by corals and coralline algae. Water moving across this area is not sufficiently warmed or evaporated to induce calcium carbonate precipitation. Therefore non-skeletal constituents are relatively rare in the sediment of this area. The lack of dilution of skeletal debris by non-skeletal grains results in a predominantly skeletal sediment, character-

ized chiefly by the remains of corals and coralline algae derived from the adjacent rock bottom. Bankward of the outer platform the progressive decrease in water depth causes an increase in current velocity in the same direction, maximum bottom agitation being attained at the barrier rim. As turbulence increases, there is a concomitant warming and evaporation of the bankward moving water masses. These conditions decrease the solubility of calcium carbonate, which precipitates as oölitic coatings around nuclei. The optimal precipitating conditions at the barrier rim cause the oölite accumulations to be built up to an intertidal level. Because of the phase lag in tidal oscillations between the shelf lagoon and the barrier rim, tidal current velocities decrease toward the center of the bank. Consequently sand-sized grains tend to remain in contact with each other for longer and longer periods bankward of the barrier rim and tend to become cemented together to form grapestone. With further bankward decrease in current velocity, abundant silt- and clay-sized carbonate particles are deposited. Many of these grains are pelleted subsequently by organisms Partial disaggregation of the initially friable fecal pellets results in the formation of mud aggregates. The precipitation of a calcium carbonate cement within the pellets and mud aggregates causes them to become indurated. Unknown environmental factors probably account for the relative lack of indurated pellets in the mud facies as compared to the pellet-mud deposits.

The evaporation and heating of calcium carbonate saturated bank waters causes precipitation to occur, but the degree of current intensity determines the relative abundance of the various non-skeletal grain types.

REFERENCES CITED

BAVENDAMM, W., 1931, On the possible role of microorganisms in the precipitation of calcium carbonate in tropical seas: Science, v. 73, p. 597–598.

———— 1932, Die microbiologische Kalkfallung in der tropischen See: Archiv. für Microbiol., v. 3, p. 205–276.

CLOUD, P. E., JR., 1959, Geology of Saipan, Mariana Islands, part 4, Submarine topography and shoalwater ecology: U.S. Geol. Survey Prof. Paper, 208-K, p. 361–445.

———— and BARNES, V. E., 1957, Early Ordovician

sea in central Texas: Geol. Soc. America Mem., no. 67, v. 2, p. 163–214.

DREW, G. H., 1914, On the precipitation of calcium carbonate in the sea by marine bacteria, and on the action of denitrifying bacteria in tropical and temperate seas: Papers Tortugas Lab., v. 5, Carnegie Inst. Washington Pub. 182, p. 7–45.

FIELD, R. M., 1931, Geology of the Bahamas: Geol. Soc. America Bull., v. 42, p. 759–784.

GARRELS, R. M., THOMPSON, M. E., and SIEVER, R., 1959, Solubility of carbonates in sea water; control by carbonate complexes (Abstr.): Geol. Soc. America Bull., v. 70, p. 1608.

GEE, H., MOBERG, E. G., GREENBERG, D. M., and REVELLE, R., 1932, Calcium equilibrium in sea water: Scripps Inst. Oceanogr., Univ. of California Bull. Tech. Ser., v. 3, p. 145–200.

GINSBURG, R. N., 1953, Carbonate sediments: A.P.I. Proj. 51, Rpt. 8, S10 Ref., p. 26–53.

—— 1956, Environmental relationships of grain size and constituent particles in some South Florida carbonate sediments: Am. Assoc. Petroleum Geologists Bull., v. 40, no. 10, p. 2384–2427.

—— 1957, Early diagenesis and lithification of shallow-water carbonate sediments in South Florida, In Regional aspects of carbonate deposition: Soc. Econ. Paleontology and Mineralogy, Special Pub., no. 5, p. 80–100.

—— and LOWENSTAM, H. A., 1958, The Influence of marine bottom communities on the depositional environment of sediments: Jour. Geology, v. 66, p. 310–318.

ILLING, L. V., 1954, Bahaman calcareous sands: Am. Assoc. Petroleum Geologists Bull., v. 38, no. 1. p. 1–95.

IMBRIE, J., and PURDY, E. G., 1962, Classification of modern Bahaman carbonate sediments, Classification of carbonate rocks: Am. Assoc. Petroleum Geologists, Mem. 1.

KELLERMAN, K. F., and SMITH, N. R., 1914, Bacterial precipitation of calcium carbonate: Jour. Washington Acad. Sci., v. 4, p. 400–402.

KORNICKER, L. S., 1959, Ecology and taxonomy of Recent marine ostracodes in the Bimini area, Great Bahama Bank: Pub. Inst. Marine Sci., v. 5, p. 194–300.

—— and PURDY, E. G., 1957, A Bahamian faecal-pellet sediment: Jour. Sed. Petrology, v. 27, p. 126–128.

LIPMAN, C. B., 1922, Does denitrification occur in sea water? Science, v. 56, p. 501–503.

—— 1924, A critical and experimental study of Drew's bacterial hypothesis on CaCO₃ precipita-

tion in the sea: Papers Tortugas Lab., v. 19, Carnegie Inst. Washington Pub. 340, p. 179–191.

—— 1929, Further studies on marine bacteria with special references to the Drew hypothesis on $CaCO_3$ precipitation in the sea: Papers Tortugas Lab., v. 26, Carnegie Inst. Washington Pub. 391, p. 231–248.

LOWENSTAM, H. A., 1955, Aragonite needles secreted by algae and some sedimentary implications: Jour. Sed. Petrology, v. 25, p. 270–272.

—— and EPSTEIN, S., 1957, On the origin of sedimentary aragonite needles of the Great Bahama Bank: Jour. Geology, v. 65, p. 364–375.

NEWELL, N. D., and RIGBY, J. KEITH, 1957, Geologic studies on the Great Bahama Bank, In Regional aspects of carbonate deposition: Soc. Econ. Paleontology and Mineralogy, Special Pub., no. 5, p. 15–79.

—— IMBRIE, J., PURDY, E. G., and THURBER, D. T., 1959, Organism communities and bottom facies, Great Bahama Bank: American Mus. Nat. Hist. Bull. v. 117, art. 4, p. 117–228.

—— PURDY, E. G., and IMBRIE, J., 1960, Bahamian oölitic sand: Jour. Geology, v. 68, p. 481–497.

PURDY, E. G., 1961, Bahamian oölite shoals, In Geometry of sandstone bodies: Special Vol., Am. Assoc. Petroleum Geologists, p. 53–62.

—— 1963, Recent calcium carbonate facies of the Great Bahama Bank 1. Petrography and reaction groups: Jour. Geology, v. 71, p. 334–355.

REVELLE, R., and FLEMING, R. H., 1933, The solubility product constant of calcium carbonate in sea-water: Proc. 5th Pacific Sci. Cong., v. 3, p. 2089–2092.

RUBIN, M., and ALEXANDER, C., 1958, U.S. Geological Survey radiocarbon dates: Science, v. 127, p. 1476–1487.

SMITH, C. L., 1940, The Great Bahama Bank: Jour. Marine Res., v. 3, p. 147–189.

SMITH, N. R., 1926, Report on a bacteriological examination of "chalky mud" and sea-water from the Bahama Banks: Papers Tortugas Lab., v. 23, Carnegie Inst. Washington Pub. 344, p. 67–72.

SORBY, H. C., 1879, The structure and origin of lime stones: Geol. Soc. London Proc., v. 25, p. 56–95.

VAUGHAN, T. W., 1917, Chemical and organic deposits of the sea: Geol. Soc. America Bull., v. 28, p. 933–944.

UNITED STATES HYDROGRAPHIC OFFICE, 1867, Great Bahama Bank, northwestern part: 47th ed., 1932, Washington, chart 26a, from British surveys between 1836 and 1848.

4

Reprinted from pages 742–743 and 748–783 of Geol. Rundschau **61:**742–783 (1972)

RECENT ALGAL STROMATOLITIC DEPOSITS, ANDROS ISLAND, BAHAMAS. PRELIMINARY REPORT

Claude L. V. Monty

Abstract

After some introductory comments on fresh water bluegreen algal calcareous formations, the general features and the ruling mecanisms of which are of fundamental importance for the interpretation of stromatolites, the distribution and general features of algal deposits across Andros Id, Bahama, and adjacent marine platforms are briefly reported.

Morphological differenciations characterizing Recent stromatolitic biostromes in process of formation over the flats of seasonal lakes (N-Eastern Andros) are then briefly analyzed. The interpretation of structures relies on the various data gathered during an observation period of three months encompassing the phase of prolific stromatolitic growth during the complete flooding of the lakes, and ending with the progressive drying up and exposure of the algal flat at the beginning of the dry season.

[*Editors' Note:* Material has been omitted at this point.]

113

Bird's eye view of the distribution of blue-green algal structures across Andros Id and adjacent marine banks.

West of Andros, i. e. in the lee of the island, the proximal Great Bahama bank is covered by highly saline to hypersaline shallow waters; the generally muddy bottom supports algal lumps and algal limestone flakes (PURDY, 1963 a, 1963 b; ILLING, 1954) whereas important areas are covered with mats of *Lyngbya aestuarii* (BATHURST, 1967) [4]; this halophylic alga acts as a bottom stabilizer and traps grains in the baffle-like carpet of its long flexible filaments; as it grows almost continuously and faster than the depositional rate of the constantly shifted detrital grains, it does not yield banded stromatolitic-like deposits but rather accumulations similar to the ones associated with the turtle grass. Furthermore, beneath the living mat, in which the filaments can grow flat or erected according to the local conditions, many slender filaments permeate the substrate in all direction; seasonally, when the alga blooms, it can peel off the substrate, as was observed by the author, and carry away the whole mat and underlying sediments; this process also prevents the formation of stromatolitic laminations that would result from the piling up of seasonal layers.

The *intertidal zone proper* is rather limited in extent with respect to the Bahama Bank and the supratidal flats [5]. Life conditions change abruptly, as organisms have to withstand daily exposures and floodings whereas the tides *periodically* wash detrital grains over the mats. The most conspicuous alga in the lower half of the intertidal zone is probably *Schizothrix calcicola* a very tiny blue-green alga which agglutinates the grains in a mesh of filaments and mucilage; generally, it does not originate laminated structures there, for these tiny algae are constantly swamped with grains during the flood and cannot add much growth during the ebb when they suffer severe conditions (see MONTY, 1967, p. 86). On somewhat exposed headlands, these mats may be interspersed with cushions and growths of *Rivularia* which has much bigger tufted filaments

---

[4]) The interested reader will find penetrating studies on the structure and sand stabilizing potentials of the *Lyngbia* mats in SCOFFIN, 1970, and NEWMAN et al., 1970.

[5]) A very interesting description of the morphology of these inter-to supratidal settings has been recently published by SHINN et al., 1969.

between which many grains remains trapped; as opposed to the fresh water *Rivularia haematites*, these *Rivularia biasolettiana* do not build laminated nor calcified structures. In the upper part of the intertidal zone (SHINN et al., 1969) as well as in many brackish tidal marshes appear conspicuous laminated growths of *Scytonema* stat. *crustaceum;* this alga, which on the West as well as on the East coast is characteristic of transitional environments between the sea and the land, has deeply stained sheaths which absorb much of the noxious solar radiations; the thickness of the mat and the prominence of the lamination varies according to local conditions; precipitation of fine-grained carbonate and local cementation by tiny carbonate crystals appears in the mats (SHINN et al., 1969, Fig. 12, p. 11) and seem to increase up flat (MONTY, 1965 b, 1967).

The *supratidal zone* comprizes a variety of environments among which the flats and the marshes are the more interesting ones to our purpose. The phases of algal growth, of sedimentary influx and of dessication become now strongly separated in time with respect to their periodicity in the intertidal zone, whereas their pattern changes also: periodical floodings of the flats are now due to exceptionnally high tides and more generally to seasonal storms or hurricanes; there is no more continuous or subcontinuous washing of particles over the mats but a sudden and brutal influx of sediment-laden waters which deposit a continuous sheet of detrital grains over the algal mats; according to the conditions of substratal humidity, the algal filaments will either permeate the storm deposit more or less rapidly to re-establish a new mat on top, or remain in a "dormant" state just beneath the surface. During the rainy reason, or when westerly winds pile up waters onto the flats, a second seasonal phenomenon occurs: the suddenly re-juvenated filaments grow and bloom to originate a significant algal layer. These particular processes, generally widely separated in time, give birth to the typical algal laminated sediments found in this supratidal zone. They are however complicated by several secondary factors: in such harsh conditions the surficial algal films may become dormant or just survive without adding any growth to the structure; they then act as a deeply stained screen which shades the underlying filaments from noxious radiations and reduce evaporation of the substratal water; at this stage, most of the growth is *internal* and the mat thickens *from the inside* (as discussed later in this paper). As far as the mineral matter is concerned, strong diagenetic phenomena may affect the deposits and end up with the formation of dolomite-rich layers or crusts (SHINN et al., 1969).

The marine supratidal flats and associated structures pass, sometimes progressively, sometimes very rapidly, to *fresh water environments* characterized by the "terrestrial" mat building alga *Scytonema myochrous* which is everywhere on the island; the dominant sediment is a very fine-grained white calcitic mud which floors the lake bottom or forms a very thin layer over the Pleistocene basement in low lying areas; fine grained calcite is also abundantly precipitated in the *Scytonema* mats which however remain generally unlithified. This general environment will be described in the following pages (see also BLACK, 1933; MONTY, 1965 b, 1967).

The distributionnal pattern of algal mats from the fresh water settings to the windgard lagoon and reef have been described in MONTY (1965 b, part II, 1967) and will be rapidly summarized here.

According to the local coastal geomorphology, fresh water environments may pass directly to the typical marine intertidal zone of the East coast, or pass progressively to brackish water settings which characterize the "swash" (that is a system of protected tidal marshes which develope just behind the coastal ridges and are more or less connected to the sea by a system of channels and creeks). There where a continuous and regular increase in salinity can be observed, the algal mats show a progressive decrease in the participation of *Sc. myochrous,* while the concurrent euryhaline alga *Schizothrix* becomes more and more abundant; this is reflected in the progressive transformation of the organization of mats which pass progressively from type 1 to type 3 and then to type 2 of MONTY, 1967 (p. 70—72).

The *swash* proper is characterized by salinities averaging $25^0/_{00}$, the replacement of *Sc. myochrous* by *Sc. statt. crustaceum,* a net decrease in the precipitated calcite, and a strong contamination of mats by debris of marine organisms and a resulting increase in the complexity of the mineralogy of mats. Here are found

(1) laminated mats and crusts built by *Sc. statt. myochrous* (together with *Schizothrix calcicola* and *Entophysalis deusta)* which are characterized by an alternation of rather plain algal layers with layers of trapped detrital grains (MONTY, 1967, p. 80—82);

(2) crusty flakes, built by *Schizothrix* and unicells (Id. p. 77—80), which are very reminiscent of the "limestone flakes" that PURDY (1963 a, 1963 b) described from the Great Bahama Bank, and

(3) various types of algal lumps (MONTY, id. p. 82—84).

Reaching the East coast, we hit true *intertidal settings* which are no more (or very rarely) hypersaline as was the case on the West coast, but face the windward lagoon and reef. The entrapment of detrital particles becomes the leading process in the mineralization of mats and very little precipitation has been observed by the author. Non laminated mats are built by *Rivularia biasolettiana* (MONTY, 1967, p. 85, pl. 16), *Dichothrix bornetiana* (id. pl. 15), *Schizothrix calcicola* (id. pl. 16, Fig. 2, pl. 17, Figs. 1 and 2). There where local conditions are not too severe, i. e. where the influx of bioclastic material is not too heavy and where the moisture is sufficient at low tide (zone of splash, or on muddy substrate), *Schizothrix calciccla* grows in pure cultures and forms laminated domes and mats in which the lamination is daily (MONTY, 1965 a, 1967 b, p. 88—92). BLACK (1933) also reported intertidal laminated deposits where all of the mineral matter is similarly composed of detrital grains: his "type A", which is built by the filamentous blue-greens *Symploca laete-viridis* and *Phormidium tenue,* appears as rather flat mats growing in the uppermost intertidal zone; the sedimentary layers show evident traces of flood deposition such as a positive grain sorting from base to top (these structures should probably be put together with the supratidal ones); his "type B", which forms at about high water mark, occurs as conspicuous laminated domes built by a rich community of unicells (*Gloeocapsa and Entophysalis*) permeated by filamentous algae.

The most abundant stromatolites found in the *windward lagoon* (the depth of which never exceeds 10 feet) are undoubtedly the laminated domes built by pure cultures of *Schizothrix calcicola* (MONTY, 1965 a, 1967, p. 88—92); these domes proliferate seasonally and are most conspicuous in Summer. Sand sheets binded by *Lyngbya aestuarii* and similar to the ones found on the Great Bahama Bank may also be locally abundant.

Some *Schizothrix* laminated domes are still found on the reef proper but most of the structures become generally loose as they are invaded by their cousin competitor, the blue-green *Lyngbya*, the long and slender filaments of which disorganize the compact structure of the basic *Schizothrix* domes.

This bird's eye view of the general distributional pattern of blue-green algal structures across Andros Id. and adjacent marine platforms shows significant differenciation that would still be more prominent if the characteristics of the associated fauna, flora and sediments were considered. The purpose of the author's actual project is now to consider and study in detail each particular settings surveyed previously; the present preliminary report will focuss on fresh water settings and will tend to describe rapidly the general features of vast stromatolitic flats which crowd many inland lakes. Detailed relationships and microstructures of mats will be exposed elsewhere.

General setting of fresh water stromatolitic flats
(Fig. 1)

West of Andros Town, passed a belt of densely vegetated oolitic coastal ridges, the island shows an alternation of wooded area (pine barren) or of strands and patches of intricate bush characterized by *Sabal bahamensis*, with desolate, low lying savanah-like regions crowded with sawgrass and some small *Avicennia;* the pitted oolitic basement may outcrop at the margin of these very shallow depressions but it is most of the time covered by no more than 30 cm. of white, sticky and calcitic mud (the thickness of this mud is the greatest in the central "deepest" part of the depression).

Much of these "plains" are completely flooded during the rainy season but dry up rapidly when rain recesses, leaving a ground covered with shrunk, contorted and dried up "polygons" (BLACK's type D).

Other depressions, a little bit lower still in altitude, remain flooded during a much longer period of time and originate this lake system that characterizes Andros Island (Fig. 2). Most of these lakes are not directly connected with the sea by tidal channels or creeks as occurs on the N.W. half of the island; they are accordingly fresh; strickingly enough some of them may show small tidal fluctuations, which means that they would rest on a lense of sea water through an intricate pattern of underground pits and holes which open in the Andros lagoon or even on the upper wall of the tongue of the ocean.

In many instances the lakes show 2 distinct physiographic parts (Fig. 3, 6):

a) A broad eulitoral zone, flooded at least twice a year (late spring and late fall-winter) and characterized by a rather regular floor, sometimes subhorizontal, sometimes sloping very gently toward the central part of the lake. We shall refer to it as the s t r o m a t o l i t i c f l a t.

b) One (sometimes two) central permanent lakelet or pond, 1 to 10 feet deep during the rainy season. These ponds are generally floored with a white, very clotty mud, appearing like unconsolidated bird's eye limestone; each constitutive clot, ranging from a few hundred microns to about 2 mm in diameter, is formed of an agglomerate of diatoms and blue-

117

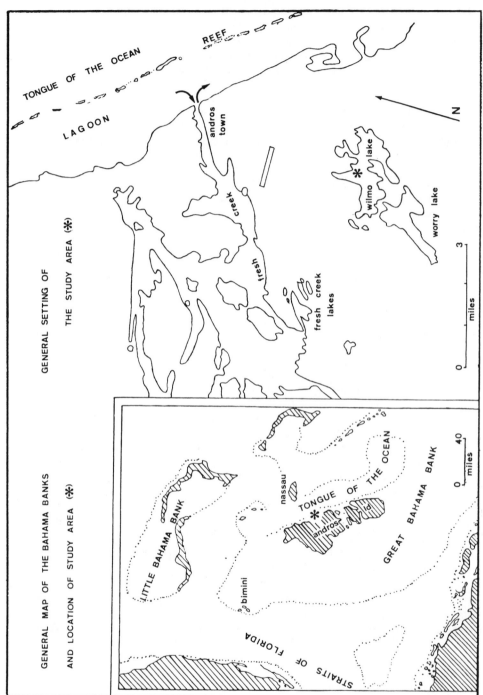

Fig. 1. Location of study area.

green algae surrounded by an aboundant mucilage loaded with crystals of calcite; such substrate composed of about 70% water and 30% mineral and organic matter, is generally too soft for the establishment of stromatolites; I have however seen instances in central Andros where tiny films of *Schizothrix* first grow over the deposit, "floating" initially over it, then thicken progressively to originate a subcontinuous buoyant algal mat up to 10 cm thick stretching over the substrate which progressively compacts.

The morphology as well as the chemistry of these lakes (tab. 1) fall into that of eutrophic "cyanophycean lakes", were it not for the phosphate content which apparently is too low here and the ratio Na/K which is too high.

Particular attention was given to Wilmo and adjacent lakes (Fig. 6), 3 to 4 miles South of Andros town (Fig. 1).

Table 1. Routine water analysis of 3 fresh water lakes. Sampling was done during the mid water level (January 1970). Complete alkaline titre (T.A.C.) measured at pH 4.6, and given in m eq.gr./1.

	pH field	pH lab	T.A.C.	Ca mgr/l	Mg mgr/l	Na mgr/l	K mgr/l	Cl- mgr/l	PO$_4$--
Wilmo lake (Fig. 1)	8	7.1	1.93	29.8	6	28.8	2.6	57	traces
Worry lake (Fig. 1)	8	7.3	1.55	25.6	5.2	25.3	1.8	55	not detected
Unnamed lake, centre of N. E. Andros	8.1	7.1	1.52	26	3.6	20.6	1.9	48	not detected

The lakes and their flats were studied at the end of the winter period of flooding, that is from January to early March; the water depth over the flats was about 1—2 feet (Figs. 4, 6) when the survey was undertaken; stromatolites were fully growing at that time. Observations were carried until the lake dries out almost completely and the stromatolite growth recesses (Fig. 7). Because of the poor quality of underwater photographs, only terrestrial photographs, taken during the shallowing up of the lake, will be given.

Zonation of the algal flat

The description of the algal flat will be presented under the form of an ideal cross section from the margin of the flat toward its central deepest part, the idea being to show the variation of stromatolite morphologies and behaviour in a very quiet and uniform environment.

The general morphological zonation originates in depths variations of but a few centimeters due to the general slope of the flat and local irregularities of its floor; these centimeters have a fundamental importance on such somewhat flat-bottomed temporary lakes in that they determine:

(1) The length of time during which prolific algal growth will last underwater. The greater the original depth of water, the longer the underwater period of growth before the lakes dries out.

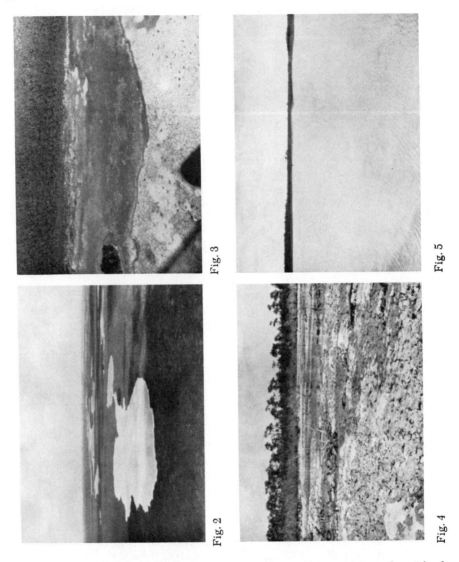

Fig. 2. Fresh water lakes at the beginning of the dry season, N.E. Andros Island. Strands of pines or bush patches appear in dark whereas low lying "plains" covered with contorted algal polygons appear much lighter.

Fig. 3. Vertical view of the margin of a fresh water lake; from top to base of the photograph: pine barren; algal flat stretching over the eulittoral zone; permanent residual lakelet floored with calcitic mud.

Fig. 4. Wilmo Lake during high water level (late December).

Fig. 5. Wilmo Lake; uppermost eulittoral zone. Shrunk algal polygons over Pleistocene substrate; mid low water level.

Fig. 6. View toward the North of 2 water bodies of Wilmo lake separated by a small transverse ridge — photograph taken during mid water level. The eulittoral zone covers the whole algal flat; in the Northern arm, the flat shows a central depression floored with white calcite mud, depression which constitutes a residual lakelet during the dry season. A drainage channel appearing as a white line connects the two arms.

Fig. 7. Wilmo Lake — mid eulittoral zone. Stromatolite flat proper during very low water level (early March); the wide flat shows a pavement of well developed algal domes 10 to 20 cm in diameter.

(2) Slight and/or local variations in the depth of the phreatic water when, during the progressive drying up of the lake, stromatolites are confined to grow subaerially on a more or less humid substrate. Other important superimposed factors will be mentionned when necessary.

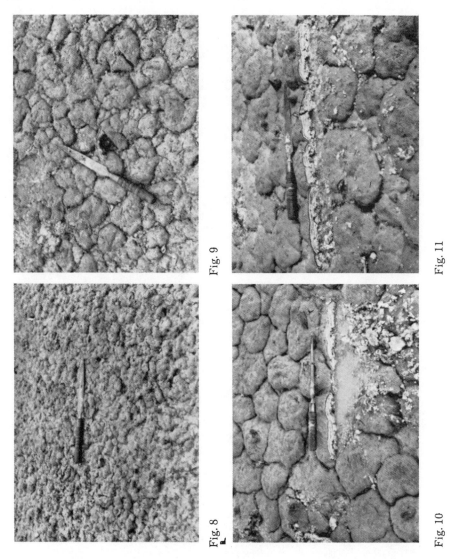

Fig. 9

Fig. 11

Fig. 8

Fig. 10

1. Upper shore

The highest parts of the shore are the least proper to the development of significant stromatolitic deposits as this belt is flooded during the shortest period of time; the Blue-Green algal mats, that grow hastingly during the very brief flood, dry up and shrink rapidly to form sorts of torn down draperies around the rock edges, irregular cracked films scattered over the drying mud flooring eventual depressions, or yet turn to sorts of polygons identical to the ones that cover huge areas of Andros Island (Fig. 5).

Locally, putrid and reducing areas of the shore are covered by a thick algal grit overlying a dark brownish mud; this greenish-brown grit, which was originally considered as reworked fragments of algal mats, is in fact composed of organic particles up to $^1/_2$ cm in size, exclusively made of huge colonies of the Blue-Green alga *Gloeocapsa* surrounded by very thick protective mucilaginous deposits.

2. Belt of smooth stromatolites

The main stromatolitic deposit starts in the mid eulittoral zone and develops downflat toward the central residual lakes (Figs. 6, 7). The upper belt shows subcontinuous irregular structures (Fig. 8) resting on rock or on algal grit; they are about 2 mm thick, very irregular in vertical section and happen to show 2 poorly defined laminae. No doubt these structures are destroyed seasonally.

A transitionnal zone (Fig. 9) brings about somewhat larger domes, about 5—10 cm in diameter and 3 mm in thickness, showing four to five laminae; they no not form coherent structures and are easily disrupted when the water cover disappears; a reason may be that they generally rest on coarse algal grit which is an unstable substrate, the more that anaerobic processes in the grit originate gaz production which disrupts the overlying domes. This transitionnal zone passes to the stromatolitic flat proper, which may be several hundred meters wide (Fig. 7) and displays a variety of rather contiguous perfectly individualized domes.

The first series (Fig. 10) is characterized by subcircular, rather flat and smooth-topped domes, about 10—15 cm in diameter. They show a prominent lamination made of alternating micritic, calcitic layers and highly organic somewhat soft layers; the general microstructure is very similar to the one I described from algal domes growing in sinkholes and originating in the combined growth of *Schizothrix calcicola* and *Scytonema myochrous* (MONTY, 1965, 1967). These, stromatolites remain rather thin (2—3 cm) when growing in rock, but usually get much thicker when on mud (Fig. 22) in which the lamination extends (Fig. 13).

Down the flat, i.e. but a few cm "deeper", the prolonged underwater growth of these subcircular domes originates amoeboid stromatolites which spread in every directions and may overgrow each other; the eventual interstices or depressions between adjacent domes are rapidly filled by lateral expansions of some prosperous domes or the rapid development of interstitial mats (Fig. 11). This

---

Fig. 8. Wilmo Lake; upper mid eulittoral zone. Irregular and very thin stromatolitic deposit at the margin of the stromatolitic flat proper: length of knife 35 cm.

Fig. 9. Transitionnal zone toward the stromatolitic flat proper; deposit made of rather small and thin algal domes, though much more clearly individualized than Fig. 8.

Fig. 10. Belt of smooth stromatolites, mid eulittoral zone. Domes subcircular in outline, about 15 cm in diameter, characterized by a very regular and smooth upper surface; prominent and persistant lamination appears in vertical section. Length of knife: 35 cm.

Fig. 11. Very closely spaced, elongated to amoeboid algal domes, which may locally overgrow each other; in the lower half, note the filling up of eventual interstices between algal domes by lateral stromatolitic expansions. Length of knife: 35 cm.

Fig. 12   Fig. 13

Fig. 12. Lower mid eulittoral zone: transition downflat of algal domes covered with small pinnacles to column-bearing stromatolites.

Fig. 13. Vertical section through a core taken under the stromatolite deposit; shows the piling up of highly compacted stromatolitic domes resting on white calcitic, strongly reworked mud.

over and lateral growth may bring about the fusion of individual stromatolites to form larger compound structures.

3. Belt of stromatolites with pinnacles and columns

The first phase of differenciation, that we have followed up to now, was rather straightforward and monotonous; it consisted primarily in the succession downflat of larger and thicker stromatolites, most of which were characterized by a rather regular and smooth upper surface. The prolonged underwater growth toward the center of the temporary lake (i.e. the actual flat) is now going to originate new structures superimposed on the previous ones.

Columns. The appearance of erected columns, up to 5—6 cm high and 3—4 cm wide (Fig. 14) overgrowing stromatolitic domes, indidualizes a prominent belt. The formation of these columns is twofolds: active clones of *Scytonema/Lyngbya* have time to bloom, as a result of a longer underwater growth; they grow hairy tufts, a few cm high, composed of juxtaposed flexuous filaments grouped in vertical bundles. Once formed, these tufts constitute a core, a new substrate, that will be overgrown by successive mats, as any other substrate, to originate a rigid column (Fig. 15).

Such columns can only form underwater and at an appropriate depth. In the initial stages indeed, the growth of the tufts results from the irregular piling up of slightly interwooven and flexuous successive bundles of filaments; these structures, which collapse at once when taken experimentally out of the water, stand vertically because of their buoyancy and of the support provided by the water mass. They become a little bit more coherent after encrustation by thin calcified films of *Schizothrix* which drape themselves around this initial substrate; the successive blue-green algal communities that can now settle on a firmer substrate will consolidate the structure, and originate a concentric banding (in cross section) similar to the one which they form around pieces of wood (Fig. 31).

By their very structure, the initial tufts are very fragile and are easily destroyed or "reaped" by heavy rains; that is why they cannot form at the periphery of the eulittoral zone where the original water depth does not afford the required protection at the time of their formation. As for the tufts which appear in the right zoneographic belt, but too late in the season (that is when the water level gets low and drops more and more rapidly), they will not be preserved either; they will collapse as the lake dries out, or will be destroyed by the atmospheric agents: this is the fate of most of the tufts shown on Fig. 14.

The base of beheaded tufts may remain firmly attached to the underlying stromatolitic dome and become encrusted by seasonal mats to form small rounded projecting knobs such as shown on Fig. 16.

These colums, composed of a core surrounded by concentric cones formed by successive encrustations may be quite reminiscent of "mini-Conophyton". Furthermore, as shown on Fig. 14, some columns are vertical, others variously inclined while some did collapse on the stromatolitic substrate in which they are reincorporated by successive overgrowths.

Pinnacles. Quite often, between the littoral belt crowded with rather smooth domes and the belt where columns do form, may appear a transitional zone, characterized by stromatolites overgrown by small pinnacles (Fig. 12). The first pinnacles to differenciate are only a few millimeters high but they become rapidly more conspicuous farther offshore (Fig. 18) i.e. toward the zone with columns in the ideal scheme.

The pinnacles appear as incipient tufts of *Scytonema* that would not have had time nor chances to grow into prominent columns, during the span of the seasonal growth, and would have been readily encrusted by successive mats. This is justified by their being confined to an intermediate belt, a little bit higher on the flat than the zone where columns do form, or by their capping local reliefs protruding through the latter zone; this means that the incipient tufts were subjected to a shorter period of underwater growth and during the drying out of the lake, were more rapidly exposed to surface turbulence.

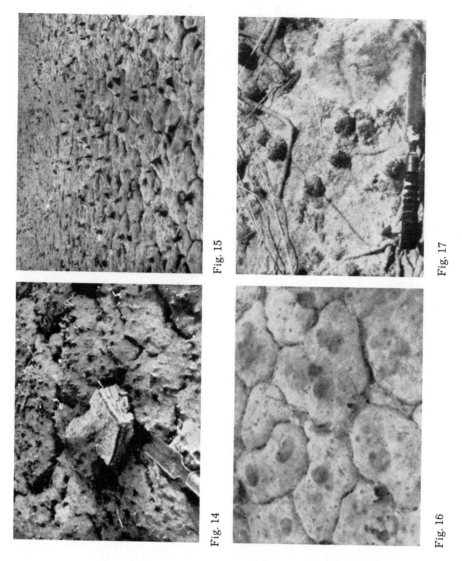

Fig. 15

Fig. 17

Fig. 14

Fig. 16

The influence of the wind rippling the surface somewhat alters the overall morphology of these pinnacles: in settings directly exposed to the trade winds, pinnacles are indeed deflected downwind and grow flat on the stromatolites, originating a lineation parallel to the wind direction (Fig. 19). If they are sufficiently close to each other, the deflected pinnacles may coalesce laterally to form subcontinuous sets of ridges perpendicular to the trade winds.

According to the slope and the local geomorphology, these 2 types of pinnacled stromatolites may be by far the most abundant on wide flats exposed to the trade winds. This may be accounted for by the fact that the general overall conditions are favorable to a prolific growth, hence the "littoral stage" represented by the flat smooth stromatolites is by-passed, but the water depth is too small and the surface turbulence too high to permit the formation and/or the preservation of columns.

4. Belt of competition

A third set of modifications, found toward the central part of the flats or in local pools, occurs when the original water depth is such that the floor remains submerged a longer period of time still, than in the precedent zone with columns.

In general, the stromatolite deposit becomes more continous (Figs. 17, 20) and the individuality of the domes is lost as a result of lateral fusion and accretion. The reeds that grow in this belt are encrusted on about 30—40 cm of their height; this gives an idea of the minimum water depth during the flood; when water recesses, they collapse over the stromatolite deposits and become part of it (Fig. 17).

Stromatolites also become abundantly covered with small rod-shaped coprolithes of gastropods and fishes that continuously feed on the algae (Fig. 17). The deposit may furthermore be invaded by green algae and a variety of hard water phanerogames (Fig. 20) to the point that its surfical features become much less apparent; in vertical section however, the classical lamination still clearly appears.

This complex community which now developes (Fig. 20) defines areas where the relative stability of the water cover makes it possible for less resistant plant to settle and compete with blue green algae; they generally start their growth in the eventual small valleys and depressions cutting through the stromatolitic deposit and spread over the floor. The upper surfaces of the leaves of these higher plants (such as *Elodea*) is generally covered with a brittle scaly layer of calcium carbonate; this shows, beside microscopic examination and the presence of whitings, that active precipitation of calcite goes on in the lake. This community generally leads to the rim of the permanent residual lakelet floored, as seen above, with "bird's eye" type of mud.

Fig. 14. Zone of columns, lower mid eulittoral zone. Algal domes are overgrown by tufts of *Scytonema* partly encrusted by films of *Schizotrix;* encrusted tufts appear light gray with rounded tips. The non, or partially, encrusted tufts (incipient columns) look much darker, with acute tips and seem to protrude through the algal dome. Note the various orientations of the tufts and columns, some of which collapsed on the underlying domes. Most of the non encrusted tufts will be destroyed by rain; they formed late in the season and became exposed before being consolidated by sufficient overgrowth (see Fig. 15).

Fig. 15. Tuft of *Scytonema* that has been encrusted by *Schizothrix* and various algal communities to form a coherent column, about 5 cm. high. Note the continuity of the uppermost layer of the fragment of basal algal dome, layer which completely envelopes the column.

Fig. 16. Algal domes with beheaded columns, the base of which has been overgrown by seasonal mats and projects as small rounded knobs. Diameter of domes, about 15 cm.

Fig. 17. Lower eulittoral zone. Subcontinuous stromatolitic deposit, with columns and knobs, invaded by reeds and covered with abundant rod-shaped slightly curved coprolites from grazing fishes. Length of knife: 35 cm. Water depth about 5 cm. Late February.

Fig. 18. Algal domes with well developed pinnacles, some of which may coalesce laterally to form small irregular ridges. The erected pattern of the pinnacles and ridges shows that they formed in a microenvironment protected from the trade winds (compare with Fig. 19).

Fig. 19. Development of lineation on pinnacle-bearing stromatolites subjected to persistant trade winds. The pinnacles and small ridges, such as shown on Fig. 18 are deflected downwind and keep on growing flat on the algal domes. Trade wind blow from left to right. Water depth about 3 cm., late February. Length of knife: 35 cm.

Microstructure of the mats and origin of the carbonate

The microstructure of these deposits, although very similar to that of algal heads described from sinkholes (Monty, 1965, p. 217 on) will be considered and illustrated elsewhere. The ever changing features of this microstructure, according to the season and time of sampling, make it difficult to provide an objective and synthetic description. During the period of successful underwater growth for instance, the surficial blooming lamina thickens to 1—1.5 cm. It is composed of long vertical filaments of *Scytonema*, generally calcified below their growing tips (Monty, 1967, pl. 6) and permeating horizontal thin and calcified daily films of *Schizothrix calcicola;* the overall pattern yields a sort of palissade type of structure. The upper layers are furthermore expanded by important gazeous vacuoles and the development of large colonies of *Entophysalis* (Monty, 1967, pl. 3, Fig. 3; pl. 8, Fig. 1 and 2).

As growth recesses, when the water level drops below the upper surface of the algal dome, this surficial layer shrinks and collapses considerably (it will

Fig. 20. Lower eulittoral zone: the stromatolitic deposit becomes invaded by competitive communities of green algae and phanerogames. Note white scaly deposit of calcium carbonate on the upper surface of the leaves of Elodea — depth 5—8 cm., late February. Length of knife: 35 cm.

be reduced to but 1—2 mm); many algal cells become dormant (Monty, 1971, p. 220) while very short, deeply stained filaments of *Scytonema* form a sort of surficial close-cropped lawn. This stained and compact surficial layer acts now as a screen that protects the underlying filaments; the latter, shaded from noxious radiations and enjoying a sufficient amount of water coming from the substrate, start growing, the more that the capillary water is trapped by the watertight upper film; these internally growing filaments are no more deeply stained, they look whitish or greenish, rarely yellow or brown like the surficial ones. This internal growth, thickens the stromatolites f r o m   t h e   i n s i d e and originates new uncalcified layers, whereas the overall microstructure is disturbed. Things change in surface too, for the general conditions are such now that only the resistant but slow growing alga *Scytonema* can grow, whereas *Schizothrix* recesses; accordingly, the palissade structure disappears. All these metabolical and structural changes are accompanied by a succession of different micro-environments within the mats, causing solution of carbonate here, reprecipitation there, consumption of mucilagenous matter and polysaccharides by anaerobic communities, etc. (see Monty, 1965, pp. 226—253).

The result of these successive and/or superimposed phenomena, occurring at different scales, acting in the same or in opposite directions, including physical, chemical, biological (micro-) processes, hinders considerably the investigation and understanding of the microstructure of these mats.

Things are not more simple when deciphering the origin of the carbonate as precipitation occurs at different scales and follows different pathways:

1. Calcite is precipitated in the lake and on the plants as a result of a chemical desiquilibrium which is reached sometimes after the filling up of the lake: while the "original" rain water gets more and more enriched in $Ca(HCO_3)_2$, by solution of the karstic substrate, the rejuvenated filaments enter an ever increasing growth phase leading to the bloom; accordingly, a point of desiquilibrium is reached sooner or later due to the increasing demand for CO_2.

2. The resulting small crystals floating in the whiting water *may* become a substrate for the tiny filaments of *Schizothrix calcicola;* the latter develop around the grains, bind some of them together and originate small loose flocs; the very active growth of *Schizothrix* and the high but very localized demand for CO_2 enhance precipitation around, or, in the floc.

3. Precipitation abundantly occurs around the filaments and within the mat as described in MONTY, 1965 (pp. 232—253), and 1967 (pp. 72—76).

4. Precipitation occurs still when the lake dries out; at that time the supersaturation of the residual carbonate may originate conspicuous whitings.

The carbonate phase involved in these mats is a low magnesian calcite [6]) appearing sometimes as perfect rhombs sometimes under a variety of crystallographic shapes which may shows rounded angles according to the site of precipitation; this latter factor also influences the crystal size which passes from one micron or less in the algal sheaths to 2—8 microns in the interstitial mucilages. The resultant stromatolite will accordingly be characterized by a micritic laminated texture zoned with brownish organic layers and showing bundles of nicely calcified filaments; the latter strongly contrast with the matrix as a result of the smaller grain size of their constitutive crystals. The ordonnance and the continuity of the tubes becomes much poorer in the lower part of the structures because of the general compaction and collapse of the individual layers (dead laminae are no more supported by the positive pressure of metabolical gazes, nor by the abundantly produced and water laden mucilages which characterize the living ones; eventual dessication furthermore enhances the contraction of individual layers); except in cases of very favorable conditions of preservation, only fragments of tubes will hence be preserved by fossilization. Finally, as described in MONTY (1965 b, p. 199, 201; 1967, p. 73, pl. 8), internally formed pellets due to bacterial action and calcification of colonies of unicells will somewhat disturb the monotonous lamination of micritic and thin organic layers.

Although BLACK noted (1933, p. 176) that "in the widespread algal deposits of the interior of Northern Andros Id., no trace of mechanical lamination was discernible and the sediment is extremely fine in texture" he nevertheless thought that the fine-grained calcitic particles were supplied to the mats by a process of flooding which supposedly took place in two different ways (id., p. 177):

(a) by heavy rainfall, the transportation of sediment consists in "a slight

---

[6]) Though locally it may contain up to 8 mole % $MgCO_3$ in lakes close to tidal creeks.

washing of material from the lands into the lakes and creeks"; he added "there is possibly also a little transference of sediment from the lake floors onto submerged parts of the algal flats, when the bottom deposits are stirred up by waves and drifted by wind-blown surface currents onto flooded areas bordering the lakes". I however believe with BLACK that during such heavy rains "the solution effects are probably more important". I do not think that washing of material from the land into the lakes is significant, for there is almost no soil on Andros and the very karstic nature of the Island prevents any runoff.

When the lakes are wide enough to present a significant residual central pond floored with fine grained calcite, there may be some transfer of material onto the algal flats as stated by BLACK. I nevertheless believe that the process is very limited for (1) the floor of many inland residual lakelets is covered with subcontinuous algal mats (2) if not, the carbonate has a clotty texture due to aggregation of calcite crystals by *Schizothrix* or unicells and diatoms; such flocs cannot account for the diversified and very particular features of the carbonate within the mats (3) during the 3 months observation period, calcification of mats kept on proceeding although no washing of fine grained material from the central residual lakelet onto the algal flat was observed (4) I found no significant differences between the microstructure of the algal heads I described from small isolated and well circumscribed sinkholes (MONTY, 1965 b, 1967) and that of the deposits reported here

(b) the second type of flooding invoked by BLACK, to account for the calcite found in mats, occurs on a grand scale during hurricanes: "... vast quantities of sea water are piled up on the shoals, and are swept over the low lying parts on Andros as a result of violent westerly winds. During such storms, the water of the banks is laden with churned up sediments and the whole of the flooded area is liable to be smoothered under a film of fine white mud, which is left behind when the flood retreats" (p. 177); he adds (id.). "The sediment brought into the center of the island by such flood water is extremely fine-grained, since all the coarser constituents are deposited before the water has travelled far in from the West coast. This uniformity of grain size prevents any mechanical lamination or any noticeable difference in the size between the particles agglutinated round the mucilage of the algae and these particles which settled out under gravity". This process is incompatible with the carbonate found in mats — a low magnesian calcite — as such catastrophic flood would bring in almost pure aragonitic mud which has never been detected here; these X-ray results are confirmed by the striking difference which exists between the morphology of the microcrystals composing the aragonitic muds of the Western banks (neddles) and that of the calcite found in fresh water lakes as revealed by the Electron microscope (NEWELL and RIGBY, 1957, pl. XVI, Figs. 1, 2). Furthermore such huge floods that would cross the island from West to East and find their way" ... through the Eastern creeks into the lagoon behind the barrier reef — a cross-country journey of some forty miles" (ibid. p. 177) are really exceptionnal at the scale of the decade and their frequency cannot account for the observed constant mineralization of the reported stromatolites. Such a process is only effective on the supratidal flats of the Western coast where it originates the well known algal laminated sediments.

Finally, the microstructure and the characteristics relationships between the

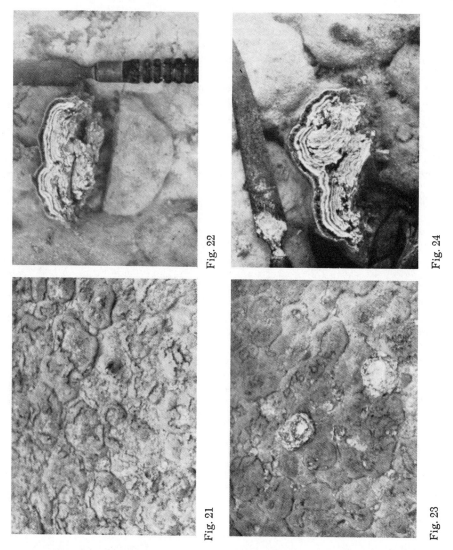

Fig. 22

Fig. 24

Fig. 21

Fig. 23

fine grained calcite and the algal material (filaments, mucilages, colonies of
unicells) as has been lengthlily described in MONTY (1965 b, 1967) show clearly
enough that the two processes invoked by BLACK cannot account for the bulk
of the observed mineralization of the biostromal deposits reported in this paper.

Alteration and reworking of the stromatolite deposit

a) Physical factors

They include stormy rains which may disrupt the marginal fragile domes
resting on algal grit, or destroy the algal columns when the water level drops.

Prolonged drought is the most important factor which completely alters the stromatolites at the periphery of the flat where the mats shrink, crack and may become brittle or very contorted. The action is less drastic in the lower eulittoral zone; the algal dome becomes more compact, some surficial layers, which are about 1 cm. thick during full growth period, shrink to about 1 mm.; however the moisture of the underlying substrate and the formation of tough water-tight algal cover makes it possible for these domes to survive.

b) Metabolical factors

Small scale peeling: Active internal growth of long slender filaments, protected from noxious radiations by a brownish calcified and watertight film of *Schizothrix* and *Scytonema*, originates the accumulation of gaz bubbles in between the fragile filaments; this weak zone is easily disrupted under the pressure of gases and the upper layer peels off (Figs. 21, 22). The peeled fragments float away and/or accumulate in the depressions separating the domes.

Large scale peeling: During the full growth period, complete mats or algal domes may become buoyant. At that time the surficial lamina, covered by a thin film of *Schizothrix,* expands up to 1 or 2 cm. (Fig. 29) and is loaded with gaz bubbles trapped in the overall mucilage. As a result of the strong positive buoyancy, whole algal mats peel off the substrate (Fig. 25) and float away, leaving sorts of big scars in the stromatolitic deposit (Fig. 26). Similarly, whole sets of buoyant algal domes detach from the substrate (Fig. 27) move around, and resediment when the bloom is over; they will be reincorporated into complex and irregular deposits by further overgrowths (Fig. 28). Some individual dome may float over for some distance before being redeposited right side up or upside down on the stromatolitic bed (Fig. 23); they will also be overgrown by seasonal mats to originate compound structures (Fig. 24).

Discussion and final comments

Comparisons

This natural history of significant non marine blue-green algal flats revealed the formation of stromatolitic deposits reminiscent of many occurences in the geological column; it also gives us an interesting counterpart of the abundant algal

---

Fig. 21. Small scale peeling in stromatolites from upper mid eulittoral zone; peeling is due to accumulation of gaz underneath the upper fragile lamina (see Fig. 22) where a weak zone progressively forms and originates the disruption of the uppermost layer.

Fig. 22. Belt of smooth stromatolites, mid eulittoral zone. Vertical section through a stromatolitic dome about 10 cm in diameter; note prominent lamination which extends in the underlying mud represented on Fig. 13. Blooming and development of gaz vacuoles in the upper laminae originates a weak zone and the peeling of top layer (cfr. Fig. 21).

Fig. 23. Two algal domes that peeled of their substrate, drifted away and have been redeposited upside down over the stromatolitic bed, at the end of their bloom; the diameter of the domes is about 10 cm.

Fig. 24. Vertical section through a complex stromatolite; a drifted and upside down resedimented algal dome (such illustrated on Fig. 23) overlies another in life position; both of them have been later overgrown by seasonal mat. This figure illustrates the way in which they situation shown on Fig. 23 will evolve.

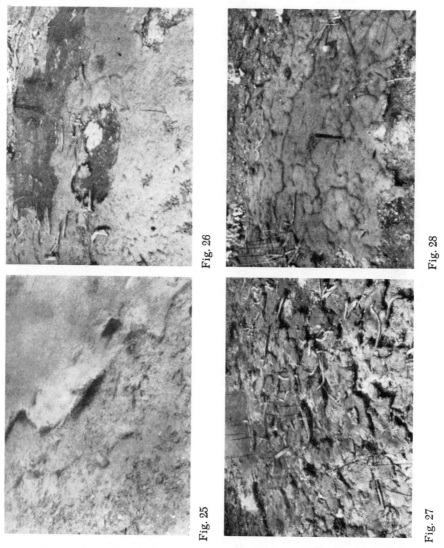

Fig. 26

Fig. 28

Fig. 25

Fig. 27

mats or laminated sediments reported from saline or hypersaline waters of inter- to supratidal settings (GINSBURG et al., 1954; DAVIES, 1969; KENDAL et al., 1968, etc.). The presently reported formations differ strongly from algal laminated sediments by their mode of formation, their microstructure, their mineralogy and associated fauna.

(1) The Wilmo Lake stromatolites are entirely controlled by algae; their over- all features do not primarily result from the periodic influx of clastic grains over algal films or mats, but from the seasonal or environmental differenciation of an algal coenose which originate most of the mineralization of the resulting structures.

(2) They show no mechanically deposited graded-bedded layers (like in Black's type A, 1933) nor storm layers (like in Shinn et al., 1969).

(3) There is no differenciation of grain sizes into distinct laminae (inter-beddings of silts and muds a.s.o.) and no variation in the granularity of the particles from the proximal to the distal end of the flat, all features which are typical of inter-to supratidal algal laminated sediments as listed in Ginsburg et al. (1954, p. 30—31) and in Davies (1969, p. 201—202).

(4) The carbonate grains are the same across the flat and from base to the top of the stromatolite; when structural or dimensionnal modications occur, they are generally bound to algal or bacterial activity, which accounts for the diffe-rences between the crystals of the sheaths and the crystals of the intersitial mucilages, for the formation of pellets in colonies of bacteria or of unicells, etc.

(5) The structure and mineralogy of the finely but thouroughly calcified filaments looks very different from what is commonly found in mats from inter-tidal settings where molds filaments are more frequent (Monty, 1965 b, pl. 58; Davies, 1969, Figs. A, B.) [7]).

(6) The carbonate mineralogy — a low to very low magnesian calcite — is very indicative of the geochimical environment and would favor a rather simple diagenetic history as opposed to what is found in association with inter-supratidal mats of Western Andros Id., the Persian Gulf, Shark Bay, etc.

(7) The Wilmo Lake stromatolites of course do not include any fragments of marine organisms (there may be some limited contamination in the deposits bordering important tidal creeks though) and the associated preservable fauna is very poor: fresh water gastropods, ostracods, thekamoebians, etc.

Such fine-grained and homogeneously laminated stromatolites permeated by remnants or fragments of calcified algal filaments, associated with a very poor fauna dominated by gastropods, forming laterally linked or spaced domes

---

[7]) See however Shinn et al., 1969, Fig. 12 where calcified filaments have apparently be found in supratidal marsh sediments.

Fig. 25. Large scale peeling in the lower eulittoral zone. Active algal growth and accumulation of gaz in blooming algal mats (see Fig. 29) endow them with a strong positive buoyancy; the whole mat peels of the substrate and gets disrupted; the resultant fragments then drift and float away. Thickness of mat 4 cm. Depth of water, about 15 cm.

Fig. 26. "Scars" left in the continuous stromatolitic deposit of the lower eulittoral zone after peeling of large fragments of mats. Note how the remaining mat floats over the substrate at the periphery of the scar and opens the way to further peeling. Depth of water: about 15 cm.

Fig. 27. Large scale peeling in the lower eulittoral zone. Floating algal domes during period of bloom; when the phenomenon occurs in region exposed to the trade winds, the whole set of detached domes will drift away to pile up some place when the bloom is over. The importance of algal encrustation around the reeds can be appreciat-ed from comparing the vertical young reeds with the dead encrusted ones lying flat over the domes.

Fig. 28. "Resedimented" domes that have been gathered into a complex deposit by seasonal overgrowth. The result of the processes illustrated in the Fig. 27 may yield a type of biohermal accumulations due to flotation, drift and local piling up of the structures rather than to plain constructional phenomenon.

capped or not with various knobs or columns, are undoubtedly features which we are used to meet in the geological column.

BLACK (1933) described algal heads from the flats of lake Forsyth and Stafford Lake (North Andros); he did not however analyze the algal zonation of the flat which he reported as a surface where the development of mats was constantly opposed by dessication of the substrate; that is why the heads he rapidly described (BLACK's type C) form individual discs standing well apart from each other (BLACK, id. Figs. 19—20) whereas the lamination is concave upward (id. Fig. 8, p. 183); this originates an internal structure very different from the one shown here on Figs. 10, 11, 22, 32 where the lamination is constantly convex upward, a typical feature of encrusting communities. The descriptions that BLACK gives of his type C heads and of their internal structure suggests however that he studied them in a pretty dessicated stage, which shows up on his photographs, by the very dark colour of the mats. Anyhow, these two "facies" or "phases" (BLACK's and ours) of fresh water algal deposits built by an identical coenose are pretty good replicates of some pre-cenozoic stromatolites.

Lack of significant lithification

One may wonder why these algal formations do not originate hard lithified structures instead of unconsolidated earthy deposits; the problem was already felt by BLACK when, comparing his Andros stromatolites with other Recent cyanophycean structures, he wrote that the Bahamian algal deposits "stand apart from all other recorded examples in being completely unlithified" (p. 184); he probably had in mind here the many concretions found in fresh water lakes or streams and the famous "Biscuit Flat" (South Australia) where MAWSON (1929) had found important accumulations of lithified algal biscuits growing on temporarily submerged flats. In fact such a type of question — how come androsian stromatolites do not get lithified? — is always a dangerous one that may have no answer that our reason can understand (*how come* some chlorophytes secrete an internal aragonitic framework, while other remain uncalcified?); we may however try to sort out some elements to think about:

(1) Androsian mats (such as BLACK's type C, D, MONTY, 1967, types 1, 2, 3, pp. 68—76; and the deposits reported here), thriving at very shallow depths and having to stand eventual periods of severe exposure, are dominated by a fresh water alga well adapted to the harsh conditions of subaerial life: *Scytonema myochrous;* its individual filaments are surrounded by a thick lamellated sheath, which generally becomes deeply stained with protective pigments and which encases itself into a thin and brittle encrustation of small calcitic crystals (MONTY, 1967, pl. 6); pure cultures of *Sc. myochrous* never go any further than this stage of calcification of the individual sheaths, as is confirmed by KANN (1941 b); so that no hard calcareous object can result.

(2) In the reported mats, *Scytonema* is accompanied by the tiny blue-green *Schizothrix calcicola* (MONTY ibid., BLACK, ibid.) which, in fresh waters originates strongly lithified structures. As exposed in MONTY (1965 b, 1967), *Sch. calcicola* is a subaquatic alga which consolidates and thickens the mats, during the phases of submergence or of increased subtratal humidity, by stretching calcified films between the *Scytonema* filaments or by binding bundles of *Scytonema* together; however being given the interference of many other algae which

somewhat disturb the framework (*Plectonema, Entophysalis, Johannobaptista,* diatoms, etc.), the abundance of released mucilage, the rather slow rate or precipitation of calcium carbonate and the difficulty of ionic migration (due to density of mucilage, eventual complexion, etc.), *Schizothrix* can originate but loose calcified structures and not the hard calcareous objects like the ones found in the lake of Constance or the tufa illustrated by Irion and Müller (1968, Figs. 6, 7).

(3) Other environmental chemical factors may also interfere: these shallow warm waters are much less agressive with respect to the karstic substrate than the waters of our temperate and cold lakes or streams; hence the ratio $CO_2/Ca(HCO_3)_2$ is rather different. Furthermore, the sudden escape of the eventual excess of CO_2, which in our european calcareous lakes and streams considerably enhances the carbonate precipitation, is very limited in the protected settings characterizing many androsian lakes. During windy days, when the surface turbulence facilitates the release of the CO_2 trapped in the water lens, the author has however observed local conspicuous whitings in residual ponds. The process is however too slow and unfrequent to add signifiant amount of carbonate to the algal structures.

General features of stromatolites vs. environment

The description of fresh water blue-green algal flats, reported here in very general terms, illustrated some aspect of the differenciation of one given stromatolitic community in response to very small variations of environmental conditions. As is the case with most studies on algal flats we are left with the impression that the algae themselves have little to do in the overall morphology of the resulting structures, but that the environmental conditions are the ruling factors in the shaping of heads and associated structures; we can indeed summarize the observations as follows:

(1) The stability and location of the final water level during the dry season originates two main ecological and morphological zones: on the one hand, these wide flats liable to be exposed to drought and characterized by pure stromatolitic deposits (Figs. 7, 18); on the other, the distal portion of the flats, or the eventual ponding areas, where stromatolites are invaded by competitive plants (Fig. 20).

(2) The eventuality of a more or less complete dessication, as well as its duration, segregates three main belts: that of thin, cracked, poorly laminated polygons, that of various algal heads and that of subcontinuous mats.

(3) The length of underwater growth is a factor of differenciation of the morphology of heads and separates smooth domes, domes with pinnacles, domes with columns (Figs. 10, 14, 18).

(4) The intervention of dynamic environmental factors may alter the deposit: the action of the trade winds intervene in the modeling of original pinnacles into striking lineations (Fig. 19), whereas heavy rains may destroy algal columns and leave residual knobs on top of the stromatolites (Fig. 16).

(5) If the nature of the substrate, hence its moisture retention, has a fundamental influence on the algal growth when the water level drops, its morphological features may directly influence the overall shape of the stromatolites, be

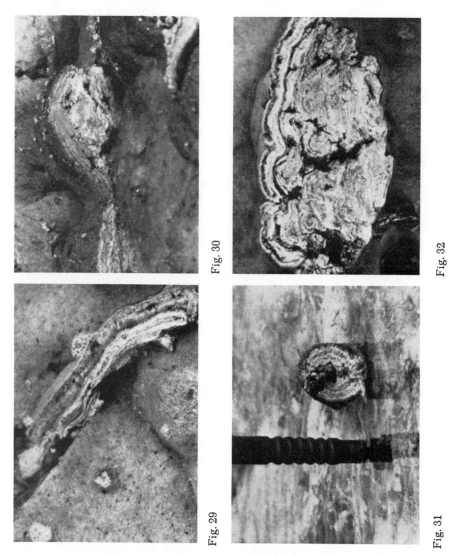

Fig. 30

Fig. 32

Fig. 29

Fig. 31

Fig. 29. Cross section through a peeling mat of the lower eulittoral zone; the upper layer (about 2 cm thick) is fully expanded as a result of algal bloom and accumulation of gaz bubbles in between the filaments.

Fig. 30. Cross section through stromatolitic deposit overgrowing the pitted Pleistocene substrate; the apparent "domes" formed here by the blue green algal mat are imposed by the presence of rocky projections which appear beneath the mat.

Fig. 31. Piece of wood encrusted by stromatolitic algae; successive encrustations originate a laminated sheath encasing the branch. Diameter of cross section is about 5 cm.

Fig. 32. Vertical section through a complex algal dome; the irregularities of the surface are here dictated by the presence of underlying discrete domes that were later overgrown.

it in the case of protruding substratal rocks (Fig. 30), of former stromatolites (Fig. 32) or even of pieces of wood (Fig. 31).

Before stating that blue-green algae are always passive agents building up structures which are primarily shaped by environmental conditions, I would like to try to frame the problem properly by enlarging the field of discussion. To this purpose let us rapidly oppose the zonation of blue-green algal crusts in the eulittoral zone of two main types of lakes, i. e. the oligotrophic and the eutrophic lakes. In the oligotrophic lakes, where the waters are generally deeper and clearer than in eutrophic ones, the shores are quite often steep and therefore narrow; among other things, this steepness of the eulittoral zone originates a set of successive and well delineated habitats, each of which closely fits the ecological requirements of a given algal coenose that consequently will become the dominant one in a given belt; the algal cover of the eulittoral zone will not consequently be uniform, but will distribute itself into successive well individualized belts leading for instance to the succession downshore of a *Tolypothrix* zone, *Rivularia* zone, *Schizothrix* zone, etc. This originates, in the same direction, a typical series of very different types of crusts or mats, each one showing its individual characteristics and growth form. Accordingly in the case of rather steep shores, the environmental pressure favors a distinct and rather brutal distribution of particular algal coenoses which i n t h e i r t u r n originate a clear distribution of *proper growth forms* (see Kann, 1941 b).

The situation is quite different in eutrophic lakes which are generally bordered by wide and flat benches. In this case, the greater part of the subhorizontal and drying up shores constitute a rather uniform habitat characterized all over by important but similar yearly environmental changes. This will favor the establishment of o n e complex basic community whose range of adaptation and tolerance comprizes the general set of environmental and seasonal conditions; accordingly the eulittoral zone will be invaded by an almost uniform algal cover, and the fundamental type of mat or crust will tend to be the same over the greatest part of the flat. The adaptational properties of the association will generally cope with environmental pressure along to main pathways:

(1) Physically induced morphological changes: resulting in the formation of thin algal polygons upshore (when growth is very intermittent), of nice and healthy domes downflat, of oriented structures in equilibrium with the local dynamic agents, etc.

(2) Morphological changes due to variations in the respective abundance of partipating algae. As opposed to the various structures distributed down the steep shores and which are built each by a definite alga, accompanied or not by a very small number of by-goers, the associations found on subhorizontal flats, submitted to rather important and continuous environmental changes, are generally complex and include a great number of participating algae [8]; among these, are generally two or three basic ones adapted to the seasonal extreme conditions to be endured by the community so that the mat can keep on growing whatever be the conditions. This organisation confers a great plasticity

---

[8]) It is however evident that when seasonal changes include extremely hard conditions, such as concentration of brines, the algal association will be restricted to one or perhaps two algae, i. e. the more resistant ones (case of *Lyngbya stratum;* see p. 775).

to the mat and opens the way to an easy differenciation both in time and in space; *in time* because the basic algae will lead each in its turn according to seasonal changes, originating that way a conspicuous laminated structure but also an ever changing external aspect of the stromatolite; differenciation *in space*, because small lateral variation in the environment (with respect to tolerance of the coenose, not to our appreciation) will favor such or such alga of the community; the balance of the participating algae will be changed, hence the properties and features of the mat; this is for instance the case for the mats described by GINSBURG et al., 1954: their types I and II are built by the same basic community but with different proportion of participating algae.

Accordingly, if in the first case (illustrated by the steep shores of oligotrophic lakes) we end up with different mats, specifically built by the different algae in different specific habitats, the second (illustrated by the flat shores of eutrophic lakes) shows different stromatolitic morphologies built by an identical but highly flexible basic algal community that copes with environmental pressure; as the latter case sums up conditions which are generally found on tidal flats where most geological studies are carried on algal mats, it is not surprizing that the "concept of passivity" of blue-green algal structures shaped by nothing else than environmental conditions opened the way to a general but misleading theory.

In view of what preceeds, we may however have a better understanding yet of the significance of these algal flats and of their stromatolitic cover. Indeed, we should always keep in mind that these wide and monotonous flats are but a mixed and intermediate habitat, accordingly colonized by a mixed and intermediate blue-green algal association (an equivalent intermediate belt can never take any significant development around oligotrophic lakes where the shores are much too steep). The resulting stromatolites and mats are then generally built up — beside many ubiquitous forms — by a collection of various specific blue greens whose ecological ranges overlap on the flats; the resulting structures appears thus as a comprimise. However, when we trace this association towards the specific biotopes of each of the leading component alga, we generally end up with rather pure cultures of them, showing each its specific growth form. For instance: tracing landward the stromatolitic deposits of Wilmo Lake, we finally reach structures built by pure growths of *Sc. myochrous* (type 1 MONTY, 1967, p. 70, Fig. 5, pl. 5), while tracing them seaward we find the pure stromatolitic domes built by *Schizothrix calcicola* (MONTY, 1965 a, 1967, p. 89—92, pls. 16-3, 17-3, 18, 19). In such wide scope, algae have surely something to do with the general features of the resulting stromatolites, and undoubtedly much more than the local hydrodynamical agents.

Competition and individualization of stromatolitic flats

When surveying marine and non marine environments, one is stricken by the fact that the most important and successful blue-green algal colonization is found on periodically exposed flats i. e. the inter to supratidal zone in the first case, the eulittoral zone of lakes in the second one; in both instances, there is of course a similarity in the type of habitat, were it not for a factor time: the frequency of exposure and submergence is indeed daily in the case of intertidal flats, whereas it is seasonal in the case of supratidal flats and of the eulittoral zone of lakes.

It has been shown elsewhere (Monty, 1972) that inter- and supratidal flats are but one of the many settings that stromatolites did colonize during their long geological history and that they had finally been confined there as a result of the strong ecological pressures developed by competitive subtidal communities. If we consider the situation in Wilmo Lake and match it with what has been reported from european lakes we find that similar competitive relationships between stromatolitic blue-green algae and other plant communities are also found in non marine settings. Kann (1941 b) discussed the problem in the case of eutrophic lakes; the abundance of nutrients found in these lakes results in a great development of all algal groups (greens, browns, diatoms, reds) as well as of various phanerogames in the upper layers of the lakes (such a situation may be analogous to the one that characterized Paleozoic and surely Mesozoic shallow seas). The settling of such a rich and varied colonization originates strong competition for space and nutrients between the different algal groups and it appears that benthic blue-green algae are poor competitors with respect to more powerful ones like the *Cladophora*. Algal mats that might form will be loose unlaminated structures in which *Cyanophyta* grow interspersed with various types of other disturbing algae. However, says Kann (id.), if the water level varies greatly in such a way that important portions of the benches become seasonally exposed to drought and intense radiation, a new habitat is created, habitat that only one algal group — the blue-greens — will be able to colonize uncompeted, because of its various morphological, cytological, physiological properties (Monty, 1971); Kann (ibid.) concludes that this is the main reason why blue-green algal crusts most characteristically colonize a well defined portion of the lakes, i. e. their eulittoral zone. Worochin (1936) described similar situations in the case of highly alkaline lakes saturated with Na_2SO_4 and containing significant amounts of iodine and bromide; for instance, the Great Tambukan lake (a shallow lake of Western Caucasus) is colonized at depths of 0.5 to 1m. by a thick growth of the filamentous green alga *Rhizoclonium hieroglyphum* (a genus close to *Cladophora*) associated with some outnumbered blue greens. As the chances of drought and exposure increase shoreward, the *Rhizoclonium* community recesses while the bottom is floored with an almost continuous layer of Cyanophyta (*Chroococcus, Lyngbya, Phormidium, Oscillatroia,* etc.) which passes, near the desicated upper shore, to leathery mats composed of an almost pure culture of *Lyngbya aesturii*. In the case of lakes which, flooded in Spring and Fall, dry up completely during the Summer (Worochin, id.) *Rhizoclonium* forms but scattered patches while blue-green algal mats, dominated by *Lyngbya*, colonize most of the flat; as Worochin states it (ibid.) the periodical dessication of the lake maintain the purity of the *Lyngbya* stratum.

All these trends aggree with what has been reported here from Wilmo Lake: pure stromatolitic deposits and heads develope on the periodically exposed flats; downflat, that is toward sites that remain submerged a longer and longer period of time, the mats become invaded by a variety of green algae and phanerogames (such as *Elodea*) which not only compete for space and nutrients with blue-green algae, but overgrow them and disrupt the stromatolitic structure.

All these examples illustrated a well known ecological "law" of population dynamics, i. e. the passage from communities with rather high diversity and low dominance in (relatively) favorable habitats, to communities with very low diversity but very high dominance in very particular or severe habitats. As has been discussed in MONTY (1972) the present status of Cyanophyta results from a long history of competition between blue-green algae and various marine encrusting communities and the confinment of significant algal mats or stromatolites on periodically exposed flats, where they can grow undisturbed and uncompeted.

One might think that predation, the intervention of boring animals, etc.... may prevent blue-green algae from growing successfully or stromatolitic structures from forming outside of these periodically exposed flats. Such a view has recently been developed by GARRETT (1970). I do not think that such a type of non competitive relationship can account for the poor development of stromatolites in the sea (MONTY, 1972). Blue-green algal mats provide a habitat to many microorganisms (worms, ostracodes, crustaceans, gastropods, etc. . . . BATHURST, 1968; NEUMAN et al., 1970) which they feed together with higher animals (fishes etc. MONTY, 1965, p. 285, Fig. 32 a, 1967, p. 81, DAVIES, 1969, Fig. 18 a etc. . . . .) This does not prevent algal laminated sediment to form (DAVIES, id.) nor stromatolitic domes to grow (MONTY, id. "mats and crust built by *Scytonema crustaceum*"). Feeding on rock-building blue-green algae is also known in fresh water structures, the building up of which does not seem to be hindered by the burrowing or browsing organisms: this is illustrated by the larvae of chironomids which built feeding and living burrows in Oscillatoriacea-tufa and feed on the algae (see STIRN, 1964; IRION & MÜLLER, 1968, etc. . . . .); furthermore, most fresh water blue-green algal deposits, be they mats, crusts or tufa are generally associated with a great number of gastropods belonging to a small number of species and feeding on the growing algal crusts; when the animal dies, its shell becomes readily incrusted by successive calcareous laminae to be incorporated into the algal crust, or to form the well known fresh water oncolites or "Schneckelisand" (BAUMAN, 1911; RUTTE, 1953, pl. 14, Fig. 4, etc. . . . .). In the latter case, the predators do not eliminate blue-green algae but on the contrary, favor the formation of oncolites by providing a hard substrate for encrusting algal communities to grow.

So, projecting the data into the marine realm and in the fossil record, I believe that the observed abundance of stromatolitic deposits on periodically exposed flats is not primarily a matter of predation but before all a matter of direct intercommuinity competition in which the competitive weakness of blue-green algae has driven them to settle in habitats where their competitors could not grow (see complete statements and restrictions in MONTY, 1972).

Floating colonies

Floating colonies such as reported from Wilmo Lake, are not a unique phenomenon but have been described from other fresh water settings (PHILLIPS, 1958, etc. . . . .) as well as from marine environments (PHILLIPS, 1963). In all cases, the phenomenon is associated with the period of algal bloom and occurs at very shallow depths (1—3 feet). Once formed, these floating aggregates originate a new habitat and the whole structure of the mat is liable to suffer some important transformations. PHILLIPS (1958) observed that the process starts by a vigorous multiplication of blue-green algae [*Arthrospira, Spirula, Oscillatoria* [9]), etc. . . .] on the muddy bottom of the pond. An abundant mucilage is secreted during the growth so that the top layer becomes air tight, while the lower filaments progressively permeate the mud and bind the particles. The abundant oxygen resulting from active photosynthesis during the Summer months remains trapped in the algal film above the mud and inflates it; the mat and the adhering mud finally tear away and are buoyed up at the water surface where the algae of the top layer multiply prodigiously. Following the first two phases of b l o o m i n g and p e e l i n g, the third one, that of f l o a t i n g originates further alteration in the mat *sensu lato:* first of all, the surficial layer forms now a sort of outer integument where the filaments grow parallel to the horizontal plane of the floating mass; secondly, in older colonies, the underlying mud acquires a stratified pattern in which three main laminae can be recognized; the original mud just beneath the mat, an underlying black layer rich in bacteria and reduced organic constituents (namely of algal origin) and finally a gray layer resulting from the re-oxygenation of the lower-most part of the black layer suddenly carried into oxidizing environments after peeling.

The process stops after a few weeks or less when the algal bloom fades away; at that moment, fragments of mats and adhering mud sink and pile up on the bottom of the lake or lagoon.

When the peeling process affects large algal mats, it is most probably enhanced by the accumulation of bacterial CO_2 and/or H_2S beneath the mat; indeed, the underlying mud looks definitly black (cfr. Fig. 26); it is crowded with bacteria and releases a strong odor of sulfur. Experiments conducted by PHILLIPS (1963) show that the pH at this level and in the released bubbles is around 7 or less, whereas that of the overlying water is around 8. Such chemical conditions account for significant solution of calcium carbonate beneath the mat (and the relative enrichment of the substratal mud in organic matter) while important precipitation keeps on around the top filaments and in the mat itself [10]). There is accordingly a recycling of the carbonate in the environment, process which somewhat inhibits the rapid formation of important deposits; solution and bacterial action destroys the buried structures and prevents the preservation of laminated deposits that would result from the piling up of successive seasonal mats; flotation and disruption of continuous mats acts in the some way.

Such chemical processes seem to be most active under subcontinuous mats growing in shallow standing waters, but not on the algal flat proper, at least there where the algal heads do grow; the discontinuity between the domes probably favors a better circulation of the lake water bathing the stromatolites and their muddy basement; furthermore, if black layers would form they would probably be reoxydated during the dry season when the algal structures suffers some shrinking.

---

[9]) These are replaced by *Lyngbya* and/or *Spirulina* in marine settings, the former in salinities greater than $27^0/_{00}$, the latter favoring salinities between $20—25^0/_{00}$ (PHILLIPS, 1963).

[10]) These simultaneous phenomena are very similar to the ones described by GOLUBIĆ (1962) to account for the formation of "Fürchensteine" or pitted rocks under algal crusts in the yougoslavian Karst.

This accounts for the fact, that although at given places the stromatolitic domes rest on a brownish black mud, most of the deposit overlies a white very fine calcitic mud in which the lamination extends.

The peeling of blooming mats and the associated phenomena reveal fundamental processes that should be kept in mind in paleoecological and morphological studies of stromatolites. These processes, operating in very quiet environments under the control of plain organic activity may completely disrupt stromatolitic deposits and induce various modifications in the algal structures. The consequences of the peeling and fragmentation of subcontinuous mats are at least threefold:

(1) Big "scars", that would surely appear as erosional traces in the fossil record, are left in the "mother deposit" when important portions of mat tear away (Fig. 26).

(2) The resulting floating aggregates, when sinking to the bottom after the period of bloom, originate a chaotic piling up of "clasts" ranging in size from a few centimeters to a few decimeters; such deposits would probably be interpreted in the fossil record as resulting from important mechanical action of waves or tides, and a very different set of environmental conditions would be erroneously reconstructed.

(3) As previously said, floating aggregates acquire a new internal structure and/or lamination resulting from the new life and biochimical conditions into which the algal mat and its adhering substrate are suddenly carried; these modifications will of course increase with the duration of surficial drift. Accordingly, the formation of floating aggregates does not only disrupt and "recombine" a stromatolitic deposit but may induce a proper differenciation of the mats which is not typical of the bottom conditions where the aggregates come from or will sink.

As far as the individual heads crowding the algal flat are concerned, we have seen that small scale peeling (Fig. 21) could also originate conspicuous scars at the surface of the stromatolite; these scars will be fossilized by discordant overgrowths of forthcoming seasonal layers. The frequency of these features joined to the presence of upsidedown resedimented domes (Fig. 24) would undoubtedly lead, in the fossil record, to an erroneous reconstruction of turbulent settings.

Finally, individual algal heads, when sinking after the bloom, may pile up against any irregularity of the lake floor or on previously sunk domes: this results in the building up of small mounds that might be interpreted as biohermal, i. e. constructional, instead of "detrital" accumulations.

The facts and hypotheses reported in this note were meant (1) to describe fresh water blue-green algal biostromes in process of formation, and to frame them in the subcontinuous series of algal deposits found across Andros Island and adjacent marine platforms. (2) To widen the scope at the light of the processes known in both marine and fresh water settings so that a sort of philosophy concerning the ecological, distributional and morphological interpretation of stromatolites and mats could be tentatively proposed; these concepts will be developed elsewhere. An attempt was made, in the discussions and comments, to show the tremendous importance of life processes in the intimate organization, shaping, distribution and alteration of stromatolitic deposits and to balance these processes with environmental physical ones.

All the small-scale processes and factors responsible for the given features and behaviour of a stromatolitic flat, as well as the complex flow of biological, social, chemical . . . phenomena within a simple algal dome or mat have always frightened me when I consider the simple (too simple!) resulting laminated structures that the paleontologist has to study; I wonder then how much of the natural history of a stromatolite is left in the thin compacted residual laminae found at the base of a Recent deposit (Figs. 13, 22); how much then in a fossil stromatolite? Well, very little most probably. But we shall miss or misunderstand this "very little message" if we do not know the rules of the game, if we do not know where the message might well come from, if we do not analyze it in full detail, if we do not sharpen our concepts. To this purpose, the Present may be a key to the Past provided:

(1) we have a good critical knowledge of the present processes at every single level of organization

(2) we dig out from the r i g h t present the r i g h t key to the r i g h t, a n a l o g o u s or h o m o l o g o u s Past.

Bibliography

Baas-Becking, L. G. M., Kaplan, I. R., Moore, D.: Limits of the natural environments in terms of p_H and oxydo-reduction potentials. — Journ. of Geology, **68**, No. 3, 243—284, 1960.

Bathurst, R. G. C.: Subtidal gelatinous mat, sand stabilizer and food, Great Bahama Bank. — Journ. Geology, **75**, 736—738, 1967.

Bauman, E.: Die Vegetation des Untersees (Bodensee). Archiv Hydrobiol. und Planktonkunde Suppl.-Bd. **I**, 26—48, 1911.

Black, M.: The algal sediments of Andros Island, Bahamas. — Royal Soc. London Philos. Trans. Ser. B, **222**, 165—192, 1933.

Blum, J. L.: The ecology of River algae. — Bot. Review, **22**, 291—341, 1956.

—: Algal population in flowing waters. — In: "The ecology of algae" Sp. publ. No. 2, Pymatuning Lab. of Field Biology. Univ. of Pittsburg, 11—21, 1960.

Bradley, W. H.: Fresh water algae from the Green River formation of Colorado. — Torrey Bot. Club, **56**, p. 421—428, 1929.

—: Algal reefs and oolites of the Green River formation. — U.S.G.S. Prof. pap., **154 G**, 203—223, 1929.

Brehm, V., & Ruttner, F.: Die Biozönosen der Lunzer Gewässer. — Int. Rev. d. ges. Hydrogr. und Hydrob., 16, 1935.

Cayeux, L.: Roches carbonatées. — Masson ED., Paris 1935.

Cerdergen, G. R.: Reofilia eller det rinnande vattnets algsamnällers. — Svensk Bot. Tidskr., **32**, 362—373, 1938.

Clarke, J. M.: The water biscuits of Squaw Island, Canandaigua lake, N.Y. — Bull. N.Y. State Mus., **8**, No. 39, 195—198, 1900.

Chodat, R.: Communication préliminaire sur les algues incrustantes et perforantes. — Arch. Sc. Phys. et Nat. Génève, 512, 1897.

Colom, G.: La paléoécologie des lacs du Ludien-Stampien inférieur de l'île de Majorque. — Revue de Micropaléontologie, 4, 1, 17—29, 1961.

Davies, G. R.: Algal-laminated sediments, Gladstone Embayment, Shark Bay, Western Australia. — In: Carbonate sedimentation and environment, Shark Bay, Western Australia. A.A.P.G. Memoir, **13**, Tulsa, Oklahoma, 169—205, 1970.

Decksbach, N. K.: Zur Kenntnis einiger sub- und elitoraler Algen Assoziationen rüssicher Gewässer. — Arch. f. Hydrobiol., **17**, 3, 492—500, 1926.

Desikachary, T. V.: Cyanophyta. — Indian Council of Agricultural Research. New Delhi, 1959.

Donaldson, J. A.: Stromatolites in the Denault formation Marion Lake, coast of Labrador, Newfoundland. — Geological Survey of Canada. Dpt. of Mines and Technical Surveys. Bulletin, **102,** 28 p., 2 figs., 7 pls. 1963.

Drouet, F.: Ecophenes of *Schizothrix calcicola.* — Proceeding of the Acad. Nat. Sc. of Philadelphia, **115,** No. 9, 261—281, 1963.

Eggleton, F. E.: Limnology of a meromictic interglacial Plunge-basin lake. — Amer. Microsc. Soc. Transact., **75,** 334—378, 1956.

Echlin, P.: The Blue-Green Algae. — Scientific American, **214,** No. 6, 74—81, 1966.

Forel, F. A.: Le Léman. — Monographie limnologique, 3, Lausanne — F. Rouge Edit., 1902.

Freytet, P., & Plaziat, J.-C.: Importance des constructions algaires dues à des Cyanophycées dans les formations continentales du Crétacé supérieur et de l'Eocène du Languedoc. — Bull. Soc. Géol de France (7), VII, 679—694, 1965.

Fritsch, F. E.: The encrusting algal communities of certain fast flowing streams. — New Phytol., **28,** 3, 165—196, 1929.

—: The lime encrusted Phormidium community of British streams. — Int. Verein für theor. und angewandte Limnologie, Verhandlungen, **10,** 141—144, 1949.

—: Algae and calcareous rocks. — The advancement of Sci., **7,** 25, 57—62, 1950.

Fritsch, F. E., & Pantin, C. F. A.: Calcareous concretions in a Cambridgeshire stream. — Nature, **156,** No. 3987, 397—398, 1946.

Garrett, P.: Phanerozoic stromatolithes: Noncompetitive ecologic restriction by grazing and burrowing animals. — Science, **169,** 171—173, 1970.

Geitler, L.: Cyanophyceae (Blaualgen) Deutschlands, Österreichs und der Schweiz mit Berücksichtigung der übrigen Länder Europas sowie der angrenzenden Meeresgebiete. — Rabenhorst's Kryptogamen — Flora — XIV. Band. Akademische Verlagsgesellschaft m. b. H., Leipzig.

Gebelein, C. D.: Distribution, morphology and accretion rate of recent subtidal algal stromatolites. — Bermuda. Jour. Sedim. Petr., **39,** 32—49, 1969.

Glazek, J.: Recent oncolites in streams of North Vietnam and of the polish Tatra Mounts. — Ann. Soc. Geol. Pol., **35,** 2, 221—242, 4 pls., 3 figs., 1965.

Ginsburg, R. N.: Ancient analogues of recent stromatolites. — Int. Geol. Congress. XXI Session, Norden, Part XXII, 26—35, 1960.

Ginsburg, R. N., Isham, L. B., Bein, S. J., & Kupferberg, J.: Laminated algal sediments of South Florida and their recognition in the fossil record. — The marine laboratory, University of Miami (unpublished) 1954.

Golubič, S.: Zur Kenntnis der Kalkinkrustation und Kalkkorrosion im See Litoral. — Schweiz. Zeitschr. f. Hydrol., **24,** 2, 229—243, 1962.

—: Algenvegetation der Felsen. — Die Binnengewässer. Bd. XXIII. — E. Schweizerbart'sche Verlagsbuchhandlung (Nägeller und Obermiller) Stuttgart, 183 p., 1967.

Gruninger, W.: Recent Kalktuffbildung im Bereich der Uracher Wasserfälle. — Abh. Karst- u. Höhlenk. — München, Reihe E (Bot), **2,** 113 S., 1965.

Howe, M. A.: The geologic importance of the lime secreting algae. — U.S.G.S. Professionnal Paper, **170,** 57—64, 1931.

Hurter, E.: Beobachtungen an Litoralgen des Vierwaldstätter Sees. — Mitt. a. d. Hydrobiol. Lab. Kaataneienbaum b. Luzern, X. Heft, 1928.

Irion, G., & Müller, G.: Mineralogy, Petrology and chemical composition of some calcareous Tufa from the Schwäbische Alb, Germany. — In: "Carbonate sedimentation in Central Europe", Müller, G., and Friedman, G. M. Ed., Berlin, Heidelberg, New-York (Springer), 157—171, 1968.

Illing, L. V.: Bahama calcareous sands. — Am. Ass. Petrol. Geol. Bull., **38,** 1—95, 1954.

JOHNSON, J. H.: Algae as rocks builders with notes on some algal limestones from Colorado. — Univ. of Colorado Studies, **23**, No. 3, Colorado (Boulder) 1936.

—: Algae and algal limestone from the Oligocene of South Park, Colorado. — Bull. of the Geol. Soc. of America, **48**, 1227—1236, 2 pls., 1 fig., 1937.

KANN, E.: Zur Ökologie des litoralen Algenaufwuchses im Lunzer Untersee. — Int. Rev. d. ges. Hydrogr. Hydrobiologie, **28**, 1933.

—: Ökologische Untersuchungen an Litoralgen ostholsteinischer Seen. — Arch. f. Hydrobiol., **37**, 2, 178—269, 1940.

—: Cyanophyceenkrusten aus einem Teich bei Abisko (Swedisch Lappland). — Arch. Hydrobiol., **37**, 4, 495—503, 1941 a.

—: Krustensteine in Seen. — Arch. f. Hydrobiol., **37**, 4, 504—532, 1941 b.

KENDAL, C. G. St. C., & SKIPWITH, Sir PATRICK A. d'E. Bt.: Recent algal mats of a Persian Gulf Lagoon. — Journ. of Sedim. Petrol., **38**, 4, 1040—1058, 1968.

KORDE, N. V.: Deposits of Blue-Green algae in sapropel lakes. — Dokladi Akad. Nauk. SSSR, **58**, 1947.

—: Blue-green algae as sources of sapropel deposits. Pt. 1: Blue-green algae in sediments of Transural Lakes. — Trudy Lab. Sapropel. Otlozhenii, 1950.

—: Biostratification and classification of Russian sapropels. — Izdatel'stvo Akademiya Nauk. SSSR. Moscow 1960.

LAUTERBORN, R.: Die Kalksinterbildungen an den unterseeischen Felswänden des Bodensees und ihre Biologie. — Mih. Bad. Landesver. Naturk. u. Naturschutz in Freiburg, **1**, H. 8, 1922.

MAWSON, D.: Some South Australian algal limestones in process of formation. — Quart. Jour. Geol. Soc. London, **85**, 4, 613—620, 1929.

MONTY, C. L.: Biostromes stromatolithiques dans le Viséen moyen de la Belgique. — C. R. Acad. Sc., **256**, 5603—5606, 1 fig., 1963.

—: Recent algal stromatolites in the Windward Lagoon, Andros Island, Bahama. — Ann. Soc. Géol. Belgique, Bull., **88**, 6, B, 269—276, 1965 a.

—: Geological and environmental significance of Cyanophyta. — Ph. D. Thesis, 429 p., 89 pls., 42 figs. Princeton University, N.J., USA, Microfilm No. 66 — 5003 Univ. Microfilm Inc. Ann Arbor, Michigan 1965 b.

—: Distribution and structure of recent stromatolitic algal mats, Eastern Andros Island, Bahama. — Ann. Soc. Géol. Belgique, **90**, 3, 55—100, 19 pls., 13 figs., 1967.

—: An autoecological approach of intertidal and deep water stromatolites. — Ann. Soc. Géol. Belgique, **94**, 265—276, 4 figs., 1971.

—: The Phanerozoic history of stromatolite communities. — Ann. Soc. Géol. Belgique, 1972 (in press).

MÜLLER, G.: Exceptionally high strontium concentrations in fresh water onkolites and mollusk shells of Lake Constance. — In: "Carbonate sedimentology in Central Europe. — Müller, G., and Friedman, G. M., Ed., Berlin, Heidelberg, New York (Springer) 1968, 116—127.

MURRAY, G.: Calcareous peebles formed by algae. — Phycological Memoirs, pt. III, 73—77, London 1895.

NAUMAN, E.: Untersuchungen über einige sub- und elitorale Algen-Assoziationen unserer Seen. — Archiv f. Botanik, **19**, 16, 1—30, 1925.

NEUMANN, A. C., GEBELEIN, C. D., & SCOFFIN, T. P.: The Composition, structure, and erodability of subtidal mats, Abaco, Bahamas. — Journ. of Sedim. Petrol., **40**, 1, 274—297, figs. 1—8, 1970.

NEWELL, N. D., & RIGBY, J. K.: Geological studies on the Great Bahama Banks. — p. 15—73 in: Le Blanc, R. S., and Breeding, J. G., eds., Regional aspects of carbonate sedimentation. Soc. Econ. Paleontologists Mineralogists, Spec. Pub. 5, 178 p., 1957.

OBERDORFER, E.: Lichtverhältnisse und Algenberiedlung im Bodensee. — ZS. Bot., 20, 10/11, 1928.

PHILLIPS, R. C.: Floating communities of algae in a North Carolina pond. — Ecol. 49, 465—766, 1958.

—: Ecology of floating algal communities in Florida. — Quart. Jour. Flor. Acad. Sces., 26, 4, 329—334, 1963.

PIA, J.: Pflanzen als Gesteinsbildner. — 355 S., Berlin (Gebrüder Bornträger) 1926.

—: Die rezenten Kalksteine. — Z. f. Krist. Mineral. u. Petrol. Mitteil. Ergänzungsbd. Abt. B, 420 S., 1933.

POLLOCK, J. B.: Blue-green algae as agent of deposition of Marl in Michigan Lakes. — Michigan Academy of Sciences XXth report, 247—260, 1918.

PURDY, E.: Recent calcium carbonate facies of the Great Bahama Bank. 1. Petrography and reaction groups. — J. of Geol., 71, 3, 334—355, 1963 a.

—: Recent calcium carbonate facies of the Great Bahama Bank. 2. Sedimentary Facies. — J. of Geol., 71, 4, 472—497, 1963 b.

RODDY, H. J.: Concretions in streams formed by the agency of Blue-green algae and related plants. — Proc. Amer. Phil. Soc., 54, 246—258, Philadelphia 1917.

RUTTE, E.: Die Algenkalke aus dem Miozän von Engelwies in Baden. — Neues Jb. Geol. u. Paläontol., Abh., 98, 2, 149—174, 1953.

SCHMIDLE, W.: Postglaziale Ablagerungen im nordwestlichen Bodenseegebiet. — Neues Jb. Mineral., II, 104—122, 1910.

SCHOLL, D. W., & TAFT, W. H.: Algal contributors to the formation of calcareous Tufa, Mono Lake, California. — Journ. Sed. Petr., 34, 2, 309—319, 1964.

SCHÖTTLE, M.: Die Sedimente des Gnadensees. Ein Beitrag zur Sedimentbildung im Bodensee. — 104 p., Diss., Heidelberg.

SCHÖTTLE, M., & MÜLLER, G.: Recent carbonate sedimentation in the Gnadensee (Lake Constance), Germany. — In: "Carbonate sedimentology in Central Europe". Müller, G., and Friedman, G. M., Ed., Berlin-Heidelberg-New York (Springer) 1968, 148—156.

SCOFFIN, T. P.: The trapping and binding of subtidal carbonate sediments by marine vegetation in Bimini Lagoon, Bahamas. — Journ. of Sedim. Petrol., 40, 249—273, figs. 1—28, 1970.

SEWARD, A. C.: Fossil plant. — (Quoted by Cayeux, 1935.)

SHINN, E. A., LLOYD, R. M., & GINSBURG, R. M.: Anatomy of a modern carbonate tidal-flat, Andros Island Bahamas. — Journ. of Sedim. Petrol., 39, 3, 1202—1228, figs. 1—37, 1969.

STIRN, A.: Kalktuffvorkommen und Kalktufftypen der Schwäbischen Alb. — Abh. Karst- u. Höhlenk., München, Reihe E (Bot.), 1, 92 p., 1964.

SYMOENS, J. J.: Note sur des formations de tuf calcaire dans le bois d'Hautmont (Wauthier-Braine). — Bull. Soc. Roy. Bot. Belgique, 82, 81—95, 1949.

SYMOENS, J. J., DUVIGNEAUD, P., VAN DEN BERGEN, C: Aperçu sur la végétation des tufs calcaires de la Belgique. — Bull. Soc. Roy. Bot. Belgique, 83, 329—352, 1951.

SYMOENS, J. J., & MALAISSE, F.: Sur la formation de tuf calcaire observée sur le versant Est du plateau des Kundulungu. — Acad. Roy. des Sces d'Outre-Mer. des Séances, 1967-6, p. 1148—1151, 1967.

TILDEN, J. H.: Some new species of Minnessota algae which live in a calcareous or siliceous matrix. — Bot. gaz., 23, 7—9, 95—104, 1897.

ULRICH, FR.: Über die Wachstumsform des organogenen abgeschiedenen Kalkspates und ihre Beeinflussung durch das Kristallizationsmedium. — Z. f. Kristallographie, 66, 513—514, 1927.

WALCOTT, C. D.: Pre-Cambrian algonkian algal flora. — Smithsonian Misc. Coll., 64, 77—156, pls. 4—23, 1914.

Wallner, J.: Zur Kenntnis ces unter pflanzlichem Einfluß gebildeten Kalkspates. — Planta Archiv f. Wissenschaft. Botanik, **23**, 51—55, 1934.
—: Zur Kenntnis der Kalkbildung in der Gattung *Rivularia*. — Beih. z. Bot. Zbl., **54**, Abt. A, 1935.
Worochin, N. N.: Zur Biologie der bittersalzigen Seen in der Umgebung von Pjatigork (nördl. Kaukasus). — Arch. f. Hydr., Hydrobiol., **17**, 628—643, 1926.

5

Reprinted from *Geol. Rundschau* **61**:520–541 (1972)

INTERACTION OF GENETIC PROCESSES IN HOLOCENE REEFS OFF NORTH ELEUTHERA ISLAND, BAHAMAS

Heinrich Zankl and Johannes H. Schroeder

Abstract

The inner shelf platform off North Eleuthera consists of a series of terraces at 6 m, 8 m, 20 m, and 35 m depth, respectively. The boundary between 6 and 8 m terraces is marked by a near continuous reef ridge, landward and seaward of which isolated reef structures are found. Reefs are aggregates of pillar structures, which in turn are composed of columnar, club-shaped or mushroom-shaped pillars of corals. The dominant primary framebuilder above 3 m depth is *Acropora palmata;* below corals of the genera *Montastrea* and *Diploria, Porites astreoides* and *P. porites* also contribute to the frame, as does the hydrozoan *Millepora.* The secondary framebuilders start developing and changing the shape of the frame as soon as they find suitable substrates; most important are coralline red algae, which encrust internal and external surfaces, but also may fuse the components of the frame; other organisms just add to the frame. In contrast boring and rasping organisms destroy the frame. Constructing or destructing organisms were found living on every surface, external or internal, accreting or diminishing.

The composite nature of the reef structure, fabrics of skeletal material, and borings of organisms provide numerous cavities of mm to m size. These are subject to syngenetic sedimentation and cementation, which help maintain the frame, but change composition, fabrics, and facies. Thus rates of organism growth and destruction, of sedimentation, cementation, and mechanical breakdown and the dynamic changes of these rates determine final shape and preservation of the reef.

Recognition of the relationship between these processes should help interpret fossil reef remnants.

Introduction

Reef formation is a complex process with numerous organic and inorganic agents interacting. The final product of this process is preserved in fossil reefs; the steps leading to this product are more or less clearly documented in the fabrics of the rocks. Such genetic sequences in reef formation have been studied in a fossil example of the Alpine Triassic (ZANKL, 1969). Functional interrelationships between different agents remain largely obscure, but these can be studied in Recent reefs. In Holocene reefs of Bermuda the close interdependence between constructing and destructing organisms on the one hand and internal sedimentation and cementation on the other was recognized for the first time (GINSBURG et al., 1967, 1969, 1971; SCHROEDER, 1972 a, b). The purpose of this study is to investigate on the functional interrelationships between the agents and processes involved and thus to facilitate interpretation and understanding of the information contained in the fabrics of fossil reefs. The present paper contains the results of field observations augmented by preliminary laboratory work; the concept outlined will be elaborated and documented further after completion of the laboratory analyses.

Methods

Reef distributions were studied in aerial photographs. To characterize the phenomena seen, two N—S traverses across the inner portions of the shelf-platform, one 800 m E of Ridley's Head, the other 1000 m E of Hawk Point were investigated by skin and scuba diving. In addition numerous spot checks were taken to obtain an idea of the representative value of features seen. Two bottom profiles were taken by means of recording fathometer (ELAC Belatrix).

Selected reef structures or 20 m wide transects across such structures were measured, and their morphology was mapped. Organism distribution was first established by visual inspection, and then quantitatively verified by counts of selected squares of $^1/_4$ to several square meters in size depending on the organism counted and the respective densities. Internal reef features were studied in fresh submarine outcrops produced by small charges of explosives; from these outcrops oriented samples were taken. Sediments were sampled from cavity floors inside and from outside the reef, but as yet only field observations are available. Laboratory work so far concentrated on analysis of sectioned surfaces and thin sections.

BAHAMA ISLANDS

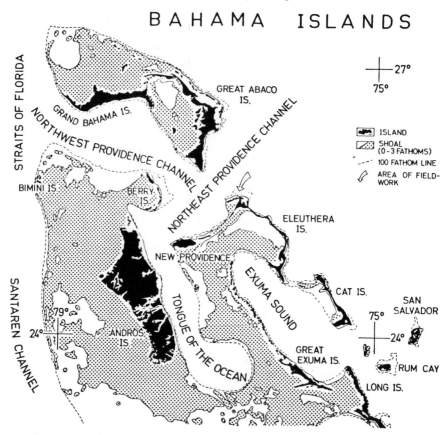

Fig. 1. Map of the northern portion of the Bahama Islands indicating the area of field work.

Physiography and general characteristics

The northern tip of Eleuthera Island is surrounded by a shallow-water platform, which drops off to the deep-sea on three sides (Figs. 1, 2). The platform descends in several steps toward the marginal escarpment. The step closest to shore is rimmed by coral reefs. Following the terminology applied to the shelf of Andros Island by NEWELL & RIGBY (1957), the area between shore and reefs is called "shelf lagoon", the area seaward of the reefs "outer platform".

The width of the shelf lagoon varies: Off Royal Island it is 3000 m, off Harbour Island the reefs abut to the shore leaving only locally a shelf lagoon of several tens of meters width. The shelf lagoon is up to 6 m deep; its floor is covered by sediment everywhere. According to field observations, sediments consist of coarse sand-sized particles close to shore, but of fine sand to silt-sized reef debris elsewhere. Large areas are covered by turtle grass. Toward the seaward margin of the shelf lagoon, patch reefs are found in increasing number, reaching their maximum density immediately lee of the reef ridge.

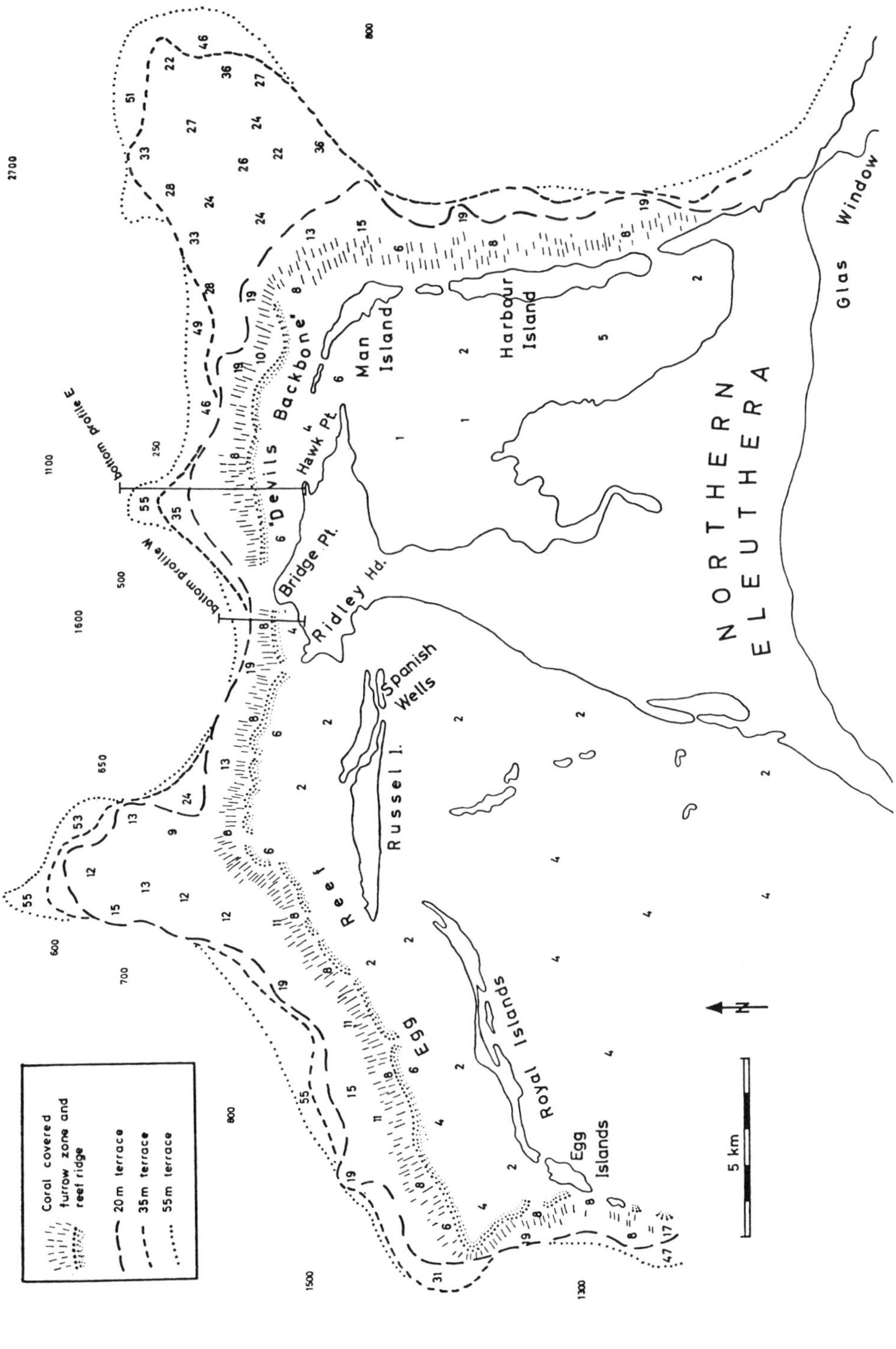

Fig. 2. The shelf platform off N. Eleuthera (waterdepths in m, taken and recalculated from Admirality Chart 2077, ed. 1958).

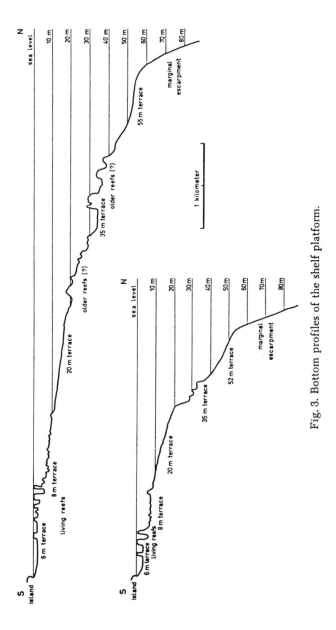

Fig. 3. Bottom profiles of the shelf platform.

The reef ridge parallels the north coast for more than 25 km; there are some small breaks in Egg Reef and in Devils Backbone. Large breaks, which are important for navigation, occur where the ridge is transposed horizontally toward the shore; Wide Opening off Bridge Point is an example of such a break.

The outer platform consists of an 8 m, a 20 m and a 35 m terrace (figs. 2, 3). The relief of the 8 m terrace is characterized by bar-shaped elevations separated by furrows [1]) and basins, which reach a depth of 8—10 m. The depressions are filled with coarse reef debris, while on the bars pillars of corals, mostly dead, are found. In general, relatively few living corals are found on the bar, but many encrusting coralline red algae and, on the floor, calcareous green algae. Near the reef ridge some mounds reach sea-level; their upper portions are covered by living corals and encrusting coralline algae. Because these mounds break the waves, they are readily visible at sea; locally they are called "boilers". Seaward the number of corals found on bars decreases, and at the expense of basins the furrows dominate the morphology forming a pattern of parallels running normal to shore. The 8 m terrace, called coral covered furrow zone, can be traced in aerial photograph as a continuous stripe along the inner margin of the outer platform (fig. 4). Similar furrows have been reported by NEWELL & RIGBY (1957) from the shelf east of Andros Island; these authors interpreted the furrows as results of Pleistocene erosion because they found furrows also in Pleistocene oolitic rock. The seaward margin of the 8 m terrace varies in pronounciation; mostly the transition to the 20 m terrace consists of a gentle slope.

The 20 m terrace covers the major portion of the outer platform. Its relief also is defined by bars and furrows, however, in comparison to the 8 m terrace the bars are relatively bare. This zone ist called bar-and-furrow-zone. The bars consist of a flat top and gentle slopes dipping toward the furrows; the difference between top and furrow generally amounts to 4—5 m. Few corals, calcareous green algae, alcyonarians, sponges are the dominant organisms on the bars. The furrows are 0.5—1 m deep; they are filled with coarse sand or gravel sized sediments. They are oriented normal to shore; individual furrows can be traced for several hundred meters; some forking furrows have been observed. STORR (1964) interpreted similar furrows off Abaco Island as surge channels. At the seaward margin of the 20 m terrace extensive mounds 100—300 m in diameter and 4—5 m in height were observed. Their irregular shape and surface suggests that these mounds may have been older reefs; further study is needed to support that suggestion. Similar, even more pronounced rises, up to 10 m high, were found at the outer margin of the 35 m terrace.

The 35 m terrace varies in width; in the area of the NE bank it reaches a width of 6 km, but at the slope of the western portion of the platform it is reduced to a mere knick in slope. The observation with respect to morphology is limited to few dives, during which the bar-and-furrow pattern was recognized to continue on this terrace.

The outer margin of the platform varies: In its NE portion a 1000 m wide gentle slope extends seaward toward the break of the marginal escarpment at a depth of 50 to 55 m. In other portions of the platform this slope is lacking, and the marginal escarpment begins next to the 35 m or even the 20 m terrace. The marginal escarpment commonly slopes with 15° seaward, but SCHALK (1946) reported dips as high as 30°—45° from this area.

---

[1]) In this paper ther terms "bar" and "furrow" describe elongate morphologically positive and negative features, respectively, which are formed by erosion. They are distinguished from the "spur" and "groove" (SHINN, 1963) which result from biogenic growth.

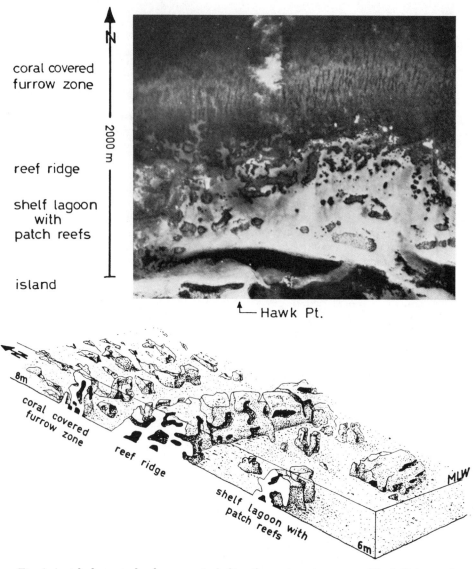

coral covered
furrow zone

N

2000 m

reef ridge

shelf lagoon
with
patch reefs

island

Hawk Pt.

8m

coral covered
furrow zone

reef ridge

shelf lagoon with
patch reefs

MLW

6m

Fig. 4. Aerial photograph of an area including the eastern traverse at Hawk Point, and
its evaluation in a block diagram.

Reef morphology

In the reef complex off North Eleuthera three types of reefs are distinguished
(figs. 4, 8):

1. The patch reefs of the shelf lagoon which vary in outline and dimension.
   They may assume block or table shape; in plan view they are subcircular
   to elongate ellipsoid. Their long dimension varies between a few meters
   and 300 meters.

2. The reef ridge marking the boundary between shelf lagoon and outer platform. The ridge consists of steep seaward and shoreward flanks with an extensive reef flat in between. The reef flat may slope gently shoreward, thus
giving rise to a somewhat wedge-shaped profile, or else it may be replaced
by a central depression separating two narrow subparallel ridges. The reef
ridge commonly is 60—80 m wide.

3. The isolated reef structures of the coral covered furrow zone. They are
knob- to block-shaped; their long dimension ranges from 20 to 150 m.

Except for their distribution and consequently their height above sea-floor
these reef types have more features in common than not. For example all of
them reach the lower portion of the intertidal and are awash at neap tides. More
important, however, construction of the reefs of all types is governed by the
same principle.

The basic structural unit is a pillar built up by corals. Depending on the coral
species involved and on the location, pillars vary in shape and size. Tall and
slender columnar pillars (common diameter 30—60 cm) frequently are formed
by a single coral species such as *Diploria strigosa*, *D. labyrinthiformis*, *Montastrea annularis*, or *M. cavernosa* (fig. 5 a). Increasing lateral growth is reflected
in club-shaped pillars which widen upward (fig. 5 b); such shapes are predominantly developed by the *Montastrea* species mentioned, subordinately also by
the *Diploria* species or by *Porites astreoides*. From club shape there is a transition
to mushroom shape (figs. 5 c, d, e). Mushroom shapes result from continuous
lateral growth and slowing or stopping vertical growth. *Diploria* and *Montastrea*
form relatively massive mushroom tops (fig. 5 c), while *Acropora palmata* forms
farther extending umbrella-like tops (fig. 5 d). The distribution of *Acropora palmata* growth forms within the reef (compact flat forms near sea level; branched
tall forms at lower levels) accounts for the corresponding distribution of mushroom-shaped pillars. *Montastrea*, especialy *M. annularis* constructs mushroom-
shaped pillars with large overhangs exhibiting downward and inward turned lips
(fig. 5 e). Evidently size and shape of pillars are governed primarily by growth
forms and ecology of the corals involved, thus indirectly also by their location
with respect to sea-level and by the substrate.

Pillars are combined to pillar structures, the next larger structural unit (figs. 5 f,
g, h, i, j); pillar structures may assume plate shape, block shape, club shape, or
mushroom shape. Of course their shape again depends on growth form and
ecology of the corals. The depth zonation of corals results in vertical variation
of pillar shapes within a pillar structure. For example, a block structure may
be composed of simple pillars at the bottom, of club-shaped pillars farther up,
both composed of *Montastrea*, *Diploria* and *Porites;* it then may be capped by
a plate of fused mushroom pillars formed by *Acropora palmata* near sea level.

The reefs off N. Eleuthera are composites of various pillar structures. The
combination of club-shaped, mushroom-shaped and plate-shaped pillar structures
results in large caves and tunnel systems (up to 5 m high, several 10 m in
length). Patch reefs and reef ridge were found to be based on few relatively
thin pillar structures which near sea-level combine to form a continuous reef
flat with large lateral overhangs (up to 6 m wide). Caves and tunnels are
accessible from the flanks, and occasionally from the top; their floors are
covered by sand- and gravel-sized rippled sediment.

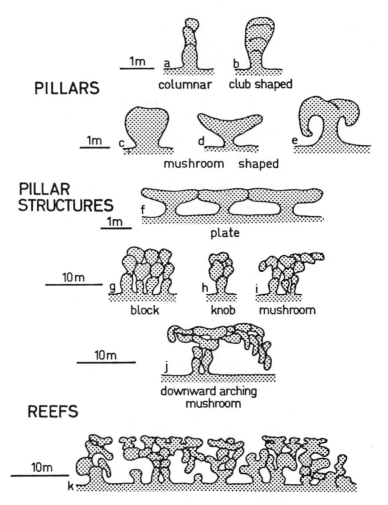

Fig. 5. Schematic illustration of coral pillars, their combination in pillar structures, and of reefs, which are aggregates of pillar structures.

Shaping the frame

While construction of the primary framework is progressing, the shape of the reef is developed further by four processes:

1. Construction by secondary frame building organisms.
2. Destruction by organisms.
3. Internal sedimentation and cementation.
4. Mechanical breakdown.

Construction by organisms

Dead portions of the coral frame are rapidly encrusted by coralline red algae and to a lesser extent by *Millepora*, both of which have also been observed encroaching on living corals. Coralline red algae may form thick crusts which eventually constitute a larger portion of the frame than the corals skeletons. Red algal encrustation fuses individual pillars and thus is instrumental in forming continuous pillar structures; for example individual mushroom pillars of *Acropora* have been observed frequently to be fused to a plate structure by these algae. This function of the red algae has been recognized by WALTHER (1888), GINSBURG & LOWENSTAM (1958), and LOGAN (1969). Algal growth is not limited to the external surface but has been observed on virtually every surface, internal or external, of the reefs (Fig. 6 a, b). Within caves and tunnels they plaster the original frame. Local differences in substrate, current, and light conditions lead to differential growth resulting in the bulbous shape of the internal surfaces. The most rapid growth is directed toward the light, either vertically (fig. 6 a) or horizontally (fig. 6 b). The packing of subsequent crusts appears to be a matter of current intensity; sheet cavities of up to 1 cm width were observed in calm internal cavities, but mostly crusts are packed relatively tightly. The framework binding activity of the hydrozoan *Millepora* is restricted to external surfaces. In addition *Millepora* constructs a frame by itself, a boxwork of vertical blades which may form 2–3 m wide subcylindrical aggregates.

Volumetrically less important, but abundant secondary frame builders are the colonial foraminifer *Homotrema rubrum* growing in cavities and on the underside of ledges and corals, and serpulid worms forming irregular aggregates of coiling tubes. Internal surfaces are frequently dotted by the patches made by calcisponges. The ahermatypic coral *Astrangia solitaria*, barnacles, brachiopods, and the hydrozoan *Stylasterina* are less important, but may be locally abundant. Bryozoans have been found in various habitats (CUFFEY, 1972).

Destruction by organisms

Destruction of the frame is initiated and proceeds simultaneously with secondary frame building. Quantitatively most important is the boring by various sponges *(Cliona)*, pelecypods *(Lithophaga nigra, L. bisulcata, Botula fusca, Petricola lapiscida* and others), and by polychaete as well as sipunculid worms on a macroscale, and the boring of endolithic algae on the micro-scale. The borers produce

Fig. 6. Shaping the reef frame by organism growth and boring. T o p : Sample from patchreef N of Bridge Point (Fig. 8 a) from 0.75 m below reef top; vertical section, top is up. The coral is encrusted at the top by secondary framebuilders, mainly crustose coralline red algae. The entire specimen is penetrated by various borings including those of pelecypods, sponges, and worms. At the top, the rate of encrusting was higher then the rate of boring, thus accretion resulted. At the bottom, the rate of boring exceeded the rate of encrustation, thus part of the coral was removed. B o t t o m : Sample from reef ridge N of Bridge Point (Fig. 8 c) from 7 m depth, about 6 m inside a cave. Pronounced laterial growth (toward cave entrance) of red coralline red alga formed a cone-shaped protrusion (only selected algal crusts are shown). At top and bottom boring rates are high enough to prevent substantial accretion, whereas horizontally the growth rate of the algae is well above that of boring.

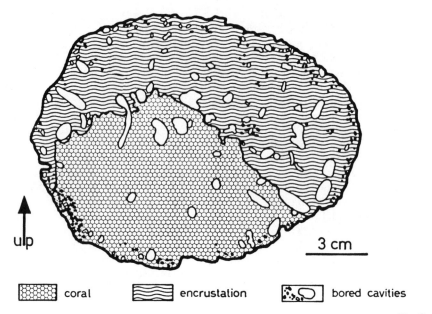

coral   encrustation   bored cavities

Fig. 6 a

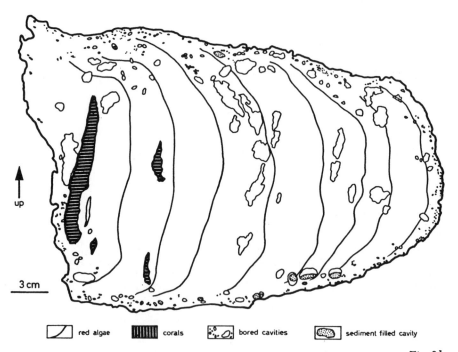

red algae   corals   bored cavities   sediment filled cavity

Fig. 6 b

additional cavities; thus their activity results in an outward increasing porosity of every structural element, inside and outside the reef.

Less significant, but by no means negligible is the external destruction by rasping fish (e.g. parrot fish), turtles, and crabs. These organisms remove considerable amounts of material (GLYNN et al., 1972), but their traces are hardly ever found beneath the surface within the reef rock.

Considering the distribution of both constructive and destructive organisms there is not a patch of surface, neither on the wave-beaten reef flat nor in the caves and cavities including boreholes within the reefs, which can be considered dead. Everywhere organisms are actively developing and changing the shape of the reef structure.

Sedimentation and cementation

Primary and secondary frame building in combination with boring result in complex reef structures enclosing a variety of cavities such as intra-skeletal, inter-skeletal and boring cavities (GINSBURG & SCHROEDER, in preparation). These cavities are filled with predominantly reef-derived sediment of mostly fine sand and silt size. The sediment is lithified internally; additional cement may line or fill remaining cavities. Cavities filled by sediment and cement are portions of the solid reef material and therefore subject to subsequent boring, renewed sedimentation, and cementation. In this way the original framework may be replaced entirely by cemented sediment and the original structure may be completely obliterated (fig. 7). From the foregoing it is evident that sedimentation and cementation maintain the framework, and therefore can be considered constructional processes. However, it is equally evident that in conjunction with boring they change composition, fabric, and facies of the reef rock.

Mechanical breakdown

Mechanical destruction of the reef frame affects primarily the external surface. As a result of periodic storm activity, protrusions are broken off the frame; coral pillars with narrow bases riddled by boring organisms are particularly prone to this breakdown. The fragments produced are rapidly encrusted by red algae. The debris is either accumulated in depressions or caves within the reef or in rubble zones (fig. 8) fringing the reef bases. Extensive rubble layers interlayered with the sand sized sediments leeward of the reef ridge document periodic maxima of this type of destruction. This process provides a lot of the material for internal sedimentation (see above).

In summary, the shape, and thus composition, fabrics, facies as well as preservation of the frame are determined by the relative rates of organic growth and boring, sedimentation, cementation, and mechanical breakdown. Since all these rates are subject to local, if not micro-environmental influences, each sample is formed in a different way, and even within a sample or thin section marked differences in formation may be found. Considering the time variable, evidently local and micro-environmental influences change. For example, current patterns, permeability, and sediment supply change with time. Thus the interplay between processes involved is dynamic, hence the variety of genetic histories within a reef is almost inconceivable.

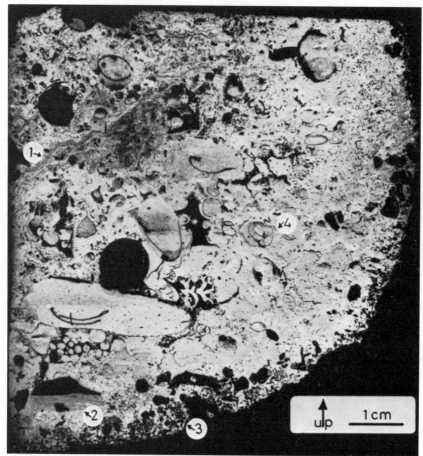

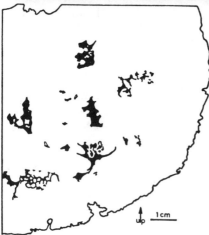

Fig. 7. Gradual replacement of organic framework by lithified sediment which in turn is bored by organisms. Top: Negative print of thin section. Except for the portions shown in sketch (left) and for the mollusc shells, black portions are cavities of various kinds mainly bored. 1. Encrusting red algae. 2. Large pelecypod boring partially filled with silt-sized laminated sediments. 3. Rim bored by sponges and settled by serpulids and other organisms. 4. "Shell-in-shell structure": Three boring pelecypods successively inhabiting the same hole. Left: Sketch taken from above section, indicating what is left over of the original coral frame (coral is black).

Ecology and structural differentiation

To illustrate structural elements and organism distributions, generalized measured sections of two reefs each of shelf lagoon, reef ridge, and coral covered furrow zone are presented (figs. 8 a–f). The variety of structural developments prohibits establishing reef types; nevertheless the examples figured show the salient features of the reef structures in this area. The following comments supplement these figures. The top of most reefs is marked by a reef flat at or near the MLW level, a bare zone which almost lacks active coral growth. This irregularly undulate surface is covered by crustose coralline algae and the crustose growth form of *Millepora*. Such surfaces occur not only on the reef top but also on shoreward (fig. 8 c) and seaward (fig. 8 e) slopes. On portions of the reef flat exposed to continuous wave and surf action reef crests developed which are characterized by the occurrence of bladed *Millepora* and the bulbous colonial sea-anemone *Polythoa mammilosa* (fig. 8 e).

The distribution of reef organisms is largely determined by water depth and degree of turbulence. With respect to corals, on reef ridge and coral covered furrow zone *Acropora palmata* is the major framebuilder at water depth above 3 m; below that depth *Monastrea annularis, M. cavernosa, Diploria labyrinthiformis* and *D. strigosa* are dominant. *Porites astreoides* and *P. porites* are encountered in both depth zones. Locally both *Porites* species are abundant and predominant at the rim of profuse coral growth (fig. 8 b) which surrounds some sheltered patch reefs; jointly with *Agaricia agaricites*, they form pronounced overhangs.

With respect to algae, in addition to the ubiquitous crustose coralline red algae, on the bare zone of the reef flat a community of brown algae such as *Turbinaria, Sargassum,* and *Dictyota* occurs accompanied by the calcareous green alga *Halimeda*. This community also is encountered on other external surfaces exposed to wave action. In more sheltered areas the brown alga *Padina* and branching red algae such as *Amphiroa* are found. On the sandfloor around the reefs calcareous green algae such as *Halimeda, Penecillus,* and *Udotea* grow sparsely.

Discussion

Distribution of coral reefs off N. Eleuthera is related to pre-existing topography; this relation has been observed in other Bahamian reefs studies. NEWELL & RIGBY (1957) found reefs east of Andros to rise from a Pleistocene substrate, and STORR (1964) found the same with respect to the reefs NE of Abaco Island. Off Abaco Island the position of the reefs is almost identical to that illustrated from N. Eleuthera: reefs are concentrated shoreward and seaward of a "reef barrier" which parallels the boundary between 4 m and 8 m terraces.

As to the structure, the reefs off Abaco Island enclose caves which in shape and dimension correspond to those observed in N. Eleuthera reefs. STORR (1964), however, draws the conclusion that the caves result from "biological erosion and collapse of loosely consolidated rock". In contrast this study shows that except for bio-erosion no additional enlargement of caves takes place, but that the caves and tunnels result from combination of coral pillars and pillar structures to reefs (fig. 5), thus are truly constructional. The basal overhang of many pillar

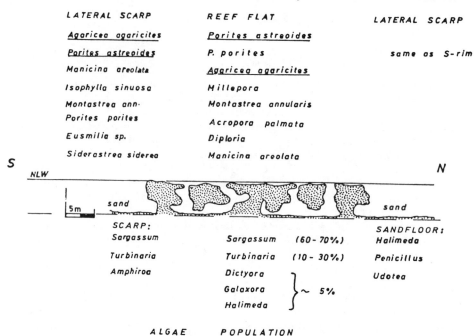

LATERAL SCARP

Agaricea agaricites

Porites astreoides

Manicina areolata

Isophylla sinuosa

Montastrea ann·

Porites porites

Eusmilia sp.

Siderastrea siderea

REEF FLAT

Porites astreoides

P. porites

Agaricea agaricites

Millepora

Montastrea annularis

Acropora palmata

Diploria

Manicina areolata

LATERAL SCARP

same as S-rim

S   N

NLW

5 m   sand   sand

SCARP:
Sargassum

Turbinaria

Amphiroa

Sargassum (60-70%)

Turbinaria (10-30%)

Dictyora

Galaxora } ~ 5%

Halimeda

SANDFLOOR:
Halimeda

Penicillus

Udotea

ALGAE POPULATION

Fig. 8 a. Generalized measured section of shelf lagoon patch reef N of Bridge Point showing distribution of organisms and structural elements (no vertical exaggeration; dominant framebuilders underlined). Note block shape, deep channels, and narrow pillar bases.

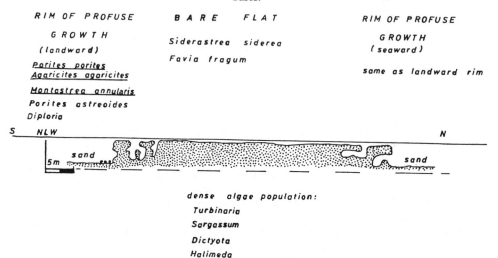

RIM OF PROFUSE

GROWTH

(landward)

Porites porites
Agaricites agaricites

Montastrea annularis
Porites astreoides
Diploria

BARE FLAT

Siderastrea siderea

Favia fragum

RIM OF PROFUSE

GROWTH

(seaward)

same as landward rim

S   NLW   N

5 m   sand   sand

dense algae population:

Turbinaria

Sargassum

Dictyota

Halimeda

Fig. 8 b. Generalized measured section of shelf lagoon patch reef N of Hawk Point showing distribution of organisms and structural elements (no vertical exaggeration; dominant framebuilders underlined). Note rim of profuse coral growth and relatively massive central portion.

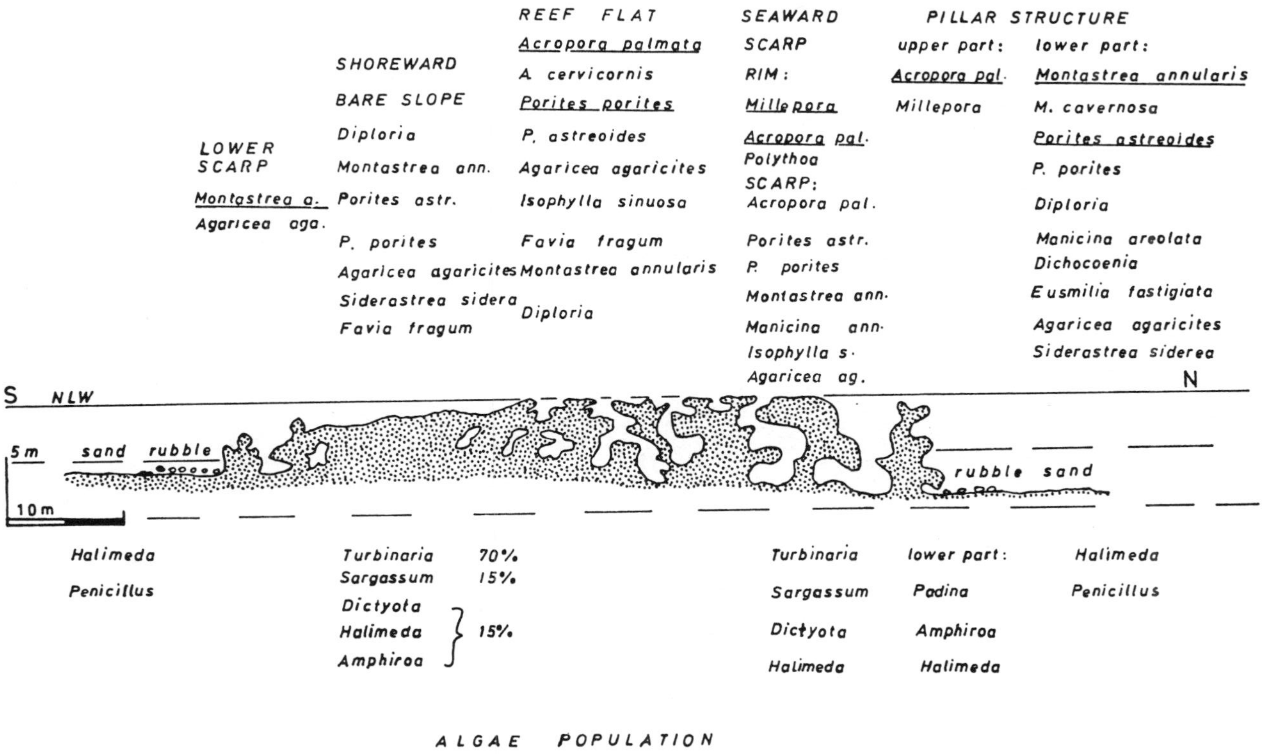

PILLAR STRUCTURE

upper part:	lower part:
Acropora pal.	*Montastrea annularis*
Millepora	M. cavernosa
	Porites astreoides
	P. porites
	Diploria
	Manicina areolata
	Dichocoenia
	Eusmilia fastigiata
	Agaricea agaricites
	Siderastrea siderea

N

SEAWARD

SCARP
RIM:
Millepora
Acropora pal.
Polythoa
SCARP:
Acropora pal.
Porites astr.
P. porites
Montastrea ann.
Manicina ann.
Isophylla s.
Agaricea ag.

rubble sand

REEF FLAT

Acropora palmata
A cervicornis
Porites porites
P. astreoides
Agaricea agaricites
Isophylla sinuosa
Favia fragum
Agaricea agaricites Montastrea annularis
Siderastrea sidera Diploria
Favia fragum

SHOREWARD

BARE SLOPE
Diploria
Montastrea ann.
Porites astr.
P. porites
Agaricea agaricites
Siderastrea sidera
Favia fragum

LOWER
SCARP
Montastrea a.
Agaricea aga.

ALGAE POPULATION

	lower part:	
Turbinaria	Padina	Halimeda
Sargassum	Amphiroa	Penicillus
Dictyota		
Halimeda		

Turbinaria	70%
Sargassum	15%
Dictyota	
Halimeda	15%
Amphiroa	

Halimeda
Penicillus

S NLW

sand rubble

5 m

10 m

Fig. 8 c. Generalized measured section of the reef ridge N of Bridge Point showing distribution of organisms and structural elements (no vertical exaggeration; dominant framebuilders underlined). Note seaward pillar structures and relatively massive shoreward slope.

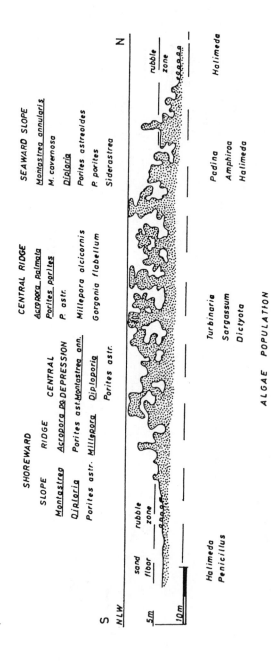

Fig. 8 d. Generalized measured section of the reef ridge N of Hawk Point showing distribution of organisms and structural elements (no vertical exaggeration; dominant framebuilders underlined). Note high columnar pillars at seaward and shoreward margins.

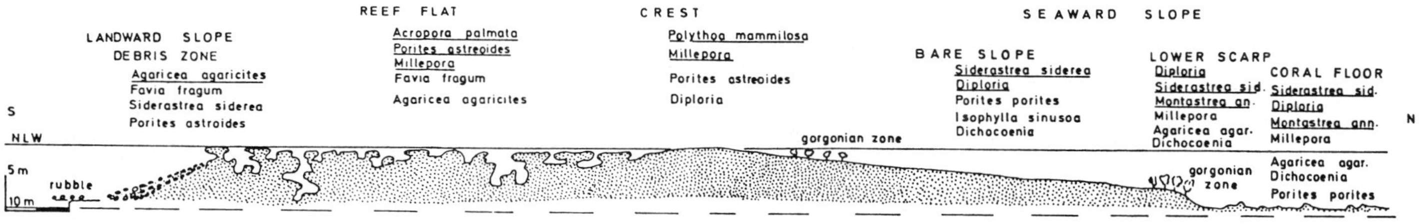

LANDWARD SLOPE
DEBRIS ZONE
Agaricea agaricites
Favia fragum
Siderastrea siderea
Porites astroides

REEF FLAT
Acropora palmata
Porites astroides
Millepora
Favia fragum
Agaricea agaricites

CREST
Polythoa mammilosa
Millepora
Porites astroides
Diploria

SEAWARD SLOPE

BARE SLOPE
Siderastrea siderea
Diploria
Porites porites
Isophylla sinusoa
Dichocoenia

LOWER SCARP
Diploria
Siderastrea sid.
Montastrea an.
Millepora
Agaricea agar.
Dichocoenia

CORAL FLOOR
Siderastrea sid.
Diploria
Montastrea ann.
Millepora
Agaricea agar.
Dichocoenia
Porites porites

gorgonian zone

Fig. 8 e. Generalized measured section of a reef of the coral covered furrow zone N of Bridge Point showing distribution of organisms and structural elements (no vertical exaggeration; dominant framebuilders underlined). Note extensive bare seaward slope and debris covered shorward slope.

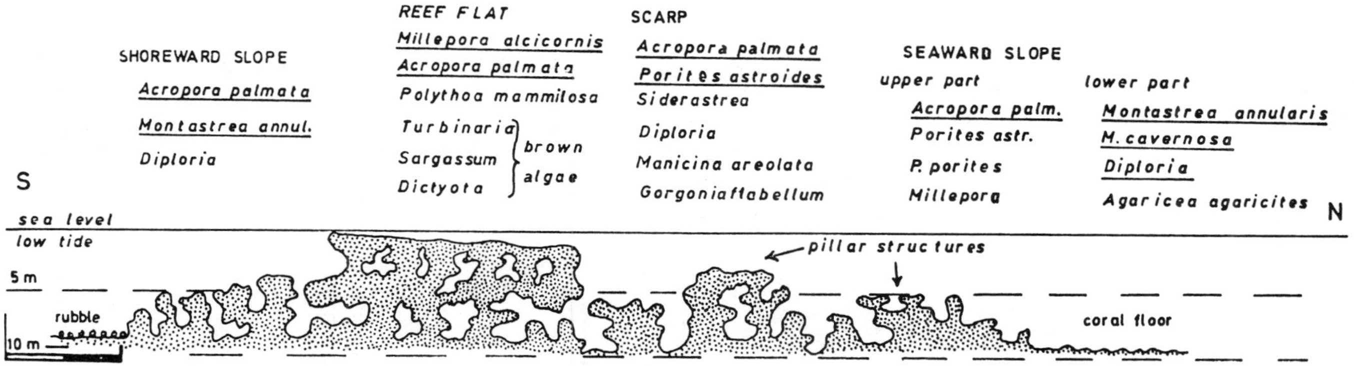

SHOREWARD SLOPE
Acropora palmata
Montastrea annul.
Diploria

REEF FLAT
Millepora alcicornis
Acropora palmata
Polythoa mammilosa
Turbinaria } brown
Sargassum } algae
Dictyota

SCARP
Acropora palmata
Porites astroides
Siderastrea
Diploria
Manicina areolata
Gorgoniaflabellum

SEAWARD SLOPE
upper part
Acropora palm.
Porites astr.
P. porites
Millepora

lower part
Montastrea annularis
M. cavernosa
Diploria
Agaricea agaricites

pillar structures

coral floor

Fig. 8 f. Generalized measured section of a reef of the coral covered furrow zone (a "boiler") N of Hawk Point showing distribution of organisms and structural elements (no vertical exaggeration; dominant framebuilders underlined). Note upward narrowing internal caves and, on the seaward side, pillar structures which do not reach MLW level.

structures was interpreted by Storr (1964) to be due to erosion; this study suggests that it results from lateral growth extending from an originally limited basis. Similarly, the ceilings and overhangs near the MLW level are interpreted to result from lateral growth which became dominant when vertical growth was curbed by sea-level. Scoffin (1972) ascribed the formation of overhangs and cavities in Recent Bermuda reefs and Silurian reefs of Shropshire (England) to frame growth, thus providing additional examples for the growth pattern suggested [2]).

Compared to pacific reef communities the number of species of reef building corals is smaller. The fauna and flora reported in the present paper from N. Eleuthera reefs is very similar to those known from other reefs of the Caribbean Sea (Newell & Rigby: 1957; Andros Island; Storr, 1964: Abaco Island; Logan, 1969: Yucatan; Shinn, 1963: Florida). The extension of vertical coral zonation differs, however, and appears to depend on exposure to turbulence, hence relative position of the reefs to the shelf margin, and on the height of the reef. The Nuevo reef off Yucatan is located near the shelf edge, with its basis at 50 m depth; Logan (1969) reported from this reef a reef flat resulting from encrusting and frame-binding activity of coralline red algae. Below that flat the *Acropora palmata* community assumes the range of 0—10 m, while the *Montastrea-Diploria-Porites* community occurs between 5—25 m. The zonation observed in the relatively shallow shelf platform off N. Eleuthera is basically identical, but vertically compressed to less than 10 m.

The interplay of organism growth, bio-erosion, sedimentation, cementation, and mechanical breakdown is certainly not restricted to the reefs off N. Eleuthera. Complex sequences of skeletal material, sediments, and cements have been observed in Holocene Bermuda reefs (Ginsburg & Schroeder, in preparation; Schroeder, 1972 b). Moreover, this interaction has been found to be an essential characteristic of fossil reefs. For example, in the Dachstein Limestone reef of the Hohe Göll (Alpine Triassic) primary and secondary framebuilders formed pillar structures enclosing growth cavities. The frame of these reefs is modified by syngenetic internal sedimentation and cementation (Zankl, 1969). Interaction of these processes was observed at various scales from the reef complex down to the micro-environment of small cavities. Krebs' (1969) study of Devonian reefs in Germany contains additional examples of growth-sedimentation-cementation sequences. These examples indicate that further delineation of these processes, their interactions, and their sequences, recognition of the resulting products, and definition of the factors affecting them are needed for comprehensive understanding of reef formation.

Conclusions

Study of the reefs off N. Eleuthera Island, Bahamas, revealed:

1. Reef distribution follows pre-existing topography.
2. The reef frame is an aggregate of coral pillars and pillar structures; it is

---

[2]) While this paper was in press, Garrett et al. (1971) presented a study of Bermuda patch reefs. Although the faunal composition as well as the dimensions of the cavities differ from those found off North Eleuthera, the formation of reef frame and cavities as described by those authors is comparable to the processes we have observed.

essentially of accretionary nature. Ecological zonation is reflected in the distribution of various structural elements.

3. The shape of the reef frame is developed further by encrusting and boring organisms which become active as soon as the frame has grown and live on every external and internal surface.

4. Internal sedimentation and cementation may fill existing cavities and thus contribute to maintaining the frame, but change its composition fabric and facies.

5. Periodic mechanical breakdown modifies the external shape and provides sediment.

6. Rates of organism growth and destruction, of sedimentation and cementation as well as mechanical breakdown and the dynamic changes of these rates determine the development of the frame and its preservation.

This study provides a comparative basis for investigation of interrelationships and relative importance of organism growth and destruction, sedimentation, cementation, and mechanical break down in fossil reefs.

Acknowledgements

We wish to thank the following agencies and officials of the Government of the Bahamas: The Ministry of External Affairs for granting permission to carry out this study, particularly Miss Patricia Rodgers of this Ministry who handled our application; the Ministry of Agriculture and Fisheries for permission, advice, encouragement, and cooperation, particularly Mr. O. S. Rusell, Permanent Secretary, and Mr. R. W. Beales, Fisheries Officer; the Ministry of Finance and the Comptroller of Customs for permitting duty-free entry of our equipment; the Ministry of Works for issuing the necessary licenses; the Lands & Survey Department for release of aerial photographs and friendly cooperation, particularly Mr. F. A. Garraway.

We extend our gratitude to the following individuals: Captain John Carpenter for continuing interest and cooperation, for experienced advice, for the use of the R. V. "Dragon Lady" which proved to be well suited for this operation; Don Zolanz, mate of the "Dragon Lady" who continuously gave his best to keep the fieldwork going; Aziel Pinder for able service as local pilot; T. H. Brzeski, R. Fiedler-Volmer, B. Foster, H. Langmann, and M. Scherer who assisted in the field; B. Dunker, B. Geringer, H. Gögdak for drafting the figures, B. Kleeberg for photographic work, and K. Zeschke for typing the manuscript.

We are indebted for financial support to: The German Research Foundation (Deutsche Forschungsgemeinschaft) which generously provided the financial basis of this project; A. Wilke of the Institut für Lagerstättenforschung, TU Berlin, who made possible purchase of additional equipment needed; the "Förderer der Berliner Fakultät für Bergbau und Hüttenwesen" who contributed toward the costs of transportation.

References

Cuffey, R. J.: The roles of bryozoans in modern coral reefs. — Geol. Rundschau, **61**, 542—550, 1972.

Garrett, P., Smith, D. L., Wilson, A. O., & Patriquin, D.: Physiography, ecology, and sediments of two Bermuda patch reefs. — Jour. Geology, 79, 647—668, 1971.

Ginsburg, R. N. & Lowenstam, H. A.: The influence of marine bottom communities on the depositional environment of sediments. — Jour. Geology, **66**, 310—318, 1958.

Ginsburg, R. N., Marszalek, D. S. & Schneidermann, N.: Ultrastructure of carbonate cements in a Holocene algal reef of Bermuda. — Jour. Sedimentary Petrology, **41**, 472—482, 1971.

Ginsburg, R. N., & Schroeder, J. H.: Recent synsedimentary cementation in subtidal Bermuda reefs. — In Bricker, P. P., Ginsburg, R. N., Land, L. S & Mackenzie, F. T.: Carbonate cements. — Bermuda Biological Station for Research, Spec. Pub. **3**, 31—34, 1969.

Ginsburg, R. N,, Shinn, E. A. & Schroeder, J. H.: Submarine cementation and internal sedimentation within Bermuda reefs (abstract). — Geol. Soc. America, Spec. Paper **115**, 78—79, 1967.

Glynn, P. W., Stewart, R. H. & McCosker, J. E.: Pacific coral reefs of Panama: distribution and predators. — Geol. Rundschau, **61**, 483—519, 1972.

Krebs, W.: Early void-filling cementation in Devonian forereef limestones (Germany). — Sedimentology, **12**, 279—299, 1969.

Logan, B. W.: Carbonate sediments and reefs, Yucatan shelf Mexico. Part 2. Coral reefs and banks, Yucatan shelf (Yucatan reef unit). — Am. Assoc. Petrol. Geologists, Mem. **11**, 129—198, 1969.

Newell, N. D. & Rigby, J. K.: Geological studies on the Great Bahama Bank. — Soc. Econ. Paleontologists and Mineralogists, Spec. Pub. **5**, 15—72, 1957.

Schalk, M.: Submarine topography off Eleuthera Island, Bahamas (abstract). — Geol. Soc. America, Bull. **58**, 1228, 1946.

Schroeder, J. H.: Calcified filaments of an endolithic alga in Recent Bermuda reefs. — N. Jb. Geol. Paläont. Mh. **1972**, 16—33, 1972 (a).

—: Fabrics and sequences of submarine carbonate cements in Holocene Bermuda cup reefs. — Geol. Rundschau, **61**, 708—730, 1972 (b).

Scoffin, T. P.: Cavities in the reefs of Wenlock Limestone (Mid-Silurian) of Shropshire, England. — Geol. Rundschau, **61**, 565—578, 1972.

Shinn, E. A.: Spur and groove formation on the Florida reef tract. — Jour. Sedimentary Petrology, **33**, 291—300, 1963.

Storr, J. F.: Ecology and oceanography of the coral reef tract, Abaco Island, Bahamas. — Geol. Soc. America, Spec. Pub. **76**, 1—98, 1964.

Walther, J.: Die Korallenriffe der Sinaihalbinsel. — Abh. Kgl. Sächs. Ges. Wiss. Abh. math.-phys. Kl., **14**, 439—484, 1888.

Zankl, H.: Der Hohe Göll. Aufbau und Lebensbild eines Dachsteinkalk-Riffes in der Obertrias der nördlichen Kalkalpen. — Abh. Senckenberg. Naturforsch. Ges. **519**. 1—123, 1969.

170

Editors' Comments
on Paper 6

6   GINSBURG
Excerpts from *Environmental Relationships of Grain Size and Constituent Particles in some South Florida Carbonate Sediments*

FLORIDA

It has been the fashion in the last several decades to study sizes, shapes, and mineralogy of terrigenous particles to determine the depositional environments and provenance (for a review see Friedman and Sanders, 1978, Chapters 2 and 3; and Friedman, 1979). The results of such studies range from frustration to promise for environments and are useful in provenance studies. However, the results are even more encouraging in the study of limestones where particles are primarily autochthonous in origin. This has been demonstrated by Ginsburg (Paper 6) in his study of South Florida carbonate sediments where he has clearly shown that particle size and nature of constituent particles reflect the different depositional environments within Florida Bay and reef tract, unless the depositional fabrics have been obliterated through postdepositional diagenetic overprint.

Ginsburg, however, was not the first in this line of study; several earlier workers made use of grain size and constituent particles in the delineation of sedimentary facies and environments, for example, Sander (1936), Henson (1950), and Bonet (1952) in their studies of Triassic carbonates, Mesozoic and Cenozoic reef tracts of the Middle East, and Cretaceous carbonates of Mexico. Pioneering work on modern carbonate sediments was initiated even earlier by Murray and Renard (1891) in their H.M.S. *Challenger* expedition. They were followed by Bathurst (1971), Bramlette (1926), Cloud (1962), Friedman (1968), Friedman et al. (1973), Goldman (1918, 1926), Illing (Paper 2), Kornicker and Purdy (1957), Logan et al. (1969), Lucia (1968), Matson (1910), Newell et al. (1957), Purdy (1963; Paper 3), Scoffin (1970), Smith, (1940), Thorp (1936), Vaughan (1916) and many others.

Ginsburg, in his study of South Florida carbonate sediments, was able to delineate certain distributional patterns of the carbonate-producing organisms and clearly demonstrated these in his Figure 8. He concluded that the zonal variances of the constituent particles are in response to differences in living abundances of the organisms that are environmental dependents. He therefore suggested that petrographic study of grain size and constituent composition of ancient limestones may provide useful data in characterizing ancient depositional environments. This type of study will be of particular help when looking into Cretaceous and younger limestones, for constituent compositions in them are closely related to modern living forms of known ecologic ranges.

REFERENCES

Bathurst, R. G. C., 1971, *Carbonate Sediments and Their Diagenesis* Developments in Sedimentology, vol. 12, Elsevier, Amsterdam, 620p.

Bonet, F., 1952, La facies Urgoniana del Cretacico Medio de la Region de Tampico, *Asoc. Mexicana Geologos Petroleros Bol.* **4:**1–259.

Bramlette, M. N., 1926, Some Marine Bottom Samples from Pago Pago Harbor, Samoa, *Carnegie Inst. Washington Pub. 344* **23:**1–36.

Cloud, P. E., 1962, Environment of Calcium Carbonate Deposition West of Andros Island, Bahamas, *U.S. Geol. Survey Prof. Paper 350,* pp. 1–138.

Friedman, G. M., 1979, Address of the Retiring President of the International Association of Sedimentologists: Differences in Size Distributions of Populations of Particles among Sands of Various Origins, *Sedimentology* **26:**3–32.

Friedman, G. M., 1968, Geology and Geochemistry of Reefs, Carbonate Sediments and Waters, Gulf of Aqaba (Elat), Red Sea, *Jour. Sed. Petrology* **38:**895–919.

Friedman, G. M., A. J. Amiel, M. Braun, and D. S. Miller, 1973, Generation of Carbonate Particles and Laminites in Algal Mats—Examples from Sea-Marginal Hypersaline Pool, Gulf of Aqaba, Red Sea, *Am. Assoc. Petroleum Geologists Bull.* **38:**541–557.

Friedman, G. M., and J. E. Sanders, 1978, *Principles of Sedimentology,* Wiley, New York-Chichester-Brisbane-Toronto, 792p.

Goldman, M. I., 1918, Composition of Two Murray Island Bottom Samples according to Source of Material, *Carnegie Inst. Washington Pub. 213* **9:**249–262.

Goldman, M. I., 1926, Proportions of Detrital Organic Calcareous Constituents and their Chemical Alteration in a Reef Sand from the Bahamas, *Carnegie Inst. Washington Pub. 344* **23:**37–66.

Henson, F. R. S., 1950, Cretaceous and Tertiary Reef Formations and Associated Sediments in Middle East, *Am. Assoc. Petroleum Geologists Bull.* **34:**215–238.

Kornicker, L. S., and E. G. Purdy, 1957, A Bahamian Fecal-Pellet Sediment, *Jour. Sed. Petrology* **27:**126–128.

Logan, B. W., J. L. Harding, W. M. Ahr, W. M. Williams, and R. G. Snead, 1969, Carbonate Sediments and Reefs, Yucatan Shelf, Mexico, *Am. Assoc. Petroleum Geologists Mem. 11,* 335p.

Lucia, F. J., 1968, Recent Sediments and Diagenesis of South Bonaire, Netherlands Antilles, *Jour. Sed. Petrology* **38:**848–858.

Matson, G. C., 1910, Report on Examination of Material from the Sea-Bottom between Miami and Key West, *Carnegie Inst. Washington Pub. 133* **4:**120–125.

Murray, J., and A. F. Renard, 1891, *Report on Deep-Sea Deposits based on Specimens Collected During the voyage of H.M.S.* Challenger, H.M. Stationary Office, London, pp. 184–248.

Newell, N. D., and J. K. Rigby, 1957, Geological Studies on the Great Bahama Bank, in *Regional Aspects of Carbonate Deposition,* R. J. LeBlanc and J. G. Breeding, eds., Soc. Econ. Paleontologists and Mineralogists Spec. Pub. 5, pp. 15–72.

Purdy, E. G., 1963, Recent Calcium Carbonate Facies of the Great Bahama Bank, I. Petrography and Reaction groups, *Jour. Geology* **71:**334–355.

Sander, B., 1936, Beitrage zur Kenntnis der Anlagerungsgefüge (Rhythmische Kalke und Dolomite aus der Trias) *Tschermaks Mineralog. u. Petrog. Mitt.,* **48:**27–209. (English translation by E. B. Knoff, 1951, Tulsa, Oklahoma, Am. Assoc. Petroleum Geologists, 207p.)

Scoffin, T. P., 1970, Trapping and Binding of Subtidal Carbonate Sediments by Marine Vegetation in Bimini Lagoon, Bahamas, *Jour. Sed. Petrology* **40:**249–273.

Smith, C.L., 1940, The Great Bahama Bank I. General Hydrographical and Chemical Features, II. Calcium Carbonate Precipitation, *Jour. Marine Research* **3:**1–31, 147–189.

Thorp, E. M., 1936, Calcareous Shallow-Water Marine Deposits of Florida and the Bahamas, *Carnegie Inst. Washington Pub. 452* **29:**37–119.

Vaughan, T. W., 1916, On Recent Madreporaria of Florida, the Bahamas, and the West Indies, and on Collection from Murray Island, Australia, *Carnegie Inst. Washington Year Book 14,* pp. 222–223, 229.

6

Reprinted from pages 2384–2415, 2419–2420, 2422–2423, and 2426–2427 of *Am. Assoc. Petroleum Geologists Bull.* **40**:2384–2427 (1956)

ENVIRONMENTAL RELATIONSHIPS OF GRAIN SIZE AND CONSTITUENT PARTICLES IN SOME SOUTH FLORIDA CARBONATE SEDIMENTS[1]

ROBERT N. GINSBURG[2]
Coral Gables, Florida

ABSTRACT

In the southern extension of the Florida peninsula variations in the submarine topography, areal geography, and hydrography which control the distribution of sediment-producing organisms are reflected in the grain size and constituent particles of the calcareous sediments being deposited. Two major environments can be recognized: (1) a curving band-shaped reef tract with good water circulation, and (2) Florida Bay, a very shallow triangular area with semi-restricted water circulation.

Florida Bay sediments have larger proportions of particles less than ⅛ mm. than the sediments of the reef tract. The constituent particle composition of the fraction larger than ⅛ mm. in Florida Bay is almost exclusively molluscan and foraminiferal, but in the same size fraction of the reef-tract sediments fragments of algae and corals are abundant. Similar distinctions in grain size and constituent particles for comparable environments can be derived from published data for the sediments around Andros Island, Bahamas.

In Florida Bay large local variations in physical environment obscure the expected effects of differences in environment from one part of the Bay to another, and no distinct sub-environments could be recognized from the gross grain size and constituent particle composition. However, in the reef tract local variations of environment are smaller, and the gradual but consistent changes in depth and water circulation effect differences in the fauna and flora, and thereby produce sediments which have recognizably different abundances of the major constituent particles as shown in Figure 1. The three sub-environments, back reef, outer reef-arc, and fore reef are indicated by progressive changes in constituent composition, and in less degree by variations in gross grain size.

Because the estimates of constituent particle composition of the reef-tract sediments were made by point counts on standard petrographic thin sections this approach can be used to analyze ancient limestones.

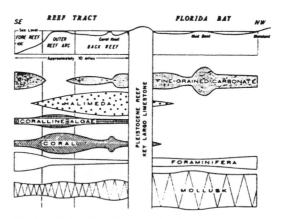

Fig. 1.—Generalized variations in sediment grain size and constituent composition of South Florida sediments.

[1] Manuscript received, March 23, 1956. Contribution No. 160 Marine Laboratory, University of Miami, Coral Gables, Florida. This investigation was supported by a grant from the American Petroleum Institute, Project 51. The Shell Oil Company provided time and facilities to complete the manuscript.

[2] Marine Laboratory, University of Miami. Present address: Shell Oil Company, 4029 Ponce de Leon Boulevard, Coral Gables, Florida.

A grant from American Petroleum Institute Project 51 provided the services of a full-time assistant, and field expenses from July, 1952, to July, 1954. Robert Fetner and Alden Jensen each served

INTRODUCTION

The increasing interest in defining the environment of deposition of ancient sedimentary rocks has focused attention on the basis of interpretation—the marine sediments forming and accumulating on the present sea floors. This paper describes the results of a study of recent carbonate sediments in South Florida which suggest how some depositional environments are reflected by the grain size and constituent[3] particles of their respective sediments.

The recently renewed interest in carbonate sediments and particularly coral reefs, which languished somewhat since the early twentieth century, is due in large part to the stimulation of the petroleum industry's search for stratigraphic traps in limestones. Analysis of the relative depositional environments of limestones is rapidly becoming a part of standard stratigraphic procedures.

There are many depositional properties of limestones which may be used for the interpretation of environment, but those which are small enough to be seen in hand specimens or well cores and are persistent laterally will be of most use to practical stratigraphy. Parameters which fulfill these requirements are those which describe the origin, chemical composition, size distribution, and arrangement of the original particles.[4] The effects of early diagenesis may also be indicative of depositional environment (Ginsburg, 1954). There are of course many carbonate rocks which have been so affected by recrystallization or dolomitization that none of the particle parameters will present an unaltered index of original environment. Excluding these rocks, there remains a large body of limestones which retain, to a very large degree, the stamp of their depositional environment.

Several workers have used grain size, constituent particles, and fabric to interpret ancient carbonates. Sander (1936, 1951) made a detailed study of Triassic carbonates, and even in dolomites was able to recognize original particles. Henson (1950) used texture, constituent particles, and other microscopic features observed in thin section, to characterize the facies of Mesozoic and Cenozoic reef tracts in the Middle East. He relied heavily on Foraminifera for ecologic inter-

as assistants for 6 months, and Paul Kruse for one year. Their industry and enthusiasm contributed much to the successful completion of this work. Gilbert Voss and Donald Moore gave valuable information on the distribution of organisms. Captain Herb Alley of Tavernier guided the writer in Florida Bay, and provided much useful information on hydrographic conditions.

Francis Shepard gave advice and encouragement during the course of the work, and his criticism of the manuscript was most helpful. Stimulating discussions with R. Michael Lloyd helped the writer clarify several problems.

The writer is indebted to several colleagues who kindly read the manuscript. Heinz Lowenstam and D. A. White read the entire paper, and their acute critique of subject matter and organization contributed much to the final draft. Leslie V. Illing criticized part of the manuscript, and his suggestions were very helpful. Wayne E. Moore gave helpful suggestions on the manuscript, and offered supporting evidence for some of the interpretations from his study of the Foraminifera. The very effective illustrations were made by Mrs. Blackwood of the Shell Development Company.

[3] "Constituents" as used in this paper refers to the origin of the sediment particles larger than $\frac{1}{8}$ mm. For example, the constituent composition of Sample I-A is Halimeda 29%, Coralline Algae 7%, Mollusk 20%, Coral 10%, Foraminifera 14%, etc.

[4] Nanz (1954, p. 100) gave texture and composition as the fundamental properties of sediment character in sandstones.

pretations (*ibid.*, p. 216). Bonet (1952) studied the facies variations of grain size and constituent particles in the Cretaceous of Mexico. Fairbridge (1954) has summarized some of the recent work on the interpretation of carbonates from thin-section examination of texture and constituent composition.

The results of the present study offer considerable encouragement to the petrographic analysis of ancient limestones. Under favorable circumstances this approach may provide useful information on relative environment, but it also has limitations, and should not be applied indiscriminately.

PREVIOUS STUDIES OF RECENT CARBONATE SEDIMENTS

The pioneer work on grain size and constituent composition of recent marine sediments is the monumental *Challenger* treatise of Murray and Renard (1891). Their facies classification of organic deep-sea deposits is based on the dominant constituents—globigerina, pteropods, diatoms, radiolarians, *et cetera*—as determined by microscopic examinations. The importance of this classification is shown by its continued acceptance and use to-day, 60 years later. In passing it is interesting to note that the sediments Murray and Renard designated as "coral mud" and "coral sand"—a distinction based on the percentage of "fine washings" in the non-calcareous residues—were composed mostly of Foraminifera, and that other organisms including coral were less abundant.

One of the earliest records of constituent composition studies of recent shallow-water calcareous sediments is the analysis by Quin of the sand in Cane Bay, St. Croix, Virgin Islands (Vaughan, 1916, p. 223), which indicated that what was ordinarily called coral sand had, in fact, only about 10 per cent coral detritus. Vaughan (1916, pp. 222–23) recognized the value of such studies in characterizing sediment types and in understanding the production and distribution of recent calcareous sediments. He initiated and encouraged most of the subsequent work along these lines. Matson (1910, pp. 120–25) made some preliminary observations on the grain size and constituent composition of a suite of samples from Florida, and Vaughan (1918, pp. 262–88) gave data on grain size and foraminiferal content of samples from Florida and the Bahamas. Goldman (1918, pp. 249–62; 1926) made detailed analyses of the constituent particles of two samples from Murray Island, Australia, and one from Andros Island, Bahamas. He determined the source organisms of each grain, and using the known composition of the organic skeleton he compared bulk chemical composition with calculated chemical composition.

Vaughan (1918, p. 287) summarized studies of the grain size and constituent particles of samples from Florida and the Bahamas. He recognized three main classes of shallow-water carbonate sediments in the Bahamas: (a) sands forming behind the reef, consisting of the fragments of reef organisms; (b) fine-grained muds in South Bight, Andros Island, and on the west side of Andros; (c) oölitic sands derived from the breakdown of oölitic rock. Vaughan (*op. cit.*) found sediments belonging to his (a) and (b) classes in Florida, and he further subdivided class (b) according to grain size.

Using grain size and constituent particles, Bramlette (1926, pp. 21–23) recognized two types of calcareous sediments in the Pago Pago Harbor, Samoa: (a) on the reef flats and slopes a poorly sorted calcarenaceous sediment about three-fourths of which is composed of fragments of algae and corals with less amounts of mollusks, Foraminifera, and other organisms; (b) in the deeper waters of the inner part of the harbor a calcilutaceous sediment (silt and clay size). In addition to a considerable percentage of inorganic material (fragments of volcanic rock from the island), Bramlette (*ibid.*, p. 22) states "Molluskan material is one of the major constituents, as are Foraminifera, pteropods, and various sponge and ascidian spicules."

Thorp (1936) gave the results of grain size and constituent particle analyses of 74 samples from the Florida-Bahama region which were collected previously by Vaughan and his colleagues. Thorp's main contribution was to show that the bulk of the sediment is of organic origin, and that the order of abundance of the major contributors is algae, mollusks, Foraminifera, and corals. Thorp (*ibid.*, p. 94) also indicated that quiet-water sediments contain more fines than those from the reef tracts in Florida and the Bahamas.

In a very significant paper, Illing (1954) gave the results of grain size and constituent particle analyses of 28 samples from the eastern Bahamas. In addition to his important discovery of the abundance of non-skeletal particles, and their mode of origin, he was able to show for the first time a distinct difference in the composition of sediments in different environments. On the margins of the banks he found the sediments predominantly organic skeletal débris, but on the banks proper, non-skeletal particles are most abundant (*ibid.*, p. 17). Furthermore, he discovered that the kind and abundance of the various non-skeletal particles are also related to variation in environment (*ibid.*, pp. 78–79).

PRESENT STUDY

During the course of the writer's studies of carbonate sedimentation and diagenesis in the South Florida area, a considerable amount of background information on sediment types and their environment was collected. Financial assistance from API Project 51 from 1952 to 1954 made it possible to continue and expand these earlier studies. Some of the results of this research have already been presented (Ginsburg, 1954). The present paper gives the results of studies of the grain size and constituent particles of the sediments.

In order to study the variations in grain size and constituent composition related to differences in environment, bottom samples were collected from the main environments. From the narrow band-shaped reef tract, 25 surface samples were collected with a bottom grab at intervals along three traverses oriented perpendicular to the trend. In the other main environment, Florida Bay, where traverses were less appropriate, 17 grab samples and piston cores were located more at random. Further discussion of methods is given in the Appendix.

In the shallow, clear water of the area, the observations of the bottom fauna and flora were made by divers. With swim fins and a face mask it is possible to examine the bottom anywhere in Florida Bay and in much of the reef tract. Two underwater breathing units for work in deeper water were provided on loan by Scripps Institution. Although all the bottom observations were of a reconnaissance nature, they suffice to give a general idea of the relative distribution and density of the floras and faunas.

During the course of this study, the writer also conducted similar investigations for the U. S. Navy in the region. The supplementary information obtained in those studies, as yet unpublished, gives support to the conclusions presented here.

REGIONAL SETTING

Figure 2 shows how peninsular Florida is located on the eastern side of a much larger submarine plateau outlined approximately by the 100-fathom curve. This platform called the Floridian Plateau by Agassiz (1888, p. 62) is surrounded by deeper water on three sides; on the east and southeast it is separated from the Bahama Banks and from Cuba by the narrow Straits of Florida, the floor of which deepens from about 2,500 feet below sea-level off Miami to 5,000 feet or more between Key West and Havana, Cuba. The Florida Current, which becomes the Gulf Stream off Cape Hatteras, flows northward through this trough.

Our knowledge of the structure and origin of the Floridian Plateau is limited to the relatively few exploratory oil wells drilled on the peninsula and the Florida Keys. From this stratigraphic information Applin (1951, p. 17) suggested that the peninsular part of the plateau was an area of non-deposition in post-Silurian Paleozoic time, and that its present structure is due to regional movements in the Mesozoic and Cenozoic.

That the southeastern part of the plateau has long been an area of non-clastic deposition which has been subsiding differentially is shown by the southward thickening of the Mesozoic and Cenozoic section from the Ocala uplift near the central part of the peninsula to near Key West (Fig. 3). Pressler (1947, p. 1856) includes this southern part of the plateau with Cuba and the Bahamas in what he called the South Florida Embayment. For further discussion of the subsurface stratigraphy see Pressler (1947) and Applin (1951).

SOUTH FLORIDA

Bottom contours in Figure 4 show the existence of a crescent-shaped platform between Miami and Key West. This surface which slopes gently from about 100 to 300 fathoms (600 to 1,800 feet) is called the Pourtales Plateau, after Count Pourtales whose studies of the flora, fauna, and bottom deposits in the 1860s are still the basic work on this part of the area.

Recent bathymetric surveys of the Pourtales Plateau have disclosed the existence of what appear to be ancient sink holes near the outer edge of a terrace-like surface about 900 feet below sea-level off Key West (Jordan, 1954, p. 1811).

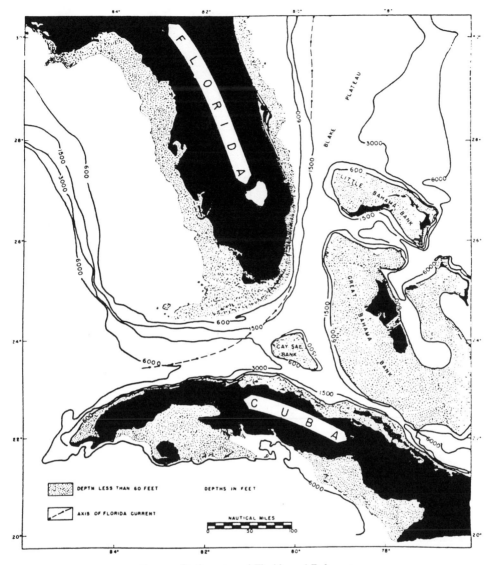

FIG. 2.—Bathymetry of Florida and Bahamas.

These holes are $\frac{1}{2}$ mile or more in diameter at the top and 450–540 feet deep. If they were formed subaerially, which is probable, the Pourtales Plateau is more probably due to subsidence or downfaulting (Shepard, 1948, p. 199) than to erosion by the Florida Current.

From the examination of numerous dredgings in the Straits of Florida, Agassiz recognized three main bottom types in the Straits (1888, pp. 286–87); (a) coral bottom or coral sand made up of the fragments of invertebrate skeletons, extend-

179

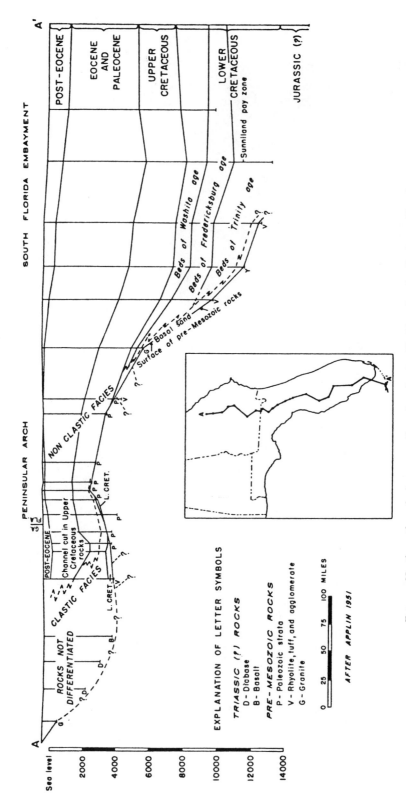

Fig. 3.—North-south geologic cross section of peninsular Florida, after Applin (1951, Fig. 3).

180

ing from the reefs to depths of about 100 fathoms; (b) the rocky bottom of the Pourtales Plateau at depths of 100–300 fathoms which consists of the débris of corals, mollusks, and echinoderms consolidated by serpulid worm tubes and coralline algae; and (c) the globigerina ooze of the deep trough of the Straits.

The upper part of the southwest-tapering shallow-water extension of the Florida peninsula (Fig. 4) is formed by two limestones. One of these, the Key Largo limestone (Sanford, 1913, pp. 184–89) is a Pleistocene coral reef, the subaerially exposed parts of which form the arc of islands known as the Upper Florida Keys between Soldier Key and Big Pine Key. From the maximum elevation of the Key Largo limestone, 18 feet at Windley Key, the exposed part has been interpreted as of Sangamon interglacial age (Parker and Cooke, 1944, pp. 68, 71). The only published figure for the thickness of the Key Largo limestone is given by Sanford (1913, p. 169) as 105 feet at Marathon (Fig. 4).

The other limestone, the Miami oölite (Cooke and Mossom, 1929, pp. 204–07) lies on the landward side, northwest, north, and west of the narrow exposed Key Largo reef. On the mainland, this typical oölite is a west-tapering wedge-shaped rock mass with maximum thickness of about 40 feet which forms a topographic ridge along the eastern and southeastern margins of the peninsula (Parker and Cooke, 1944, p. 70). This limestone also extends above sea-level as the Lower Keys and westward to Rebecca Shoal (Fig. 4), where Sanford (1913, p. 182) gave a figure of 50 feet for its thickness. These two isolated areas of exposure are probably part of a continuous rock body, as oölitic calcarenite has been collected from the rock floor of Florida Bay at several points.

Parker and Cooke (1944, p. 68) consider that the Miami oölite is contemporaneous with the upper exposed Key Largo limestone, and at Stock Island near Key West, the writer has observed that the two lithologies occur in the same rock pit below sea-level. However, the juxtaposition of the two lithologic types does not necessarily mean time equivalence, and it is still possible that the Miami oölite is younger than the Key Largo.

No direct information is available on the nature of the uppermost part of the narrow platform, east, southeast, and south of the Florida Keys on the edge of which living coral reefs form an intermittent barrier. From a well at Marathon, Sanford (1913, p. 169) reported 156 feet of limestone, including 105 feet of Key Largo lithologic type underlain by at least 545 feet of siliceous sand of Miocene age. If the siliceous sands extend southeastward the several miles to the edge of the platform, then perhaps there also the upper 150 feet are limestone. Perhaps this limestone includes reef arcs like the Key Largo, as well as calcarenites laid down under bank conditions.

MAJOR ENVIRONMENTS, FLORIDA BAY AND REEF TRACT

The discontinuous island barrier formed by the Pleistocene Key Largo reef separates the present band-shaped reef tract from a series of shallow bays and sounds (Fig. 4). Table I compares the bathymetry, areal geography, and hydrog-

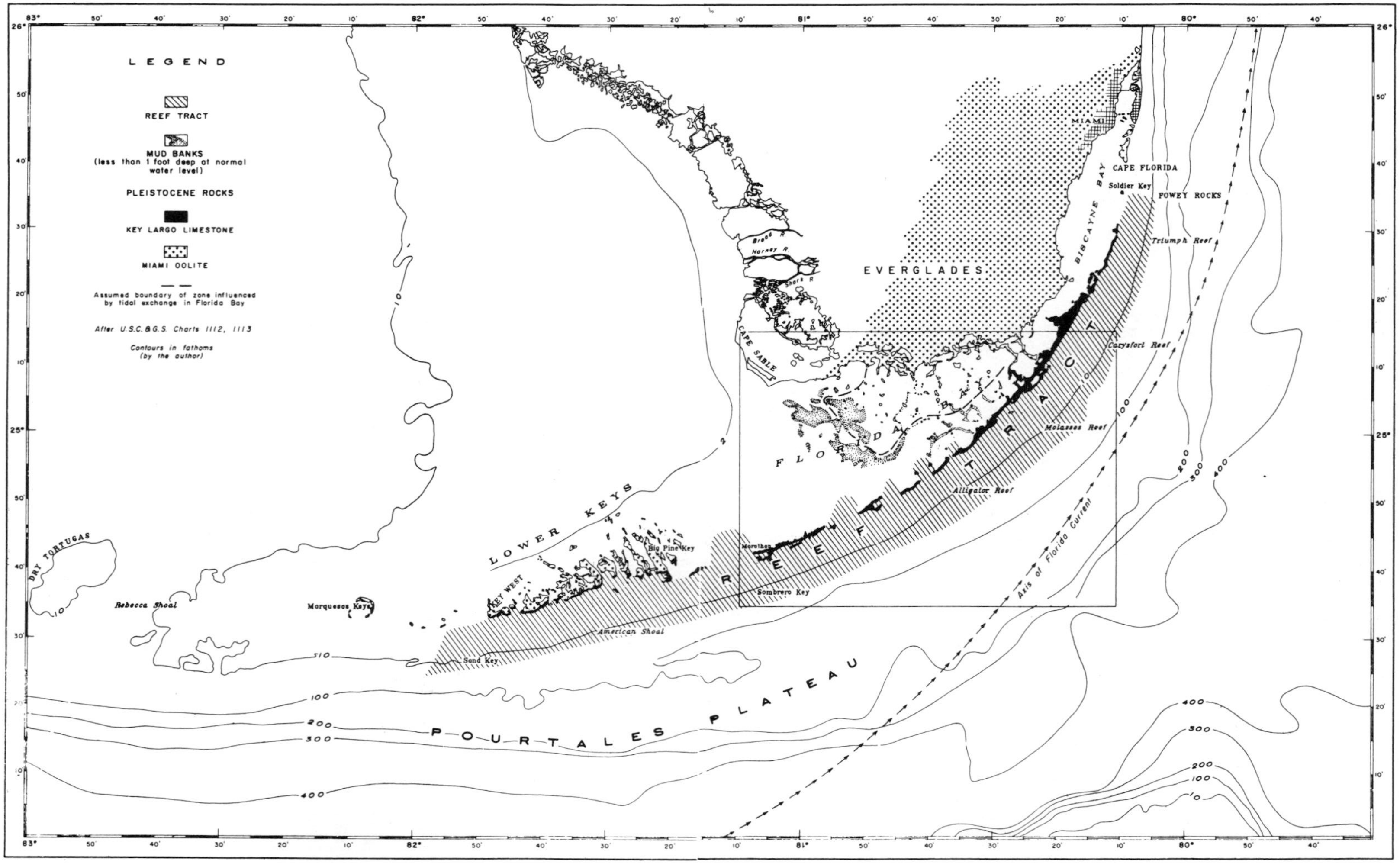

FIG. 4.—Chart of southeast margin of Floridian Plateau and Straits of Florida. Boxed area of Reef Tract is shown enlarged in Figure 5.

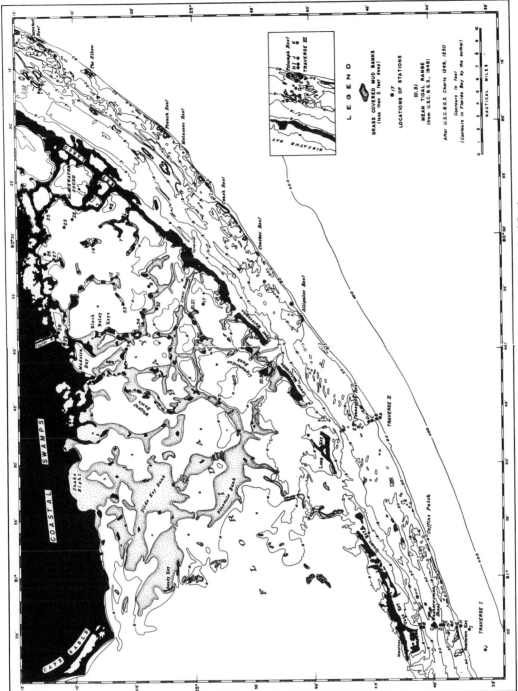

Fig. 5.—Bathymetric chart of Florida Bay and part of reef tract showing locations of all stations.

raphy of these two areas. Because only the east half of Florida Bay has been
studied, this discussion is restricted to that part.

The differences in bathymetry, areal geography, and hydrography between
Florida Bay and the reef tract, given in Table I, are considered sufficiently large
to characterize two major environments. The rapidity of the change from one
major environment to the other depends on the extent of the barrier between
them. Where there are channels through the Pleistocene limestone islands, as

TABLE I. MAJOR ENVIRONMENTS OF SOUTH FLORIDA

	Florida Bay	Reef Tract
BATHYMETRY		
Depth range	0–10 feet	0–300 feet
Maximum local relief	6 feet	30 feet
AREAL GEOGRAPHY		
Shape	Solid triangular	Arcuate band
HYDROGRAPHY		
Circulation	Restricted, periodic tides only on margins	Open, semi-diurnal tidal exchange with Florida Current
Variations in temperature and salinity		
Temperature	15°–40°C.	15°–33°C.
Salinity	10–40 °/₀₀	32–38 °/₀₀
Available plankton and nutrients	Low	Normal for tropical waters
Turbidity	Generally high	Periodically high only in lagoonal part

between the Matecumbe Keys (Fig. 5), the nearshore reef-tract environment extends tongue-like into Florida Bay, and the change to restricted conditions is
gradual. Where the barrier is more complete, as along Key Largo, the change in
conditions is more rapid, but there is still a transition zone parallelling the islands
because the extremely porous limestone permits some tidal exchange between
Bay and reef tract.

Bathymetry.—Most of Florida Bay is less than 6 feet deep, and as shown in Figure 5 there are
large areas of mud banks which are less than 2 feet deep and in many places they have grown to
the surface. These mud banks which rise from the very gently sloping rock floor produce the only
relief in Florida Bay.

The lower limit of the reef tract is taken here arbitrarily as 300 feet. Within it there is considerable
local relief in the form of elongate coral reefs, rocky shoals veneered with calcareaceous sediment,
and irregularly shaped coral knolls and pinnacles which rise as much as 30 feet from the bottom.

Areal geography.—As shown in Figure 5, Florida Bay is a massive triangular-shaped area, apex
to the south, of which the base is about 35 miles wide. It is connected to the ocean on the southeast
by the channels through the Keys, and to the Gulf of Mexico on the southwest by narrow channels
through the almost continuous mud banks. In contrast, the reef tract is a narrow curving band 5–10
miles wide which is open to the Straits of Florida.

Hydrography.—The open circulation of the reef tract is due to semi-diurnal tidal exchange with
the open oceanic water of the Florida Current, and to wind-driven water movements. The tidal
ranges along the edge of the platform where living coral reefs and shoals form an intermittent barrier
decrease southward from Miami. At Fowey Rock (Fig. 4) off Miami the mean range is 2.4 feet, the
spring range is 2.9 feet; about 40 miles southeastward at Molasses Reef the mean range is 2.2 feet,
and the spring range 2.6 feet; and about 45 miles farther along the reefs at Sombrero the mean range

is 1.6 feet and the spring range is 2.0 feet (Tide Tables, East Coast, 1955, p. 222). The prevailing easterly winds probably also contribute to the water circulation of the reef tract by moving water across the pseudo-barrier, and in the winter months northeast winds may give a general drift toward the southwest.

The Florida Current approaches the reef tract most closely off Miami where its west edge is often about 4 miles from the reefs (personal communication, Lansing Wagner); south of Fowey Rock, the north-flowing current is separated from the reefs by an intermittent countercurrent which flows along the edge of the shallow-water platform (Thorp, 1936, p. 50, and personal observations).

Tidal exchange between Florida Bay and the reef tract is limited to the southeast margin of the Bay. In the east half of Florida Bay the mean tide is less than 0.5 foot (Fig. 5). Much larger fluctuations in bay level and the accompanying water movements are produced by wind-driven water pileup, and by seasonal changes in ocean level. Strong sustained winds from any direction can cause fluctuations in bay level by piling up water against the island barrier formed by the Keys or by the mud banks, and by literally blowing the water out of the Bay into the open Gulf southwestward. Seasonal fluctuations in bay level corresponding with variation in the level of the surrounding oceans (Marmer, 1954, p. 116) are also observed in Florida Bay. Observations during 1954–1955 at Community Harbor, Tavernier, by Captain H. Alley suggest a normal maximum annual range of 21 inches, with seasonally high levels in October. Unusual fluctuations of bay level can be produced by hurricanes, as during the 1948 storm when gale force winds from the southwest raised the bay level at Tavernier 3–4 feet above normal, according to the observations of Captain Alley.

Drainage of fresh water from the mainland into Florida Bay is also a major factor in circulation, as shown by its influence on salinity.

Variations in temperature and salinity.—Relatively large seasonal and annual fluctuations of temperature and salinity occur in Florida Bay. Davis (1940, pp. 348–50) presented some data from Florida Bay which showed how the salinity of the Bay near the mainland increased $10^{\circ}/_{\circ\circ}$ during 1938 because of low rainfall on the mainland. The results of salinity determinations made during the present study are shown in Figure 6. They suggest seasonal fluctuations in salinity over a large part of the eastern Bay due to fresh-water drainage from the mainland, the mechanics of which are discussed further in a later section.

The maximum range of temperature in Florida Bay from about 40 observations was from 19° to 38°C. The upper limit was found in water less than one foot deep on a mud bank during the summer. The lower limit is also from very shallow water, and even lower temperatures are expected during the winter, probably as low as 15°C. In the deeper water around the mud banks the available data suggest a range from 20° to 30°C.

Although salinity and temperature data for the reef tract are very scattered, they indicate much more uniform conditions than in Florida Bay. Dole and Chamber (1918, p. 311) found that fresh-water drainage from the mainland after heavy rains could reduce the salinity at Fowey Rocks on the edge of the platform 10 miles away by as much as one part per thousand. However, 7 miles south of Fowey Rocks, at Triumph Reef, where the reef tract is separated from the mainland by shallow bays, Smith, Williams, and Dawson (1950, p. 123) found no effect of rainfall on salinity. These authors also found a greater range of salinity in the stations in Biscayne Bay ($34.09^{\circ}/_{\circ\circ}$ to $39.36^{\circ}/_{\circ\circ}$) than in the reef tract to the eastward ($35.25^{\circ}/_{\circ\circ}$ to $36.50^{\circ}/_{\circ\circ}$) (*ibid.*, p. 122).

The most complete data available for the reef tract are those reported by Chew (1954) from the approaches to Key West Harbor. His studies extended from the island of Key West across the major part of the reef tract to depths of 150 feet. He gave a maximum range of salinity from $33.3^{\circ}/_{\circ\circ}$ to to $37.0^{\circ}/_{\circ\circ}$ but excluding Gulf of Mexico water entering from Northwest Channel the lower limit is $35.2^{\circ}/_{\circ\circ}$ (Chew, p. 2).

Vaughan (1918, pp. 321–36) gave the result of temperature measurements between 1879 and 1899 for the surface water along the reefs. The ranges from minimum to maximum temperatures at the following locations were as follows: Sand Key Lighthouse (off Key West) 17.9° to 32.2°C.; Carysfort Reef, 18.2° to 30.3°C.; Fowey Rocks (off Miami), 15.6° to 31.2°C. Smith, Williams, and Dawson (1950, p. 122) gave some comparative data from the inner reef tract which showed that the temperature range near the Keys is larger (19.15° to 32.3°C.) than along the reefs at the edge of the platform (24.35° to 29.8°C.). Chew's data from the reef tract at Key West showed maximum annual range from 20° to 31°C.

Plankton and nutrients.—The only available data on the abundance of plankton are for stations in Biscayne Bay and the reef tract eastward where Smith, Williams, and Davis (1950, pp. 131–32) found lower plankton volumes and diatom populations in the southern part of Biscayne Bay than in the open reef tract. An even smaller plankton supply is to be expected in Florida Bay where circulation is more restricted.

The only available data on the concentration of phosphate and nitrate are those from Biscayne Bay and the reef tract eastward where Smith, Williams, and Dawson (1950, pp. 124–25) found that "dissolved mineral nutrients were extremely low or undetectable at all stations throughout the year, except at a station in the northern part of the Bay where sewage was abundant."

Turbidity.—Throughout Florida Bay steady winds of 15 m.p.h. or more are sufficient to turn the water so turbid that the bottom is not visible in depths of 6 feet. This condition exists during much of the period between November and April when stronger wind velocities are more frequent than during the remainder of the year. Schools of mullet which are abundant in Florida Bay produce unusually turbid water by expelling undigested sediment picked up from the bottom and by stirring the bottom in feeding. These streaks of white water, or fish muds as they are called locally, are a common feature of the Bay.

Estimates of the amounts of sediment in suspension after 2 days of fresh northeast winds were made by passing through a continuous-flow centrifuge 5-liter samples of water collected at the surface and 18 inches above the bottom. The amount of calcium carbonate determined by R. Michael Lloyd from acid digestion of six filtrates taken at three different stations, which are shown in Figure 5 were as follows:

Station FB-52: surface, 16.0 mg/l;   18 inches above bottom, 17.2 mg/l
Station FB-53: surface, 11.0 mg/l;   18 inches above bottom, 9.4 mg/l
Station FB-54: surface, 17.1 mg/l;   18 inches above bottom, 17.6 mg/l

In the reef tract sustained winds of about 15 m.p.h. or more also produce periods of white water by stirring up fine sediment from the bottom. Such turbid water ordinarily is restricted to the area between the reef and the Keys, and especially the parts near the Keys. Chew (1954, p. 2, Figs. 5, 6, 9, 10, 13, 14, 17, 18) presented the results of sechi disk readings across the reef tract at Key West which showed less transparent water between the reefs and the Keys than outside the Keys. Sediment-laden waters from Florida Bay can also add to the turbidity of the water in the reef tract where they are brought through the island barrier by the tidal exchange.

FLORAS AND FAUNAS

The distinct differences between the restricted circulation of shallow Florida Bay, and the open circulation of the deeper reef tract produces major differences in the abundance and variety of the sediment-producing floras and faunas of these two environments. The reef tract has an extremely varied and abundant population of organisms compared with the few numbers and species in Florida Bay.

In general the calcareous flora of the reef tract is dominated by green algae of the family *Codiaceae*, but locally red algae of the family *Corallinaceae* are very abundant. The reef tract fauna is extremely varied; its major calcareous elements are corals, mollusks, echinoderms, alcyonarians, Foraminifera, worms, bryozoans, and crustaceans. Most of Florida Bay has a non-calcareous grass flora and a molluscan-foraminiferal fauna. In the marginal zone of Florida Bay, and especially around tidal channels leading across the Keys, reef-tract organisms may be abundant. For example, the green algae, *Halimeda* sp. and *Penicillus* sp., species of which flourish in the shallow inshore waters of the reef tract, can be found as far as 5 miles back in the Bay, and the hardy corals, *Porites furcata* and *Siderastrea* sp., have been collected a few miles within the Bay. Further discussion of the flora and fauna of these two environments is presented in a later section.

GRAIN SIZE AND CONSTITUENT PARTICLES

To compare the grain size and constituent composition of the sediments from Florida Bay and from the reef tract, 50 samples were analyzed, 25 from Eastern Florida Bay, and 25 from the reef tract. The samples from the Bay included surface scoop samples taken by a diver, and parts of cores. The samples from the reef tract were taken at intervals along three traverses as shown in Figure 5 with a Peterson grab; the locations of all samples are shown in Figure 5. Further discussion of methods is given in the Appendix.

TABLE II. AVERAGE GRAIN SIZE AND CONSTITUENT COMPOSITION OF
SEDIMENTS FROM FLORIDA BAY AND REEF TRACT

| | Florida Bay (17 samples[1]) | | Reef Tract (25 samples) | |
	Average %	Range %	Average %	Range %
Grain size				
Weight percentage less than ⅛ mm.	49	10–85	17	0–68
Constituent composition of fraction greater than ⅛ mm.				
Algae	½	0–1	42	7–61
Mollusk	76	58–95	14	4–33
Coral	—	—	12	2–26
Foraminifera	11	1–32	9	3–32
Non-skeletal	3	0–3	12	3–24
Miscellaneous	½	0–4	9	2–23
Unknown	1	0–3	8	4–15
Ostracods	2	1–6	—	—
Quartz	6	0–20	—	—

[1] Only surface scoop samples and the upper parts of cores are included in this average (Table X).

Table II gives the average grain size and constituent composition of the samples from Florida Bay and from the reef tract. Complete analyses are given in Tables VIII and IX.

The average percentage of fines in Florida Bay is much larger than the percentage from the reef tract. Most of the coarse fraction (larger than ⅛ mm.) in Florida Bay consists of the remains of mollusks and Foraminifera, but the reef tract sediments have abundant algae, coral, and various minor constituents not found in the Florida Bay samples.

Although the average values for the grain size and constituent composition of the sediments from the two major environments are quite distinct, the ranges suggest that there are samples which could be assigned to either environment. Such overlapping ranges are certainly to be expected when it is recalled how the reef-tract environment extends directly into Florida Bay through the channels between the Keys. It would probably be very difficult to determine whether the sediments along such a channel as the one between Upper and Lower Matecumbe Keys are of the reef-tract facies or the Bay facies. But gradually as the sampling extended farther into the Bay and the more delicate producers of the reef-tract sediments such as algae and corals were eliminated, the constituent compositions are reduced to Mollusca and Foraminifera. Where there is no open channel—as along Key Largo—the change in composition on either side of the island will naturally be found relatively more abrupt, even though the porous Key Largo limestone allows tidal movement of reef-tract water to penetrate the Bay sufficiently to permit the growth of the more hardy reef-tract organisms. In this way the rapidity of changes in constituent composition may give information on the completeness of a barrier between environments.

FLORIDA BAY SUB-ENVIRONMENTS

There are no major trends in the depth variations in Florida Bay. Differences of a few feet do occur between the greatest depths near the Keys, 8–10 feet, and the greatest depths in the northern and northeastern parts of the Bay, 4–6 feet, but these changes are gradual. These differences are overshadowed by the large local variations in depth between the near-surface mud banks and adjacent areas of deeper water (Fig. 5).

The shapes and lateral extents of the mud banks and their associated islands, built by the sediment-trapping action of mangroves (Davis, 1940), produce some differences in geography and circulation from one part of the Bay to another. In the southeast part of the Bay the mud banks are very narrow, they have numerous channels, and they surround relatively large areas of deeper water. In the central and west parts of the Bay the banks are generally wider, more irregular in plan, and the areas of deeper water smaller than in the southeastern Bay. Along the mainland, and in the northeast corner of the Bay the banks are replaced by an almost continuous rim of mangrove islands, which may represent the final results of the island-building activities of these plants. The wider and more continuous are the mud banks and islands, the more complete is the separation of the Bay into a series of semi-isolated depressions locally called lakes.

Differences in the hydrography or circulation-salinity across Florida Bay offer the only basis for dividing the Bay into sub-environments. The data are rather limited, but they suggest that the Bay can be divided into an outer or marginal zone where there is frequent tidal exchange with the reef tract and salinities are near normal, and a larger interior or central zone of semi-restricted circulation in which there may be larger fluctuations in salinity. The estimated position of the boundary between these two zones is shown by the dashed line in Figure 4, but its precise location at any time may be quite different, depending on the specific meteorological and hydrographic conditions.

Superposed on any variations in physical environment across the east part of the Bay are the local fluctuations in conditions produced by the near-surface, grass-covered mud banks which surround areas of deeper water, 4–8 feet deep. In the deeper water the temperature range is smaller than on the banks, and wave action reaches to the bottom, judged by the absence of all but several inches of sand-sized sediments with sparse grass and algae over the irregular rock.[5] In contrast to the rock-floored deeper water, the banks are entirely sediment, mostly silt- and clay-size with a luxuriant cover of grass (*Thalassia* and *Halodule*). Although the banks are periodically exposed by wind-driven water movements, wave action is restricted to their exposed margins. Thus with the exception of salinity, the local variations in physical environment are larger than the variations from one part of the Bay to another.

[5] Shown by probing with a metal rod at about 50 different places in the east part of the Bay.

Circulation-salinity.—The location of the boundary between the marginal zone and the interior zone of Florida Bay depends on the interaction of fresh-water introduction from the mainland swamps, annual fluctuations in sea-level, the amount of tidal exchange with the reef tract, and the strength and direction of the winds. Maximum drainage of fresh water from the mainland is expected during the rainy season, May through October, August, September, and October are generally the wettest months (Local Climatological Summary for 1950, U. S. Weather Bureau, Miami, Florida). The maximum rainfall occurs during the period when sea-level is relatively high, as shown by regional studies and local observations. From the seasonal variations in the height of sea-level at Key West

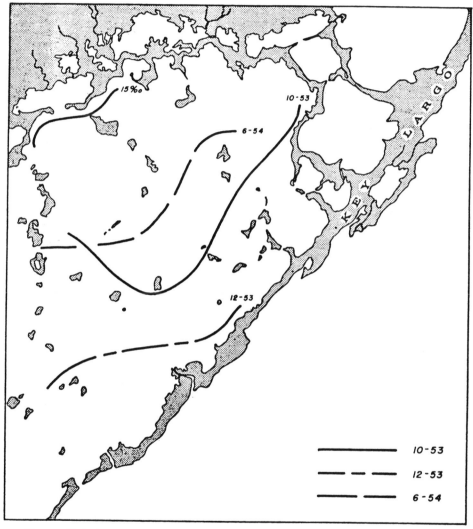

Fig. 6.—Variations in salinity of northeast part of Florida Bay, 1953–1954. Lines show approximate positions of the 30°/₀₀ isohaline at dates indicated. Numbers of stations for each date are: 16 for 10/53, 11 for 12/53, and 16 for 6/54.

between 1930–1948, Marmer (1954, p. 116) showed that sea-level is relatively high from August to December, with maximum of about ½ foot in October. This period of high sea-level is well known in Florida Bay because it permits boats to pass across banks normally too shallow (personal communication. Captain H. Alley). By November or early December, Bay level recedes to normal, which probably accelerates the drainage of fresh water into the Bay from the mainland, and the interior zone of reduced salinity may extend farther south and southeast. The movement of the 30°/₀₀ isohaline during 1953–1954 (Fig. 6) suggests that delayed release of fresh water from the mainland as described above was operating then.

Reef-tract water moves into Florida Bay through the channels across the Keys as shown in Figure 5; exchange also proceeds through the porous reef limestone of the Keys as shown by tidal fluctuations in the eastern part of the Bay far from open channels, for example, along Key Largo south of Blackwater Sound. Spring tides will tend to bring a greater than usual amount of reef-tract water into the Bay, and increase the width of the marginal zone.

As described earlier, wind-driven water movements can raise or lower Bay level considerably. Sustained and strong easterly winds can literally blow water out of Florida Bay into the open Gulf of Mexico, and the lowered level of the Bay will thus increase the amount of reef-tract water which enters across the Keys.

In similar fashion winds from the north can accelerate the introduction of mainland drainage into the northern part of the Bay, and winds from the west can move the water of the marginal zone into the interior zone.

Turbidity.—The extent of wind-produced turbidity in the Bay is more dependent on the local condition, fetch, and the orientation of the banks, than on the general circulation of the area. Some of the depressions or areas of deeper water are consistently more turbid than others whose general configurations are about the same. Perhaps this is due to the preference of bottom-feeding fish, such as mullet, for certain areas.

FLORA AND FAUNA

Reconnaissance observations by divers of the flora and fauna in the two circulation-salinity zones of the Bay suggest that for the present study the differences between them are relatively minor. In the marginal zone, and especially in the channels leading through the Keys, there are some of the typical organisms of the shallow nearshore parts of the reef tract, for example, the green algae, *Halimeda tridens*, *H. monile*, and *Penicillus* sp., and the corals, *Porites* sp. and *Siderastrea* sp. Farther in the Bay beyond the influence of tidal action, these reef-tract organisms disappear. In the marginal zone beyond the channels the quantity of reef-tract organisms is very low, and the macrofauna is dominated by mollusks as it is in the central zone. There are of course variations in the noncalcareous grass flora between the central and marginal zones, and probably differences in the accessory faunas and floras, but in general terms the similarities overshadow the differences.

No doubt there are differences in the variety and abundance of the molluscan faunas in the two zones of the Bay, but even on the generic level the large local variations between the mud banks and the deeper water may mask regional differences. As an example of the effect of local variation on the molluscan fauna, Allen's data (1942, p. 111) on the variety, abundance, and number of living specimens are given in Table III. His "marl flats," and "shoal" refer to parts of the banks. The data suggest that the banks support a molluscan fauna much more varied and abundant than the fauna in the areas of deeper water. These differences, which can occur within a distance of less than a few hundred yards on the margins of the mud banks, may be as large as changes across the entire Bay.

The most abundant mollusks found in the dredgings and collections from 14

TABLE III. MOLLUSCAN FAUNA FROM BANKS IN FLORIDA BAY
(From Allen, 1942, p. 111)

Water Depth	Character of Area	Average Number[1] of: Live Mollusks	Genera	Dominant
(Inches)				
½–5	Broad marl flat; fairly hard bottom; scattered grass	1,220	6	*Cerithium minimum*
1–2½	Narrow marl flats with soft bottom; no grass	68	6	*Modulus modulus*
2–2½	Broad marly flat; soft bottom; no grass	258	4	*Cerithium minimum*
2½–6	Broad marl flat; fairly hard bottom; scattered grass	150	9	*Modulus modulus*
3–5	Semi-enclosed slough; soft bottom	1,072	5	*Anomalocardia cuneimeris*
4½–5½	Broad marl flat; fairly hard bottom; no grass	129	8	*Cerithium minimum*
4½–5½	Broad marl flat; fairly hard bottom; sparse grass	88	6	*Cerithium muscarum*
5–7	Broad marl flat; fairly hard bottom; scattered grass	227	5	*Pinctada radiata*
(Feet)				
2	Edge of shoal; heavy grass on bottom	83	7	*Pinctada radiata*
3	Edge of shoal; heavy grass on bottom	76	1	*Pinctada radiata*
6	Open bay; marly bottom with sparse grass	75	1	*Pinctada radiata*
6½	Open bay; marly bottom; no grass	0	0	

[1] The samples were standardized as to the size of area dredged.

stations, most of which are from the deeper-water areas, are *Chione, Laevicardium, Tellina, Cerithium, Pinctada,* and *Bulla;* other genera which were found at various stations are *Modulus, Marginella, Mytilus, Battilaria, Fasciolaria, Melamipus, Conus, Arca, Pecten, Cardium,* and *Turbo.*

GRAIN SIZE AND CONSTITUENT COMPOSITION

Comparison of the grain size of sediments in the two circulation-salinity zones of the Bay does not disclose any distinct differences although the number of samples is limited. Again, the large local differences in conditions on the mud banks and in deeper water produce a range of variations in grain size which is as large as that for the entire Bay. Locally on the exposed edges of the banks the sediment is commonly sand-size, as for example at Station 11 (Fig. 5) where there was 72 per cent greater than $\frac{1}{16}$ mm. Up on the banks, and on the bank margins protected by the slope of the bottom or by their orientation with respect to winds, the sediment is much finer, as for example at Stations 12 and 10 where the percentages of sand-size particles are 8 per cent and 47 per cent, respectively.

The percentage of Foraminifera is the only difference between the constituent compositions of the central and marginal zones. Samples from the central zone have 10 per cent or more Foraminifera, and those from the marginal zone have

191

less than 10 per cent (Table X and Fig. 5). Foraminifera are most abundant in samples taken near the mainland where salinity fluctuations are the rule (Stations 26, 27, 28, 29, Table X). Perhaps the relatively greater abundance of Foraminifera in brackish water is due to a large population of a few species which are adapted to the changes in salinity, and the absence of mollusks able to withstand such changes. Moore (ms., 1955) has given data on the variety of the foraminiferal faunas in the Bay.

Although a limited number of samples from each zone were analyzed, the data on constituent composition like the data on gross grain size suggest that the effects of local variations in environment within both circulation-salinity zones of Florida Bay tend to obscure any possible differences between the two zones. However, more detailed studies of the variety and abundance of the mollusks and Foraminifera may show more subtle distinctions.

REEF-TRACT SUB-ENVIRONMENTS

For the present discussion, the reef tract is defined to include the arcuate band-shaped area east, southeast, and south of the Keys between 0 and 300 feet. The reef tract is divided into a semi-protected pseudo-lagoonal part and an area of deeper water by a discontinuous chain of elongate living reefs and rocky shoals along the edge of the shallow-water part (Fig. 5). The main differences in the physical environment of these three subdivisions of the reef tract—fore reef, outer reef, and back reef—are due to depth of water and water circulation.

Some data on annual variations in temperature and salinity in the north part of the reef tract have been reported by Smith, Williams, and Davis (1950, p. 122). They found smaller fluctuations of temperature and salinity at Triumph Reef, Traverse III-Figure 5 (24.35°C.–29.8°C.; 32.25 °/oo–36.36 °/oo), than on the nearshore side of the back reef at Elliot Key 5 miles westward (Fig. 5: 19.10°C.–32.3°C.; 34.43 °/oo–37.40 °/oo). Chew's data (1954) for the reef tract off Key West indicates similar ranges; the surface water in the fore reef had a smaller range of temperature and salinity variations (23°C.–30°C.; 36 °/oo–37 °/oo) than the surface water in the back reef (20°C.–30°C.; 33 °/oo–37 °/oo). Chew found little or no stratification of temperature and salinity in the back reef, but his results show that bottom water in the shallow part of the fore reef is usually slightly cooler than surface water.

Wave action from the sea is most intense along the outer reef-arc, and it may extend up on the platform a short distance where there are no near-surface reefs or shoals. In the back reef wind-driven waves can produce considerable agitation of the surface water, especially during the winter when the winds are consistently stronger than in the summer.

Sustained winds of more than 15 m.p.h. form waves which stir the bottom in the shallower parts of the back reef and turn the water turbid with suspended sediment. Chew's data (1954, Figs. 5–18) of sechi disk observations off Key West show that water along the outer reef-arc and in the fore reef is consistently much clearer than the water in the back reef. However, because the white suspended sediment reflects and scatters light, the illumination on the bottom in the back reef observed by divers does not appear to be much less than when the water is clear.

The difference in physical conditions between the fore reef, outer reef, and back reef, together with variations in the type of bottom produce differences in the fauna and flora of these sub-environments. The following brief discussion is based on reconnaissance observations by divers at numerous localities along the reef tract, supplemented by local dredging, and the experience of the writer's colleagues.

OUTER REEF-ARC SUB-ENVIRONMENT

The outer reef-arc consists of a series of elongate living reefs and rocky shoals separated by areas of deeper water which are usually floored with ripple-marked calcarenaceous and calcirudaceous sediments. Although scattered observations by aqua-lung divers and dredging suggest that flourishing coral growth extends locally to 80 feet or more, Figure 5 shows that the reefs with distinct topographic relief rise from water shallower than 60 feet, and more commonly less than 40 feet. Vaughan and Wells (1943, p. 52) state that " . . . the maximum depth at which corals are active in building reefs is 46 meters, and most reef-building takes place in depths of 15 fathoms or less." Because the lower limit of the most flourishing growth of reef-building corals probably varies according to the slope and nature of the bottom, and the particular hydrographic conditions, no single fixed depth can be given for the lower limit of the entire outer reef-arc sub-environment. Instead the zone from 50 feet to 150 feet is designated as a transition zone separating outer reef-arc from fore reef (Fig. 8).

The inner or shoreward limit of the outer reef-arc is difficult to fix precisely, because like the lower boundary its precise location depends on the particular topography and hydrography. Here again a transition zone seems more appropriate, and its probable range is shown in Figures 8 and 9.

The number and size of reefs which form topographic barriers along the margin of the platform are actually quite small, and there are relatively large areas where there are no reefs or rocky shoals, as for example between Coffins Patch and Tennessee Reef, and between Tennessee Reef and Alligator Reef[6] (Fig. 5). In these areas the hydrographic conditions which characterize the outer reef-arc sub-environment probably extend well up on the platform, and the change to back-reef conditions is gradual.

Even where the reefs are large and more closely spaced, as off Key Largo, there is no semi-continuous reef barrier at the margin of the platform, and there are many gaps or channels between the reefs and shoals. Furthermore, the reefs

---

[6] The areas without reefs are generally opposite openings through the Keys (Fig. 5). Perhaps these openings permit water from Florida Bay, which may at times be warmer and more saline than normal ocean water, to reach the edge of the platform, thus inhibiting the growth of reef-building corals.

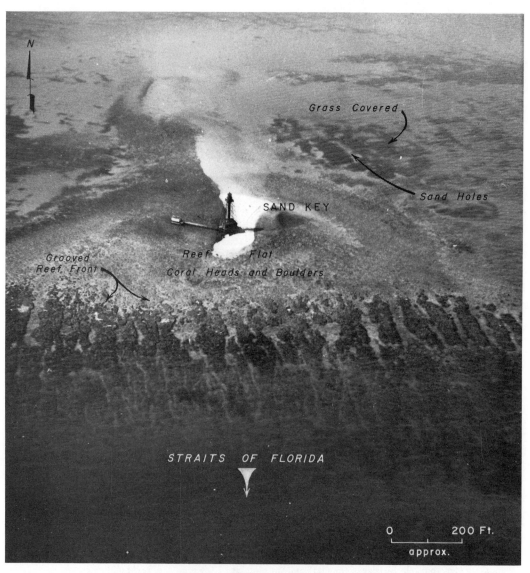

FIG. 7—Aerial photograph of Sand Key reef and lighthouse off Key West looking north across reef. Ridge and groove structure of reef clearly shown. Sand holes are grass-free calcarenaceous sediment a few feet below surrounding grass-covered bottoms.

which form topographic barriers do not everywhere occur on the seaward margin of the platform. At several places, for example, between French Reef and The Elbow and between The Elbow and Carysfort Reef, small, isolated reefs similar to those normally found at the margin of the platform occur about a mile from the edge. In these areas, the outer reef-arc sub-environment is wider than where the reefs occur on the margin of the platform. Thus, in plan, the outer reef-arc sub-environment is irregular, with small finger-like projections and larger bulges all directed toward the Keys, and the change to back-reef conditions is rapid only where there is a long, near-surface reef barrier.

The individual, actively growing reefs are nowhere longer than about a mile, although there are rocky ridges probably reef-built up to 3 miles long on the outer-arc. Most of the reefs which approach sea-level are less than $\frac{1}{2}$ mile long; and they show only on a small scale the structures and zonation which is so typical of Pacific Ocean reefs. Figure 7 shows a well developed example of one of these reefs which reaches to the low-water mark. It has the typical radiating groove-and-ridge structure extending seaward. The grooves floored with calcarenaceous sediments range from 15 to 50 feet wide and up to about 25 feet deep. The intervening ridges of honeycombed reef-rock veneered with living corals, algae, alcyonarians, *et cetera* are as much as 100 feet wide.

The primary structural element of the reef mass in Figure 7 is the moose-horned coral, *Acropora palmata*. This large branching coral, which is capable of annual increase in height of an inch per year (Vaughan, 1915, p. 229), provides both a framework, around and in which detrital material can accumulate, as well as considerable cobble- and boulder-size detritus. This entire mass is eventually stabilized by a covering of the hydrozoan *Millepora alcicornis* and incrusting algae. Although the incrusting algae bind detritus and cover coral fragments with a calcitic envelope which inhibits destruction by boring and burrowing organisms, they are nowhere in Florida as abundant as in the Pacific, where they form a continuous cover along the reef crest—the lithothamnion ridge (Ladd, Tracey, *et al.*, 1950, pp. 413–14). In Florida, behind the wall-like masses which take the full force of the waves, there are heads of *Montastrea* sp., *Siderastrea* sp., *Porites* sp., and *Diploria* sp. up to several feet in diameter and clumps of the staghorn coral, *Acropora cervicornis* and *A. prolifera*. On, around, and under these corals and their detritus there is a wide variety of smaller organisms which contribute to the total reef mass. Incrusting and branching coralline algae, *Halimeda* sp., smaller corals, alcyonarians, mollusks, Foraminifera, and bryozoans all contribute to the detritus. The areas between coral heads are floored with rudaceous and ripple-marked arenaceous detritus. On one reef in the study area, Molasses Reef, algal-covered cobbles and boulders of *Acropora* are piled up to form a ridge about 50 yards long which is entirely exposed at low water, like the shingle ramparts of the Great Barrier Reef (Fairbridge, 1948, pp. 76–78).

If the term "reef flat" is used for the more or less level surface exposed or awash at low tide, and extending lagoonward from the reef crest, then the reef

195

patches in the area of Figure 5 have only a trace of this feature. Behind the reef mass and zone of coral heads and débris, one may find a more or less regular rocky surface thinly floored with rudaceous and arenaceous débris, but it is ordinarily at depths between 5 and 15 feet below mean low water. However, the reefs in the south part of the reef tract do have a narrow reef flat, commonly with coral boulders, and small islands of rubble which may extend above the high-water mark.

The most numerous flourishing reefs which reach near the low-water mark are found south of Carysfort Reef. From Pacific Reef north to Fowey Rocks the outer reef consists of elongate rocky shoals with a relatively small amount of living coral. Vaughan (1918, p. 321) suggested that the lower winter temperatures in the north part of the reef tract are responsible for the scarcity of corals: Fowey Rocks off Miami 15.6°C.; Carysfort Reef off Key Largo 18.2°C.

FORE-REEF SUB-ENVIRONMENT

This sub-environment of the reef tract fronts the reef on the southeast and east. Its upper bathymetric limit occurs within the transition zone shown in Figures 8 and 9, and its lower limit is undefined because the present study did not extend below 300 feet.

Relatively little is known of the flora and fauna of the fore reef and the lower transition zone because the depths restrict extensive direct observation, and because dredging is hampered by rough rocky bottom. In Traverses I and III, sediment samples from the fore reef were recovered only after several attempts at some stations. The bottom sampler often came up empty, and occasionally with a rock fragment. Biologists attempting to dredge specimens in these depths have usually lost their dredges in the rock. The fine sand and silt-size sediment is probably confined to depressions, but locally, as off Key West, it may form a continuous cover more than a few feet thick over the rock.

The few successful dredgings from the shallower part of the fore reef have contained mostly small mollusks, and a few urchins. In 200–300 feet off Key West and off Miami, pebble- to cobble-size porous masses of serpulid tubes and shells of bivalves incrusted by red algae and bryozoans have been recovered. Small solitary corals were abundant in the dredgings off Key West.

Diving off Molasses Reef in 200 feet, Paul Kruse found a smooth sandy bottom with numerous living green algae; species of *Halimeda*, *Penicillus*, and *Udotea* were collected.

BACK-REEF SUB-ENVIRONMENT

The back reef includes that part of the reef tract between the inner transition zone of the outer reef-arc sub-environment and the Keys. Its greatest depths, 40 feet, occur in the south part near Sombrero, and north of Alligator Reef the greatest depths are 25 feet. In the same manner that projections of the outer reef-arc sub-environment extend northwest into the back reef, the back-reef environment may extend into Florida Bay through the wide channels between the Pleistocene reefs. Three of these embayments occur between Grassy Key and Upper Matecumbe Key (Fig. 5).

Three main types of bottom occur in the back reef: (a) rocky with thin ripple-marked sediment, commonly only in the depressions, (b) sediment with or without a thick cover of marine grasses or algae, or both, and (c) patch reefs and rocky with scattered corals. The distribution of these types is irregular and complicated, and only the general outlines can be given.

Thin, ripple-marked, medium to coarse calcarenaceous sediment is most common in the outer part of the back reef near the reef-arc, on those back shoals whose depth is less than 2 fathoms, and in the depressions or sand holes in the grass-covered areas. Examples of the sand holes are shown in Figure 7. In general they are as much as 2 feet deeper than the surrounding grassy bottoms. Near the Keys, the bottom is rocky with only a few inches of sediment in the depressions. The rock is reef rock, commonly well recrystallized like the sub-aerially exposed Key Largo limestone. Worms, crustaceans, burrowing mollusks, and urchins are the main elements of the sparse fauna in the rippled calcarenaceous sediments. A larger and more varied fauna occurs on the bare rock, consisting of corals—including *Siderastrea* sp., *Porites* sp., and *Manicina* sp.—alcyonarians, sponges, green and red algae, mollusks, and a variety of smaller organisms.

Carpets of turtle grass, *Thalassia testudinum*, cover large areas of the back reef. Its long slender leaves extend as much as a foot above the bottom, and its extensive root system penetrates the sediment to approximately the same depth. The density of grass coverage is variable, ranging from scattered plants separated by about a foot, to an almost continuous carpet through which one can not see the sediment bottom. Although turtle grass has no bathymetric limit in the back reef, the dense carpets are most abundant in water deeper than 6 feet.

The baffle-like carpet of grass provides a protected habitat which is occupied by a variety of organisms. It suffices here to mention only the more abundant sediment-producing forms. The green algae of the family *Codiaceae* are very abundant in the grass beds, but they also form the sparse cover on bare sediment bottoms. *Halimeda* sp. is abundant in the grass beds, represented by cushion-like masses up to a foot or more in diameter of *H. opuntia*, and by the erect, rooted, branching *H. tridens* and *H. monile*. Other lightly calcified *Codiaceae*, such as *Penicillus* sp. and *Udotea* sp., are equally abundant. Branching red algae, *Goniolithon* sp. and *Amphiroa*, are particularly abundant on shallow sand banks near the Keys, and beach sands commonly have a large percentage of the fragmented remains of *Goniolithon*. Branching coral, *Porites porites* and *P. furcata*, are particularly common in the inshore grass beds. The rose coral, *Manicina areolata*, is also found in the grass beds. Various mollusks occur in the grass; the most common are *Chione*, *Turbo*, *Pecten*, *Lucina*, *Fasiolaria*, *Astrea*, *Mytilus*, *Tegula*, and *Modulus*. Echinoderms are represented by *Clypeaster rosaceus*,

Lytechinus variegatus, *Tripnuestes esculenta*, and, strangely, colonies of the typical reef form, *Diadema*, which may be found in small patches of bare sediment on the edges of shallow grass banks. The grass leaves themselves are covered with a gray sediment-rich slime, and they support an epilithic microfauna and flora, including calcareous algae, Foraminifera, bryozoans, *et cetera*. Voss and Voss (1955) give a detailed account of the inshore zonation.

Patch reefs and the smaller coral knolls which rise above the surrounding bottoms occur throughout the back-reef area; those rising within 6 feet of mean low water are shown in Figure 5. Considerable variations in size, shape, relief, and distribution are indicated. The individual patches range in size from a few hundred feet to areas of several hundred yards, and where they are close together, as west of Carysfort Reef, they may cover as much as a square mile. An equant to circular shape is most common, but there are discontinuous patches off Key Largo which are a mile long and only ¼ mile wide. The relief of the patches between Alligator Reef and Fowey Rocks is 15–20 feet and in places 30 feet above the general level of the back reef. Farther southwest, off Key West and in the Dry Tortugas, where the general depth of the back reef is 40 and 60 feet, respectively, the relief of the patch reefs increases to 50 feet in the lagoon of the Dry Tortugas (*U. S. Coast and Geodetic Survey Chart 585*). Patch reefs occur throughout the back reef, but are more abundant in the seaward third than elsewhere. As shown in Figure 5, there are only a few patch reefs near shore, and nowhere in the study area is there anything approaching a fringing reef.

The amount of living corals on the patch reefs is variable, and their distribution sporadic. Hemispherical coral heads up to several feet in diameter occur in groups separated by rocky gorgonian-covered interareas. *Montastrea annularis*, *Porites asteroides*, *Siderastrea siderea*, and *Diploria* sp. form the large heads. Other smaller corals include *Porites porites*, *Agaricia* sp., *Montastrea cavernosa* and others. *Millepora alcicornis* occurs as an incrustation on gorgonians and other projections. Incrusting coralline algae are present on dead corals and on the undersides of projections but in no great abundance. The gorgonians are especially abundant, some of the larger forms being several feet high. Organic erosion of corals on these patch reefs seems very intense, and several examples of coral heads have been observed which have toppled over or split in half probably because their bases were undermined by boring and burrowing organisms.

Just as the changes between the physical environments of the three subdivisions of the reef tract are gradual and difficult to separate sharply, so are the floras and faunas transitional from one subdivision to the other. Despite these gradual changes, and the limited bottom observations, especially in the fore reef, the following pattern of distribution for the living major sediment-producing organisms is suggested.

Halimeda—Most abundant in back reef and in outer reef-arc
Coralline algae—Most abundant in outer reef-arc and fore reef; locally abundant on patch reefs and in shallow, nearshore parts of back reef

Coral—Most abundant in outer reef-arc and on patch reefs in back reef; may be locally abundant in shallow, nearshore parts of back reef

Foraminifera—No observations

Mollusks—Difficult to evaluate because many species burrow or live under rocks

GRAIN SIZE AND CONSTITUENT COMPOSITION

In order to indicate the extent to which the sub-environments are reflected in the grain size and constituent composition of their sediments, the analyses of the samples from each of the three traverses given in Table VIII are portrayed graphically in Figure 8 and 9.

Grain size.—The weight percentage of the total sample finer than $\frac{1}{8}$ mm. is used as a measure of grain size. Although this type of measure gives no idea of the particle-size distribution it does have more practical value than measures of the size distribution based on a complete sieve analysis because a comparable measure can be determined for ancient unrecrystallized limestones. Quartile parameters of the size distribution are given in Table VIII. Figure 8 shows the variations in the percentage finer than $\frac{1}{8}$ mm. across the reef tract. In all three traverses there is a general increase in the percentage of fines in the inner part of the back reef, especially in those parts of the area which are slightly deeper than their surrounding bottom (Stations D and C in Traverse I, for example). In Traverse I and III the percentage of fines increases as the water deepens from the outer reef into the fore reef. This relationship is not shown in Traverse II which did not extend into the fore reef. In all three traverses the outer reef-arc and the near-by part of the back reef have little or no fines.

Constituent composition.—Constituent composition of the fraction larger than $\frac{1}{8}$ mm. was determined by point counts on thin sections of representative sub-samples. The particles were assigned to one of the following eight categories: *Halimeda*, coralline algae (all calcareous Rhodophyceae), mollusks, coral, Foraminifera, non-skeletal (physical aggregates, fecal pellets, oöliths, cemented aggregates and rock fragments), miscellaneous (echinoids, worm tubes, ostracods, and *Millepora*, bryozoans, *et cetera*), and unknown. The methods of compositional analysis, and the precision of the estimates are discussed in the Appendix. Only those variations in the estimates of volume percentage which are greater than 5 per cent should be considered significant, and then only when there is a consistent trend in the variations.

The results of these compositional analyses for the 24 samples from the three reef-tract traverses are given in Table VIII. Variations in the percentage of the five major constituents—*Halimeda*, coralline algae, mollusks, coral, and Foraminifera—along each of the three traverses are shown graphically in Figure 9 together with bottom profiles.

Traverses I and III, which both crossed the reef tract where there are distinct topographic reef barriers, show similar variations in the abundances of the major constituents.

1. *Halimeda* is low or absent in the fore reef below 200 feet. Its abundance

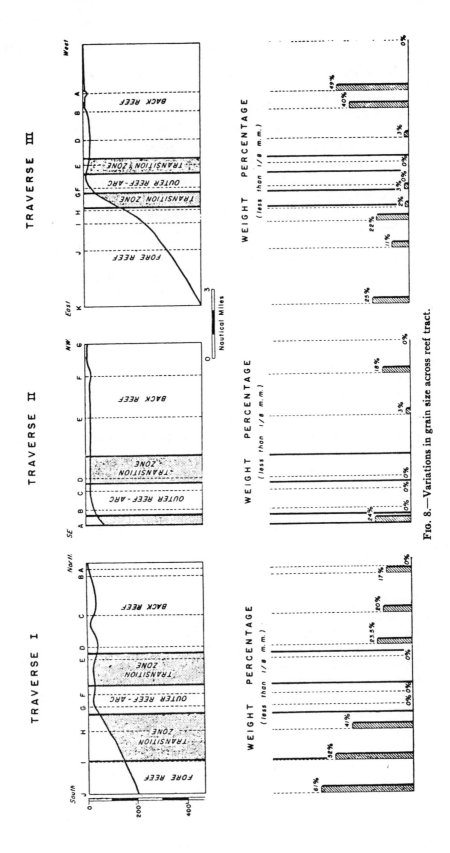

FIG. 8.—Variations in grain size across reef tract.

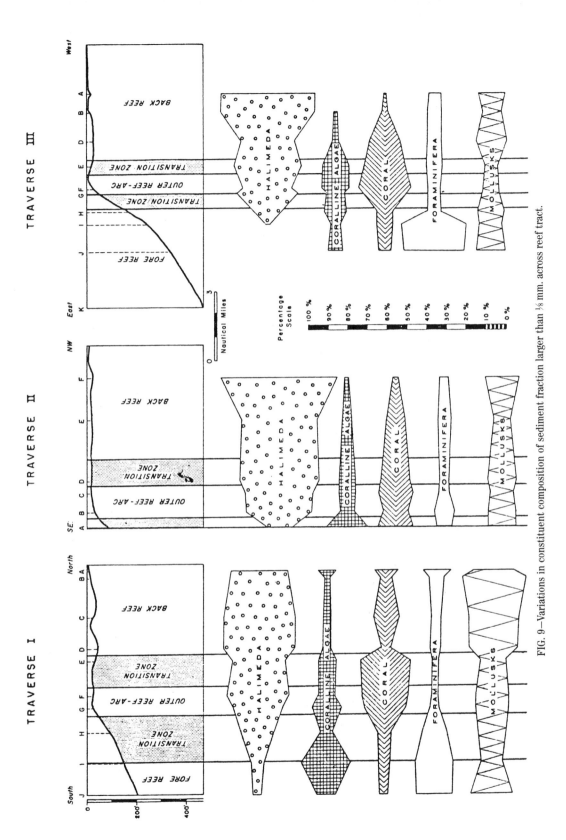

FIG. 9—Variations in constituent composition of sediment fraction larger than ⅛ mm. across reef tract.

201

rapidly increases in the lower transition zone, and high values extend from the outer reef-arc across the entire back reef. Maximum percentages of *Halimeda* are found in the nearshore parts of the back reef in Traverse III, and in the central part of the back reef in Traverse I.

2. The percentage of coralline algae is low in the back reef except nearshore, as at Station A in Traverse I (Fig. 9). It increases rapidly in the upper transition zone, and reaches maximum values in the outer reef-arc and adjacent lower transition zone. Judged by the results of dredging in this region, increasing percentages of coralline algae are expected at greater depths.

3. Coral reaches its maximum in the outer reef-arc and in parts of the adjacent transition zones. It decreases rapidly in the lower transition zone. Coral also decreases between the outer reef-arc and the central part of the back reef. The decrease is very rapid in Traverse I, but somewhat less so in Traverse III. A patch reef at Station C of Traverse I (Fig. 9) produces a local increase of coral in the back reef.[7]

4. Foraminifera increase seaward of the outer reef-arc (Traverse I), or seaward of the lower transition zone (Traverse III).

5. The percentage of mollusks is lower in the outer reef-arc and adjacent transition zones than in the fore reef and back reef.

As Traverse II did not extend into the fore reef, it does not give a complete section across the reef tract. However, the variations in composition between the outer reef-arc and back reef in Traverse II resemble those of Traverses I and III, but they are much more gradual.

The kinds and relative abundances of the minor sediment constituents are also expected to show variations in the sub-environments of the reef tract. However, because these variations are probably relatively small, and will require more detailed examinations of the sediments than time permitted, they were not attempted in the present study.[8]

Because the estimates of abundance of the constituents are relative, the variations shown in Figure 9 are not necessarily the result of differences in the

[7] The lack of samples around this patch reef, and the method of construction of Figure 9, exaggerate the extent of increased coral. Bottom observations around other patch reefs in the area, and examination of aerial photographs suggest that sediment derived from them does not extend more than a few hundred feet beyond their growth limits. Thus, the maximum at Station C in Traverse I is probably much more local than the gradual increase shown in Figure 9.

[8] One usually minor constituent, the non-skeletal category, does attain major significance in several samples. The term is used by Illing (1954, pp. 24–45) for all particles whose existence is due to the binding and cementing action of precipitated aragonite. As the percentage of non-skeletal particles is generally small (Table IX), it was not broken down into the sub-categories used by Illing. However, at three stations in the fore reef where there is a considerable proportion of non-skeletal grains, examinations show that it is mostly rock fragments, ooliths, and pellets (Traverse I, Stations H and I; Traverse III, Station H). The rock fragments are oolitic limestone with a calcite cement, and are indistinguishable from the Pleistocene Miami oolite of the mainland. Because all these stations are in water at least 100 feet deep, it is unlikely that such limestone is forming there now. The fragments may be derived from a submarine outcrop of oolitic limestone, or they may have been brought to their present position from the mainland by streams during a low-water stage of the Late Pleistocene.

amount of the living constituents or differences in their growth rates. For example, the increase in Foraminifera in the fore reef may not indicate a greater rate of production of Foraminifera per unit area, but instead may be the result of relatively large decreases in the other constituents. However, in Traverses I and III, it appears that the variations of the constituent compositions do reflect differences in the living abundances of the organisms that are produced by differences in physical environment, because the major variations in the abundances of particles of *Halimeda*, coral, and coralline algae correspond with the observed growth densities.

Transportation of débris produced in one sub-environment to another would be expected to obscure the relationship between the sediments and their source areas. But because transportation like the growth of organisms is controlled by physical conditions it may also reflect differences in environment. For example, the absence of a topographic reef barrier in Traverse II permits wave action and tidal currents to extend much farther up on the platform than in Traverse I or III. In the back reef of Traverse II there is only a small amount of the stabilizing grass cover that is common in the other traverses, and the mega-ripples seen from the air indicate that there is movement of bottom sediment. Probably then, the very gradual decreases in coralline algae and coral of Traverse II (Fig. 9) are produced by transportation of sediment from the outer reef-arc to the back reef. Thus, the changes and the rates of change in the traverses provide indications of differences in environment.

The constituent composition of the size fraction smaller than $\frac{1}{8}$ mm. was not estimated for practical reasons, and probably these finer particles can be transported much farther than the sand-size particles. Therefore, fine sediment produced in one sub-environment may accumulate in another. If such movement occurs, about 20 per cent of several back-reef samples, and up to 60 per cent of some fore samples, could have been produced outside the sub-environment where they now occur (Fig. 8).

[*Editors' Note:* Material has been omitted at this point.]

CONCLUSIONS

1. Variations in the physical environment (bathymetry, areal geography, and hydrography) of carbonate depositional environments in South Florida are reflected in the grain size and constituent particle composition of their sediments. Major differences in environment, as between the reef tract and Florida Bay, produce sediments with quite different grain size and constituent composition, but where the changes in environment are small and gradual, as within the reef tract, the sediments are less distinctive, and their differentiation requires recognition of gradual changes in relative abundances of the major constituent particles.

2. Local variations within a major environment may be so large that their effects on the grain size and constituent composition of the sediments obscure small and subtle variations produced by gradual changes in the physical conditions across a large area.

3. The bulk of the reef-tract sediment larger than $\frac{1}{8}$ mm. accumulates in the same sub-environment in which it is produced, and the reefs are not major sources of sand-size sediment for the adjacent areas.

4. Petrographic study of the grain size and constituent composition of ancient limestones may provide valuable data on paleo-environments.

DISCUSSION

It is perhaps not surprising that major differences in physical environment between Florida Bay and the reef tract (Table I) produce sediments of quite distinct grain size and constituent composition (Table II). Either property, grain size or constituent composition, differentiates the average sediments of these major environments. However, because the ranges of variation of these parameters overlap, there are samples which could fall into either environment unless both properties are used as criteria. Even on the basis of both grain size and constituent composition, single samples from the boundary zone between Florida Bay and the reef tract could not be definitely assigned to one major environment. However, in a series of samples taken across the boundary zone, the gradual changes in grain size and constituent composition could be recognized, and they might provide very useful information on the nature of the boundary.

A second level of environmentally controlled sediment differentiation is illustrated by one of the major environments, the reef tract. It can be divided into sub-environments on the basis of changes in bathymetry and hydrography, and although these changes are gradual, they are reflected in the distribution of the major sediment-producing organisms, their fragmented remains, and the grain size of the sediments. Figures 8 and 9 show how variations in the percentages of fines (less than $\frac{1}{8}$ mm.), and changes in the relative abundance of *Halimeda*, coralline algae, coral, and Foraminifera correspond with the three sub-environments of the reef tract.

There are individual organisms or features which are limited to one major environment or sub-environment, for example, certain Foraminifera (Moore, *ms.*), corals, or the incrusting habit of coralline algae. Although such organisms and features alone could be used to recognize environments, they can not provide as much information about gradual changes in conditions as multiple indices or assemblages of organisms. Moreover, when the single index organism is absent because of some unusual local conditions, the assemblage will probably still reflect the prevailing conditions.

The extent of local variations in physical conditions must be considered when attempting to subdivide the major environments of Florida Bay and the reef tract. In Florida Bay the rapid changes in physical and biological environment from the near-surface grass-covered mud banks to the deeper water of the adjacent rock-floored depressions are so large that they produce variations in grain size and constituent composition of the sediments as large as, or larger than, those produced by progressive changes across the Bay. By a detailed study of the generic and specific composition of the faunas it may be possible to recognize the differences between the margins of the Bay which are influenced by marine circulation, and the central and north parts of the Bay in which there are larger variations in salinity. However, in terms of the present study the sediments **are** homogeneous.

There are also local variations in physical environment within the reef tract. The patch reefs and coral knolls of the back reef produce sediments whose **grain** size and constituent composition may approach those of the outer reef-arc sub-environment. Because the sediment produced by these patch reefs and knolls appears to be limited to their confines and the adjacent bottom, it only locally interrupts the normal back-reef type of sediment. Where the patches and knolls are larger and more numerous, they could obscure the progressive changes across the reef tract in the same way as do the mud banks in Florida Bay.

The general similarity between the distribution of living sediment-producing organisms in the reef tract and their fragmentary remains in the sediments suggests that there is no major transportation of sand-size sediment from one sub-environment to another. In Traverses II and III about 15 per cent of the sediment larger than $\frac{1}{8}$ mm. in the central part of the back reef may have been brought in from the outer reef-arc, but in Traverse I there is no such indication of transportation. As the constituent composition of the sediment fraction less than $\frac{1}{8}$ mm. was not determined, it is impossible to estimate the amount of the fine sediment in the fore reef and back reef that is produced *in situ*. Perhaps a large part of it is derived from the outer reef-arc where mechanical and organic erosion are intense.

As described in the Appendix, the constituent composition data for the **reef** tract were determined from standard petrographic thin sections of representa-

[*Editors' Note:* Tables VI and VII have been omitted.]

TABLE VIII. GRAIN SIZE AND CONSTITUENT COMPOSITION OF REEF-TRACT SEDIMENTS

Sample Number	Grain Size				Particle Constituents of Fraction Greater than ⅛ mm.							
	Wt. % Less than ⅛ mm.	Wt. % Less than 1/16 mm.	Md_ϕ	Qd_ϕ	Halimeda	Coralline Algae	Mollusk	Coral	Foraminifera	Non-Skeletal	Miscellaneous	Unknown
Traverse I												
I-A	0	0	1.15	0.60	29	7	20	10	13	3	8	10
I-B	16.9	9.5	1.30	1.10	30	1	33	5	7	16	3	5
I-C[1]	20.0	11.6	2.00	0.82	36	2	25	12	4	7	7	7
I-D	23.5	17.5	0.80	1.48	36	2	24	3	5	17	7	6
I-E	—	—	0.56	0.80	27	8	12	25	3	12	7	6
I-F	—	—	0.85	0.60	32	7	16	22	3	7	6	7
I-G[2]	—	—	0.36	0.44	27	13	16	21	3	8	8	4
I-H	67.7	45.5	3.85	—[3]	16	5	13	6	12	24	11	13
I-I	61.1	33.7	3.45	0.81	6	24	10	5	19	14	12	10
I-J	52.1	37.9	3.20	—[3]	4	8	20	3	18	16	16	15
Traverse II												
II-A	24.0	13.9	0.65	1.62	13	21	14	17	8	11	12	4
II-B	—	—	0.98	0.56	34	6	14	12	10	12	5	7
II-C	—	—	0.85	0.52	34	7	11	17	6	13	4	8
II-D	—	—	1.50	0.42	41	7	14	14	4	11	2	7
II-E	2.4	—	2.15	0.50	38	4	8	9	6	18	10	7
II-F	17.7	7.3	1.35	1.00	60	1	17	2	5	5	4	6
Traverse III												
III-A	49.0	—[3]	—[3]	—[3]	46	0	13	1	6	14	5	15
III-B	39.8	13.0	2.85	c 56	46	2	9	3	7	12	14	7
III-D	3.0	—[3]	1.00	0.59	26	3	15	15	7	17	7	10
III-E	—	—	1.28	0.61	33	13	6	20	8	9	7	4
III-F	—	—	0.70	0.52	24	14	8	26	6	7	11	4
III-G	3.4	—	0.93	0.76	30	9	8	22	6	8	12	5
III-H[1]	1.7	—	0.25	1.16	15	11	13	8	7	22	18	6
III-I	22.0	8.1	2.35	0.81	2	5	11	8	32	9	19	14
III-J	11.2	4.3	2.18	0.68	0	7	14	6	30	5	23	15

[1] Particle constituent compositions are averages of the estimates from two different thin sections.
[2] All values are averages of two samples collected 50 feet apart.
[3] Not calculated because of incomplete analysis.

TABLE IX. GRAIN SIZE AND CONSTITUENT PARTICLE COMPOSITIONS OF FLORIDA BAY SEDIMENTS

Sample Number[1]	Grain Size Weight Percentage							Constituent Particle Compositions[2] Percentage of Fraction Greater than ⅛ mm.							
	Depth Below Top in Cores (cm.)	Greater than 1 mm.	Greater than ½ mm.	Greater than ¼ mm.	Greater than ⅛ mm.	Less than ⅛ mm.	Less than 1/16 mm.	Mollusk	Foraminifera	Aggregates	Ostracods	Quartz	Halimeda	Miscellaneous	Unknown
1C-a*	0–6	18.7	11.5	9.4	6.6	53.8	50.0	95	2	—	1	1	—	—	1
7C-a*	0–8	0	9.5	25.6	16.1	48.8	43.1	76	14	5	2	2	—	—	1
7C-B	16–22	5.8	4.7	4.5	8.6	76.4	67.8	72	17	2	8	—	—	—	1
8S*	—	9.9	17.0	17.2	21.3	34.6	29.4	79	4	—	3	13	—	—	1
10C-a*	0–5	14.3	7.7	7.1	5.8	65.1	52.7	69	7	18	5	—	—	1	—
11S-2*	—	1.4	8.0	31.0	30.1	29.5	27.8	58	23	2	5	7	1	1	3
12C-b	11–18	2.1	1.4	0.9	1.1	92.7	91.6	79	14	—	7	—	—	—	—
14S*	—	3.6	3.1	2.5	3.0	87.8	85.1	72	11	—	2	7	—	4	4
17D*	—	2.9	7.7	7.6	12.2	69.6	52.4	82	7	—	—	5	1	2	3
20S*	—	14.3	14.5	19.9	11.6	27.4	25.0	86	6	2	1	4	1	—	—
22C-a*	0–6.5	4.1	6.1	18.2	29.6	42.0	35.3	71	3	1	2	20	1	—	2
22C-b	35–42	4.3	4.8	9.5	13.0	68.4	61.5	82	3	—	3	11	1	—	—
22C-c	47–60	6.7	2.9	2.7	3.9	83.8	80.0	92	1	—	4	2	—	1	—
22C-d	72–76	12.6	7.4	6.3	8.3	65.4	60.0	91	1	—	2	3	1	—	2
22C-r-a*	0–6	6.4	5.5	4.7	12.3	71.1	57.8	82	6	4	3	4	—	—	1
23S*	—	6.3	7.6	7.6	12.9	65.6	59.3	72	6	1	—	19	1	—	1
25C-a*	0–4	11.1	6.8	5.5	6.6	70.0	66.5	83	6	3	3	4	—	—	1
26C-a*	1–10	9.7	5.5	8.8	11.2	64.7	54.1	64	31	—	3	2	—	—	—
27C-a*	0–4	9.5	7.8	19.3	14.9	48.6	41.8	62	32	—	1	5	—	—	—
28C-a*	7–12	8.8	6.4	8.1	8.4	68.5	63.0	92	3	1	4	—	—	—	—
28C-b	20–28	8.9	5.6	5.4	4.4	75.8	72.8	76	16	2	5	1	—	—	—
28C-c	32–40	4.2	2.6	2.7	2.7	87.8	85.5	77	16	—	6	—	—	—	1
29C-a*	10–18	10.3	5.9	4.6	3.9	75.3	73.3	74	18	4	2	—	—	—	2
29C-b	35–45	5.4	4.2	3.5	2.9	84.1	81.6	79	13	6	2	6	—	—	—
34S-2*	—	68.4	11.9	5.2	3.4	11.2	9.6	86	11	2	—	—	—	—	1

* These samples were used for the average values and ranges given in Tables II and IV.
[1] The locations of the samples are shown in Figure 5. The capital letter indicates the type of sample (C = core, S = surface scoop, D = dredge) and the small letter indicates the divisions of the cores based on examination.
[2] From counts of about 300 grains in each size fraction.

tive subsamples, and the same technique could have been used for the Florida Bay samples. The measure of grain size used in this study was the weight percentage of material finer than $\frac{1}{8}$ mm. A comparable volume measure could easily be made in thin section, and Jaanusson (1952) used this type of estimate in studying the grain size of Ordovician limestones in Sweden. The recognition of the major contributors in ancient limestones may present problems, but with the help of published photographs and reference sections of identifiable fossils it will probably be possible to recognize major constituents in limestones which are not thoroughly recrystallized. Cuvillier (1951) used dominant constituents together with diagnostic Foraminifera to define Cretaceous facies in France, and Robert Terriere is determining the relative abundance of the major constituents of Pennsylvanian limestones from West Texas.

The results of this study will be of most direct applicability in Tertiary and perhaps Cretaceous limestones in which the sediment constituents are so closely related to living forms that modern ecologic ranges can be used.

For the organic constituents in the Paleozoic, it may be dangerous to use the extrapolated ranges of comparable living organisms, and initial interpretations will necessarily be based on biomechanical considerations and the implications of environmentally controlled sediment structures and textures (Ginsburg, 1954). Even without precise information on the ecology of the major constituents, studies of variations in their abundance will be helpful in characterizing ancient depositional environments, and in determining their homogeneity and relative isolation.

[*Editors' Note:* The Appendix has been omitted.]

BIBLIOGRAPHY

AGASSIZ, A., 1888, "Three Cruises of the *Blake*," *Bull. Mus. Comp. Zool., Harvard College*, Vol. 14 pp. 286–87.

ALLEN, ROBERT PORTER, 1942, "The Roseate Spoonbill," *Nat. Audubon Soc. Research Rept. 2*, p. 111.

APPLIN, PAUL L., 1951, "Preliminary Report on Buried Pre-Mesozoic Rocks in Florida and Adjacent States," *U. S. Geol. Survey Circ. 91*, pp. 1–28.

BØGGILD, O. B., 1930, "The Shell Structure of the Mollusks," *D. Kgl. Danske, Vidensk. Selsk. Skrifter*, Naturvidensk. Og Mathem. Afd., 9, Raekke, II. 2, pp. 1–325.

BONET, FEDERICO, 1952, "La Facies Urgoniana del Cretacico Medio de la Region de Tampico," *Bol. Asoc. Mex. Geol. Petrol.*, Vol. IV, Nos. 5–6, pp. 1–259.

BRAMLETTE, M. N., 1926, "Some Marine Bottom Samples from Pago Pago Harbor, Samoa," *Carnegie Inst. Washington Pub. 344, Pap. Tortugas Lab.*, Vol. 23, pp. 1–36.

CHAYES, F., 1949, "A Simple Point Counter for Thin-Section Analysis," *Amer. Min.*, Vol. 34, pp. 1–11.

CHEW, FRANK, 1954, "Physical Oceanography of Key West Harbor and Approaches," *Marine Lab., Univ. Miami Rept. 54-13*, pp. 1–18.

COOKE, C. WYTHE, AND MOSSOM, STUART, 1929, "Geology of Florida," *Florida Geol. Survey 20th Ann. Rept.*, pp. 29–288.

CUVILLIER, J., 1951, *Correlations Stratigraphiques par Microfacies en Aquitaine Occidentale.* E. J. Brill and Company, Leiden.

DAVIS, JOHN H., JR., 1940, "The Ecology and Geologic Role of Mangroves," *Carnegie Inst. Washington Pub. 517, Pap. Tortugas Lab.*, Vol. 32, pp. 307–409.

DOLE, R. B., AND CHAMBERS, A. A., 1918, "Salinity of Ocean-Water at Fowey Rocks, Florida," *ibid., Pub. 213, Pap. Tortugas Lab.*, Vol. 9, pp. 299–315.

EMERY, KENNETH O., TRACEY, J. I., JR., AND LADD, H. S., 1954, "Geology of Bikini and Nearby Atolls," *U. S. Geol. Survey Prof. Paper 260-A*, pp. 56–59, 97–99, 107, 115.

FAIRBRIDGE, RHODES W., AND TEICHERT, CURT, 1948, "The Low Isles of the Great Barrier Reef: a New Analysis," *Geogr. Jour.*, Vol. CXI, pp. 76–80.

FOLK, ROBERT L., 1954, "The Distinction between Grain Size and Mineral Composition in Sedimentary-Rock Nomenclature," *Jour. Geol.*, Vol. 62, pp. 348–49.

GINSBURG, R. N., 1954, "Early Diagenesis and Lithification of South Florida Carbonate Sediments" (abst.), *Jour. Sed. Petrology*, Vol. 24, p. 138.

GINSBURG, ROBERT, AND LLOYD, R. MICHAEL, 1956, "A Manual Piston Coring Device for Use in Shallow Water," *ibid.*, Vol. 26, No. 1, pp. 64–66; 2 figs.

GOLDMAN, MARCUS I., 1918, "Composition of Two Murray Island Bottom Samples According to Source of Material," *Carnegie Inst. Washington Pub. 213, Pap. Tortugas Lab.*, Vol. 9, pp. 249–62.

———, 1926, "Proportions of Detrital Organic Calcareous Constituents and Their Chemical Alteration in a Reef Sand from the Bahamas," *ibid., Pub. 344, Pap. Tortugas Lab.*, Vol. 23, pp. 37–66.

HENSON, F. R. S., 1950, "Cretaceous and Tertiary Reef Formations and Associated Sediments in Middle East," *Bull. Amer. Assoc. Petrol. Geol.*, Vol. 34, pp. 215–38.

HOUGH, J. L., 1939, "Bottom-Sampling Apparatus," in TRASK, P. D., *Recent Marine Sediments*, Amer. Assoc. Petrol. Geol., p. 661.

ILLING, L. V., 1954, "Bahaman Calcareous Sands," *Bull. Amer. Assoc. Petrol. Geol.*, Vol. 38, pp. 1–95.
JAANUSSON, V., 1952, "Untersuchungen uber die Korngrosse der Ordovizischen Kalksteine," *Geol. Foren. Forhandl.*, Bd. 74, H. 2, p. 124.
JORDAN, G. F., 1954, "Large Sink Holes in Straits of Florida," *Bull. Amer. Assoc. Petrol. Geol.*, Vol. 38, pp. 1810–17.
LADD, H. S., TRACEY, J. I., JR., WELLS, J. W., AND EMERY, K. O., 1950, "Organic Growth and Sedimentation on an Atoll," *Jour. Geol.*, Vol. 58, pp. 413–14.
MARMER, H. A., 1954, "Tides and Sea Level in the Gulf of Mexico," in "Gulf of Mexico: Its Origin, Waters, and Marine Life," *Bull. U. S. Fish and Wildlife Serv.*, Vol. 55, No. 89, pp. 115–16.
MATSON, G. C., 1910. "Report on Examination of Material from the Sea-Bottom between Miami and Key West," in VAUGHAN, T. W., 1910, "A Contribution to the Geologic History of the Floridian Plateau," *Carnegie Inst. Washington Pub. 133, Pap. Tortugas Lab.*, Vol. 4, pp. 120–25.
MOORE, W. E., 1955, "Recent Foraminifera from the Florida Keys," ms.
MURRAY, J., AND RENARD, A. F., 1891, *Report on Deep-Sea Deposits Based on Specimens Collected during the Voyage of HMS Challenger*, pp. 184–248. HM Stationary Office, London.
NANZ, R. H., 1954, "Genesis of Oligocene Sandstone Reservoir, Seligson Field, Jim Wells and Kleberg Counties, Texas." *Bull. Amer. Assoc. Petrol. Geol.*, Vol. 38, p. 100.
NEWELL, NORMAN D., RIGBY, J. KEITH, WHITEMAN, ARTHUR J., AND BRADLEY, JOHN S., 1951, "Shoal-Water Geology and Environments, Eastern Andros Island, Bahamas," *Bull. Amer. Mus. Nat. History*, Vol. 97, Art .1, pp. 1–30.
PARKER, G. G., AND COOKE, C. WYTHE, 1944, "Late Cenozoic Geology of Southern Florida, with a Discussion of the Ground Water," *Florida Geol. Survey Bull. 27*, pp. 67–71.
PRESSLER, E. D., 1947, "Geology and Occurrence of Oil in Florida," *Bull. Amer. Assoc. Petrol. Geol.*, Vol. 31, pp. 1851–62.
SANDER, B., 1951, *Contributions to the Study of Depositional Fabrics—Rhythmically Deposited Triassic Limestones and Dolomites*," Amer. Assoc. Petrol. Geol., Translation from original German by E. B. KNOPF.
SANFORD, SAMUEL, 1913, "Southern Florida," in MATSON, G. C., AND SANFORD, SAMUEL, "Geology and Ground Waters of Florida," *U. S. Geol. Survey Water-Supply Paper 319*, pp. 169, 180–87.
SHEPARD, F. P., 1948. *Submarine Geology*, pp. 108–09. Harper and Brothers, New York.
SMITH, C. L., 1940, "The Great Bahama Bank I, General Hydrographical and Chemical Features," *Jour. Marine Res.*, III, 2, pp. 147–69.
SMITH, F. G. WALTON, WILLIAMS, R. H., AND DAVIS, C. C., 1950, "An Ecological Survey of the Subtropical Inshore Waters Adjacent to Miami," *Ecol.*, Vol. 31, pp. 120–25, 144.
THORP, E. M., 1936, "Calcareous Shallow-Water Marine Deposits of Florida and the Bahamas," *Carnegie Inst. Washington Pub. 452, Pap. Tortugas Lab.*, Vol.29, pp. 37–119.
U. S. DEPT. OF COMMERCE, 1955, *Coast and Geodetic Survey Tide Tables East Coast North and South America*, U. S. Govt. Printing Office, pp. 222–23.
———, WEATHER, BUREAU, 1950, *Local Climatological Summary with Comparative Data for Miami, Florida*.
VAUGHAN, T. W., 1916, "On Recent Madreporaria of Florida, the Bahamas, and the West Indies, and on Collections from Murray Island, Australia," *Carnegie Inst. Washington Yr. Bk. 14 for 1915*, pp. 222–23, 229.
———, 1918, "Some Shoal-Water Bottom Samples from Murray Island, Australia, and Comparison of Them with Samples from Florida and the Bahamas," *ibid.*, Pub. 213, Pap. Tortugas Lab., Vol. 9, pp. 285–88.
———, 1918, "The Temperature of the Florida Coral Reef Tract," *ibid.*, p. 319.
———, AND WELLS, J. W., 1943, "Revision of the Suborders, Families, and Genera of the Scleractinia," *Geol. Soc. America Spec. Paper 44*.
VOSS, GILBERT L., AND NANCY A., 1955, "An Ecological Survey of Soldier Key, Biscayne Bay, Florida," *Bull. Mar. Science of Gulf and Caribbean*, Vol. 5, No. 3, pp. 203–29.

Editors' Comments
on Papers 7 and 8

7   **SHEARMAN**
      Recent Anhydrite, Gypsum, Dolomite, and Halite from the
      Coastal Flats of the Arabian Shore of the Persian Gulf.

8   **KENDALL and SKIPWITH**
      Holocene Shallow-Water Carbonate and Evaporite Sediments
      of Khor al Bazam, Abu Dhabi, Southwest Persian Gulf

PERSIAN GULF

The Bahama islands and Florida are located in a moist climate in which evaporite minerals are not precipitated among carbonate sediments. By contrast, the climate of the Persian Gulf is hot and arid and evaporite minerals form as part of the sediment.

The origin of evaporites remains a vigorously debated subject since A. G. Werner in the late 1700s without any sign of solution. So far, at least eight models have been proposed: (1) a series of shallow-water basins (Branson, 1915): (2) a partially enclosed epicontinental sea (Krumbein, 1951; Sloss, 1953); (3) a lateral salinity gradient across the basin (Scruton, 1953); (4) subsidence of saline epeiric shelf (Landes, 1960); (5) a deep-water barred basin (Schmalz, 1969); (6) a basinal density stratification (Sloss, 1969); (7) hypersaline or *sabkha* environment (Shearman, 1966; Friedman, 1972); and (8) desiccated basin model (Nurmi, 1974). Still other models have fed this controversy (see Dean and Schreiber, 1978; and Friedman and Sanders, 1978, Compliment D).

Oversimplified, the controversy boils around two points—whether evaporites represent deposition in an enclosed, starved, standing body of water younger in age than basin-marginal carbonates or whether they represent a continuous process of evaporite deposition simultaneous to basin-margin carbonates across the basin in response to lateral biological and geochemical gradients. The adherents of different models of evaporite deposition have raised so many new problems that the controversy persists.

Paper 7 by Shearman represents one of the models of evaporite formation in a sabkha environment. The sea-marginal sabkhas along the Persian Gulf represent the classical site for diagenetic origins of evaporites. As a result, the nearshore, lagoonal, and coastal plain sediments along this belt have been subjected to intensive study since 1960 resulting in a considerable volume of publications (Butler, 1969; Evans, 1966; Illing et al., 1965; Kendall and Skipwith, 1968; Kinsman, 1964; Purser, 1973; Shinn, 1971; and Wells, 1962). Shearman and his students at Imperial College, United Kingdom are among the pioneers who conducted their studies in the carbonate environments of the Persian Gulf. The preliminary report of his original findings is Paper 7. Shearman describes occurrences of dolomite, gypsum, halite, and, for the first time, anhydrite in the sabkha sediments. The anhydrite occurs as bands, lenses, and nodules, varying from an inch up to approximately a foot in thickness. He demonstrates that all these evaporite minerals are of early diagenetic origin. Shearman further suggests that occurrences of these evaporites on a regional scale in an arid, supratidal sabkha setting may well explain the thick ancient back-reef and lagoonal dolomites and evaporites rather than the classical theory of an origin through evaporation of barred, stagnant bodies of water.

Paper 8 by Kendall and Skipwith gives a more detailed account of flora, fauna, and sediment facies of a part of the southwest Persian Gulf and relates them to distinct geomorphologic units of the area that, in turn, are related to water chemistry and hydrodynamics. The sediment components are similar to those in the Bahamas and Florida Bay except for the higher incidence of bryozoan skeletal debris and terrigenous particles. The authors suggest that though each of the sediment types occupies a small area in the Persian Gulf, there is good potential of preservation in the sedimentary record because of rapid coastal progradation. The authors also relate different evaporite minerals, which are diagenetic in origin, to intertidal and sabkha environment and trace four parallel gradational belts of evaporite mineral assemblages in response to evaporation and capillary action. They further demonstrate that particles less than 1/16 mm in diameter are more sensitive environment indicators than those less than 1/8 mm as suggested by Ginsburg (1956). Moreover, blue-green algae, according to them, are probably the major agency of precipitation of fine-grained aragonite and are not products of skeletal breakdown. The intimate relationship of blue-green algae with alteration of particles to a homogeneous microcrystalline fabric has been convincingly demonstrated; this erasure of the original depositional texture by blue-green algae suggests that these changes are not late diagenetic as proposed by many (Banner and Wood, 1964; Bathurst, 1964; Folk, 1965; Orme and Brown, 1963; and Wolf, 1965a, 1965b).

REFERENCES

Banner, F. T., and G. V. Wood, 1964, Recrystallization in Microfossiliferous Limestones, *Geol. Jour.* **4:**21-34.

Bathurst, R. G. C., 1964, The Replacement of Aragonite by Calcite in the Molluscan Shell Wall, in *Approaches to Paleoecology,* J. Imbrie and N. D. Newell, eds., Wiley, New York, pp. 357-376.

Branson, E. B., 1915, Origin of Thick Gypsum and Salt Deposits, *Geol. Soc. America Bull.* **26:**231-242.

Butler, G. P., 1969, Modern Evaporite Deposition and Geochemistry of Co-existing Brines, the Sabkha, Trucial Coast, Arabian Gulf, *Jour. Sed. Petrology* **39:**70-89.

Dean, W. E., and B. C. Schreiber, 1978, *Marine Evaporites,* Soc. Econ. Paleontologists and Mineralogists, Short Course Note 4, 188p.

Evans, G., 1966, The Recent Sedimentary Facies of the Persian Gulf Region, *Royal Soc. London Philo. Trans.,* ser. A, **259:**291-298.

Folk, R. L., 1965, Some Aspects of Recrystallization in Ancient Limestones, in *Dolomitization and limestone diagenesis, a symposium,* L. C. Pray, and R. C. Murray, eds., Soc. Econ. Paleontologists and Mineralogists Spec. Pub. 13, pp. 14-48.

Friedman, G. M., 1972, Significance of Red Sea in Problem of Evaporites and Basinal Limestones, *Am. Assoc. Petroleum Geologists Bull.* **56:**1072-1086.

Friedman, G. M., and J. Sanders, 1978, *Principles of Sedimentology,* Wiley, New York-Chichester-Brisbane-Toronto, 792p.

Ginsburg, R. N., 1956, Environmental Relationships of Grain Size and Constituent Particles in some South Florida Carbonate Sediments, *Am. Assoc. Petroleum Geologists Bull.* **40:**2384-2427.

Illing, L. V., A. J. Wells, and J. C. M. Taylor, 1965, Penecontemporaneous Dolomite in the Persian Gulf, in *Dolomitization and Limestone Diagenesis—A Symposium* L. C. Pray, and R. C. Murray, eds., Soc. of Econ. Paleontologists and Mineralogists Spec. Pub. 13, pp. 89-111.

Kendall, C. G. St. C. and P. A. d'E Skipwith, 1968, Recent Algal Mats of a Persian Gulf Lagoon, *Jour. Sed. Petrology* **38:**1040-1058.

Kinsman, D. J. J., 1964, The Recent Carbonate Sediments near Halat el Bahrani, Trucial Coast, Persian Gulf, in *Deltaic and Shallow Marine Deposits,* L. M. J. U. van Straaten, ed., Developments in Sedimentology, vol. 1, Elsevier, Amsterdam, pp. 185-192.

Krumbein, W. C., 1951, Occurrence and Lithologic Associations of Evaporites in the United States, *Jour. Sed. Petrology* **21:**63-81.

Landes, K. K., 1960, The Geology of Salt Deposits (and) Salt Deposits in the United States, in *Sodium Chloride—The Production and Properties of Salt and Brine,* D. W. Kaufmann, ed., Reinhold, New York, pp. 28-95.

Nurmi, R. D., 1974, The Lower Salina (Upper Silurian) Stratigraphy in a Desiccated, Deep Michigan basin, *Ontario Petroleum Institute 13th Ann. Conference Paper 13,* 26p.

Orme, G. R., and W. W. M. Brown, 1963, Diagenetic Fabrics in the Avonian Limestones of Derbyshire and North Wales, *Yorkshire Geol. Soc. Proc.* **34,** Pt. 1, No. 3, 51-66.

Purser, B. H., ed., 1973, *The Persian Gulf: Holocene Carbonate Sedimentation and Diagenesis in a Shallow Epicontinental Sea,* Springer-Verlag, New York, 471p.

Schmalz, R. F., 1969, Deep-Water Evaporite Deposition, A Genetic Model, *Am. Assoc. Petroleum Geologists Bull.* **53:**798–823.

Scruton, P. C., Deposition of Evaporites, *Am. Assoc. Petroleum Geologists Bull.* **37:**2498–2512.

Shearman, D. J., 1966, Origin of Marine Evaporites by Diagenesis, *Inst. Mining and Metallurgy Trans.* **75B:**208–215.

Shinn, E. A., 1971, Holocene Submarine Cementation in the Persian Gulf, in *Carbonate Cements*, O. P. Bricker, ed., The Johns Hopkins University Press, Baltimore and London, pp. 63–65.

Sloss, L. L., 1953, The significance of Evaporites, *Jour. Sed. Petrology* **21:**63–81.

Sloss, L. L., 1969, Evaporite Deposition from Layered Solutions, *Am. Assoc. Petroleum Geologists Bull.* **53:**776–789.

Wells, A. J., 1962, Recent Dolomite in the Persian Gulf, *Nature* **194:**274–275.

Wolf, K. H., 1965a, Grain Diminution of Algal Colonies to Micrite, *Jour. Sed. Petrology* **35:**420–427.

Wolf, K. H., 1965b, Petrogenesis and Paleoenvironment of Devonian Algal Limestones of New South Wales, *Sedimentology* **4:**113–178.

7

Copyright ©1963 by the Geological Society of London
Reprinted from *Geol. Soc. London Proc.* **1607**:63–65 (1963)

RECENT ANHYDRITE, GYPSUM, DOLOMITE, AND HALITE FROM THE COASTAL FLATS OF THE ARABIAN SHORE OF THE PERSIAN GULF

D. J. Shearman

The islands and lagoons of the coast of the Sheikdom of Abu Dhabi, part of the Trucial Coast of the Persian Gulf, are backed by a wide coastal flat, the *sebhka*, the surface of which stands only a few feet above high-tide level. The *sebhka* is built essentially of Recent uncemented calcite and aragonite sediments, e.g. skeletal sands, pellet sands, and carbonate mud, with accessory quartz and other terrigenous minerals. These sediments are similar to those accumulating at the present day in the intertidal zone and around the islands and lagoons, and the coastal flats appear to have built up and out as a result of Recent carbonate sedimentation. The sediments of the *sebhka* have been investigated on a reconnaissance scale by short cores, and shallow pits and trenches dug to the water table (depths down to 4 ft). Anhydrite, gypsum, dolomite, and halite have all been found in relative abundance in the sediments above the present water table.

Anhydrite has been identified only in the sediments of the *sebhka* above high-tide level. It has not so far been found anywhere in the intertidal zone, or in any of the lagoons. It occurs as bands and lenses and as layers of nodules. Individual bands range from a fraction of an inch up to approximately a foot in thickness. Often, more than one layer is present at any one place: in one pit, four bands and one layer of nodules with an aggregate thickness of 18 in. occurred within a depth of 4 ft. Bands of anhydrite were found which transgress the foreset bedding of the host sediment ; others are transgressive, not only to the bedding, but also to the surface. These transgressive relationships indicate that the anhydrite is diagenetic in origin.

Gypsum crystals are common. They have been found as scattered individuals and as lenses at various depths below the surface, and locally they occur in abundance at the surface. Several distinct crystal habits have been found. Some of these are restricted to the *sebhka*, but others are found also within and beneath algal mats in the present intertidal zone. Thus, there appears to be gypsum of more than one generation, some having been inherited from the tidal-flat stage of the evolution of the coastal flats. Many of the gypsum crystals are sand crystals, which poikilitically enclose the aragonite, calcite, and quartz grains, etc., of the host-sediment. Others, however, especially those that appear to have formed in the intertidal zone, are clear or enclose only relatively small amounts of sediment.

Dolomite was identified by X-ray methods, and is widespread. Wherever the dolomite was found, the sediments of the upper parts of the *sebhka* consist of aragonite and calcite, and this passes down, either gradationally or relatively abruptly, into very fine-grained dolomite.

Dolomite has not as yet been found in the sediments of the present intertidal zone of the Abu Dhabi coast, but in view of the occurrence described by Wells (1962) from the intertidal zone around Qatar, it may well occur.

Halite is usually present in small amounts in the top half-inch or so of the *sebhka*, where it occurs as an infilling between the sand grains.

In many places the surface of the *sebhka* is strongly blistered. When observed during January 1963, most of the blisters were cavernous, but some of them contained lenses of soluble salts up to half an inch thick.

The investigations are still in a preliminary stage, but the evidence so far indicates that the anhydrite, gypsum, dolomite, and halite are all early diagenetic in origin. Curtis & others (1963) suggested that in the absence of significant rainfall the ground-waters of the *sebhka* are fed from the highly saline marine waters of the coastal lagoons which percolate into the *sebhka* through the sediments of the intertidal zone. These ground-water solutions become concentrated by evaporation at the surface of the *sebhka*, and it is probable that a strong concentration-gradient exists in the solutions of the capillary zone above the water table. It appears unlikely that the dolomite, anhydrite, and gypsum were formed simply by the evaporation of the marine ground-waters of the *sebhka*; it is more probable that reactions between the solutions of the ground and capillary waters and the aragonite and calcite of the original sediment played an important contributory role.

The field-relationships of many ancient back-reef and lagoonal dolomites and evaporites are not fully consistent with the classical theory of an origin by evaporation of an enclosed body of sea-water. The occurrence of dolomite and evaporite minerals that apparently formed on a regional scale in sediments above sea-level may well provide a more acceptable explanation of ancient deposits. Many of these may be interpreted as expressing phases of emergence of the back-reef or lagoonal areas.

Acknowledgements. The specimens exhibited were collected by a research party from Imperial College as part of a study of the Recent sediments of the Trucial Coast financed by the D.S.I.R. The group included G. Evans, C. Kendall, D. J. J. Kinsman, and Sir Patrick Skipwith.

References

CURTIS, R., & OTHERS. 1963. Association of dolomite and anhydrite in the Recent sediments of the Persian Gulf. *Nature, Lond.* **197**, 679–80.

WELLS, A. J. 1962. Recent dolomite in the Persian Gulf. *Nature, Lond.* **194**, 274–5.

DISCUSSION

Dr L. V. ILLING commented that although it had not yet been proved that the anhydrite was forming at the present time, the position of some of the thinner seams parallel and close to the *sebhka* surface suggested this. If so, it was (as far as he was aware) the first time it had been recorded.

Theoretically there would appear to be no difficulty in the formation of anhydrite (rather than gypsum) by direct precipitation from warm brines. Van't Hoff's researches at the turn of the century indicated that it should be the stable phase above 25° or 30°C, according to the nature of the water. More recent considerations of the effect of the hemihydrate suggested that the temperature separating the fields of gypsum and anhydrite might be as high as 42°C (Groves, A. W. 1958. *Gypsum and anhydrite.* Overseas Geological Surveys, Mineral Resources Division, London, p. 5).

Whatever the critical conditions might be, whether it was a question of temperature or humidity or the composition of the water, the fact remained that anhydrite had not so far been seen in the speaker's studies of very similar *sebhka* sediments around the coasts of Qatar, less than 200 miles to the north-west. Instead, there were on parts of the *sebhkas* there a surface layer composed predominantly of moderate-sized, clear crystals of gypsum, formed in the ordinary way by precipitation from the concentrating sea-water. These were in contrast to the much larger, turbid gypsum crystals, full of carbonate inclusions, which occurred a foot or two down, just as they did in the cores demonstrated by Mr Shearman.

The sequence of sedimentary changes across these *sebhkas* in Qatar had been summarized by Dr A. J. Wells of the Koninlijke/Shell Exploratorie en Produktie Laboratorium, Rijswijk, the Netherlands (*Nature, Lond.* **194**, 274–5). All the features demonstrated by Mr Shearman, except the anhydrite, were present. In particular, there was the same vertical sequence, with aragonite muds and pelletoid sediments at a depth of two to three feet passing up into a zone one or two feet below the surface in which some or all of the aragonite had been replaced by microcrystalline dolomite. This was the level at which the big turbid gypsum crystals occurred, and it seemed likely that they were forming as a by-product of the conversion of aragonite to dolomite :

$$2\,CaCO_3 + Mg^{++} + SO_4^{--} + 2H_2O \rightarrow CaMg(CO_3)_2 + CaSO_4.2H_2O$$

The dolomite had the consistency of a stiff mud, and the partial obliteration of textural relations as well as the microscopic evidence of replacement showed that it was forming during early diagenesis. The parent carbonate material was the aragonite sediment laid down in the adjacent lagoons. The source of magnesium was the hypersaline water that occasionally flooded across and permeated through the *sebhka* sediment from the lagoons, becoming more and more concentrated by evaporation at the *sebhka* surface.

The exhibit, in fact, provided proof of penecontemporaneous dolomitization, which had been prophesied and taken for granted for over fifty years.

8

Reprinted from *Am. Assoc. Petroleum Geologists Bull.* **53**:841–869 (1969)

Holocene Shallow-Water Carbonate and Evaporite Sediments of Khor al Bazam, Abu Dhabi, Southwest Persian Gulf[1]

CHRISTOPHER G. ST. C. KENDALL[2] AND SIR PATRICK A. D'E. SKIPWITH[2,3]

Sydney, Australia

INTRODUCTION

The Khor al Bazam is part of the prograding coastal complex that forms the arid southwestern shore of the Persian Gulf (Fig. 1). The complex consists of a barrier, lagoons, and supratidal flats. The sediments are predominantly shallow-water carbonates and supratidal evaporites (Fig. 2). The carbonates consist of

[1] Manuscript received, November 15, 1967; accepted, March 29, 1968.

[2] Department of Geology, University of Sydney.

[3] Ocean Mining, Malaysia.

This paper is based on Ph.D. dissertations at Imperial College, London, England. The writers express their appreciation to the Ruler of Abu Dhabi, the Department of Scientific and Industrial Research, The Iraq Petroleum Company, the Trucial Oman Scouts, D. W. Gill, Douglas Shearman, G. Evans, Henry Nelson, and Volkmar Schmidt. Also, the writers thank those who helped with their advice and editing of the manuscript—R. L. Folk, Godfrey Butler, Henry S. Chafetz, and John Gries. The University of Texas Geology Foundation provided financial support for preparation of this report.

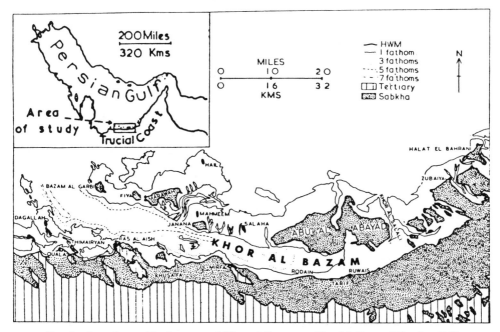

FIG. 1.—Location and bathymetry of Khor al Bazam on Trucial Coast of Persian Gulf.

217

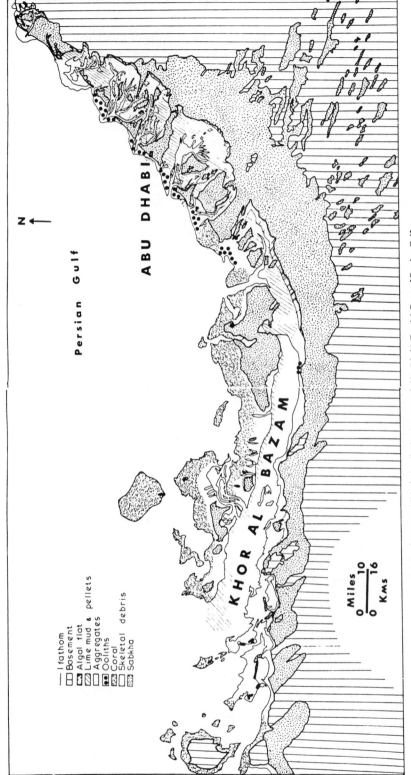

Fig. 2.—Interpreted lithofacies distribution around Abu Dhabi, Trucial Coast of Persian Gulf.

Persian Gulf

ABU DHABI

KHOR AL BAZAM

N

1 fathom
Basement
Algal flat
Lime mud & pellets
Aggregates
Ooliths
Coral
Skeletal debris
Sabkha

Miles
0   10
0   16
Kms

218

[*Editors' Note:* In the original, Figure 2 appears across the facing pages 842 and 843.]

oölites, grapestones, pellets, muds, and a variety of skeletal grains. The evaporites are anhydrite, gypsum, and dolomite. The carbonates pass landward into eolian quartz sand (Evans, 1966). Modern environments comparable to the Trucial Coast have not been reported, but analogues are recorded in the Permian rocks of West Texas, the Cretaceous of central Texas, and the Permian of northeast England.

Laboratory Techniques

Samples collected from the Khor al Bazam were sieved to facilitate identification of the component grains. A split of approximately 30 g made by coning and quartering was washed, agitated, and wet-sieved over a 0.066-mm-diameter sieve. The silt and clay grades were dried and weighed; the sand grade was dried, sieved, and weighed.

No attempt was made to remove the organic matter from the sieve samples because hydrogen peroxide disrupts the microcrystalline grains (Cloud, 1962; Ginsburg, 1956). A separate split was treated first with hydrogen peroxide and then with hydrochloric acid to find the percentage of organic matter and insoluble residue.

The percentages of the component grains were calculated for the several size grades of the samples. Visual estimation was used for most of the samples. Thus, the gross composition of the sediment was used to establish its general character (Emrich and Wobber, 1963; Folk and Robles, 1964).

Thin sections were made to establish the internal texture of the component grains. Feigel's solution and cobalt nitrate stains were used to distinguish calcite from aragonite, but the presence of organic mucilage interfered with their action.

Grain-Size Distribution

The pattern of grain-size distribution in the Khor al Bazam coincides with changes in relief of the sea floor. Sediments of the shelf areas generally are well sorted and lagoonal sediments are poorly sorted. Most of the Khor al Bazam sediments are coarse to fine sand, and are positively skewed about a mode in the medium-sand grade. However, in the lee of the offshore bank, at the western end of the Khor al Bazam opposite Bazam al Garbi (Fig. 1), and in mangrove swamps, the silt and clay content of the sediments in many places exceeds 30 percent (Fig. 3). These fine constituents are probably products of transport and deposition from the offshore bank and of direct precipitation.

Ginsburg (1956) used the percentage of material $< \frac{1}{8}$ mm as the most critical environmental indicator in Florida Bay and the adjacent reef tract. The percentage of material $< \frac{1}{16}$ mm has been found to be much more sensitive to environment in the Khor al Bazam

(Fig. 3). This distribution coincides with an area of mud mounds fixed by *Halodule*, much like the mounds described in Florida Bay by Ginsburg (1956).

SEDIMENTS

Most of the sediments found in the area of the Khor al Bazam are bioclastic. Many species but few phyla are represented. The skeletal material is composed of either aragonite or high- or low-magnesium calcite. Nonskeletal carbonates are composed of aragonite, which probably is derived biochemically but not from direct secretion by biota (Kinsman, 1964a). Using strontium content as an indicator, Kinsman found that the aragonite mud in a lagoon southwest of Abu Dhabi is not abraded skeletal material. However, he was not able to establish whether this aragonite was precipitated by evaporation and concentration of marine waters through physicochemical means or was precipitated within plant microenvironments during photosynthesis. The writers suggest that in the Khor al Bazam blue-green algae are probably the major agency of precipitation of nonskeletal aragonite.

Algally Induced Alteration of Carbonates

In common with other Holocene shallow-water carbonates, some of the sediments of the Khor al Bazam show a distinctive type of alteration that affects all carbonate grains, irrespective of their origin (Kendall *et al.*, 1966). The texture of the grains is altered to homogeneous microcrystalline fabrics of aragonite which, in the final stage, are virtually indistinguishable from each other. In many instances the external shape may be the only indication of the grains' origin. This process has been observed in the Bahamas (Illing, 1954; Purdy, 1963), Portuguese Timor (Wolf, 1965b), the Persian Gulf off Qatar (Houbolt, 1957), and in British Honduras (Purdy, 1965).

Illing (1954) attributed the alteration to bacteria or algae. In contrast, Newell *et al.* (1960) ascribed altered zones in oöids to the effects of decaying colonies of boring algae. Later Purdy (1963) concluded that organic matter trapped within the grains promotes the process. Bathurst (1966) ascribes alteration to algae.

In Holocene sediments of the Khor al

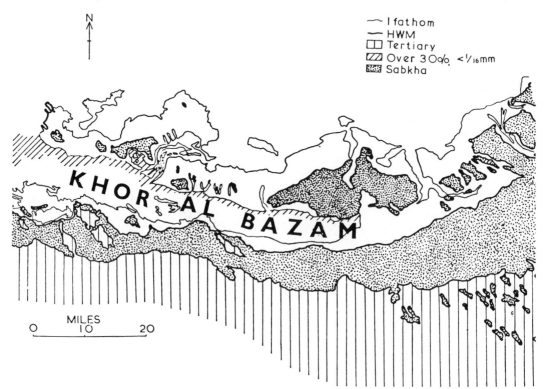

FIG. 3.—Distribution of clay and silt-size carbonate in Khor al Bazam, Persian Gulf.

Bazam, Foraminifera and red calcareous algae seem particularly susceptible to alteration, whereas thick-walled shells and shells with a more coarsely crystalline fabric are less susceptible. Alteration in the latter generally is confined to the surface, but a few lobes of fine-grained carbonate extend deep into the shell.

Alteration also occurs in aggregates of the microcrystalline carbonate that forms fecal pellets and grapestones. These aggregates are initiated as poorly bound, open-textured grains that are rapidly hardened and cemented as they alter. Oölites are similarly affected.

Crusts of radial acicular aragonite crystals form in small internal cavities of some mollusk shells, Foraminifera, and grapestone lumps. These infillings generally alter into disorganized microcrystalline mosaics of aragonite.

In ancient limestone many components have a fabric similar to that of the Holocene altered materials, but they are preserved as calcite. Banner and Wood (1964) attributed this to late diagenetic grain growth; Orme and Brown (1963) and Folk (1965), apparently describing the same phenomena, postulated a late diagenetic process of "grain diminution," confirming some of Bathurst's (1964) and Wolf's (1965a, b) findings. It is demonstrated that these changes are not exclusively late diagenetic, but are molded on alterations of depositional fabric.

The various stages in the process of alteration can be studied most usefully in peneroplids. Peneroplid Foraminifera are particularly susceptible to alteration (Murray, 1966). Detailed study of specimens collected from the Khor al Bazam showed all stages of alteration. The tests show a series of surface changes, the surfaces ranging from translucent and porcelaneous on fresh tests to opaque and saccharoidal on altered specimens (Fig. 4A). These changes are accompanied by rounding and pitting of tests. Many of the pits are seen to contain blue-green algae. Alteration may be so pronounced that tests are almost indistinguishable from aragonite ovoids of fecal and accretional origin. Both Illing (1954) and Purdy (1963) noted similar changes in the Foraminifera of the Bahamas.

Thin sections of fresh peneroplid tests under plane polarized light show a brown body color. They contain no crystal shapes apart from a faint structure parallel with the walls (Figs. 5–7) and a very faint granulation under high power. In reflected light they are milky white. Through crossed nicols the tests are seen to be composed of parallel sheaths of crystals which show low polarization colors and lie parallel with the curvature of the walls. The sheaths are embedded in a finely granular matrix of crystals which show pin-point polarization in grays and yellows of the first order.

Thin sections of fresh tests stained with Feigel's solution suggested that the tests were not aragonite. X-ray analysis by R. Curtis, Imperial College, London, showed the tests to be of high-magnesium calcite with a small amount of aragonite. The aragonite was either part of the tests or, more probably, an infilling of the chambers. If the second assumption is correct, the result of the analysis confirms the findings of Sollas (1921) and Wood (1948).

The brown body color and the low-order polarization colors of the fresh tests are not optical properties of high-magnesium calcite. Several theories have been proposed to explain these anomalies. For example, Sollas (1921) attributed the brown body color in plane polarized light and the milky color in reflected light to the reflection of light through the felted crystals of the test. He did not explain the low-order polarization colors. In contrast, Wood (1948) attributed the brown body color to small amounts of lead and iron. However, Frondel et al. (1942) were unable to trace any connection between the distribution of Fe, Cu, Mn, Al, Sr, and Mg and the morphology or color of calcite crystals. Wood (1948) also proposed that the low polarization colors are produced by small crystals optically compensating each other in the slide, but this is unlikely, because thin sections of micrite (Folk, 1962) do not show low polarization colors. Folk (personal commun., 1966) points out that chalcedony and "turbid" feldspar also are brown in transmitted light and are milky under reflected light. These optical properties are interpreted as being caused by minute water-filled vacuoles (Folk, 1955; Folk and Weaver, 1952). The same properties in the peneroplids may be of similar origin. Electron micrographs of a peneroplid test in a paper by Hay et al. (1963, Pls. 5, 6) show the test to be composed of a felt of crystals about 0.5 μ long and 0.1 μ in diameter. The disorganized arrangement of some of the crystals should leave voids at least 0.1 μ in diameter. The voids probably are filled with water which produces a brown body color as in the chalcedony and "turbid" feldspars. The presence of the water may be related to the low refractive index of the calcite perpendicular to the C axis (1.6), the low polarization colors,

221

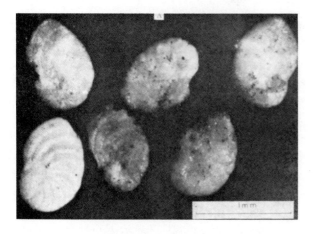

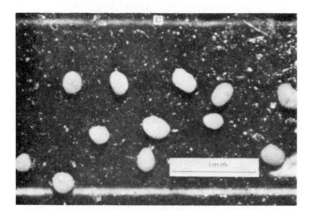

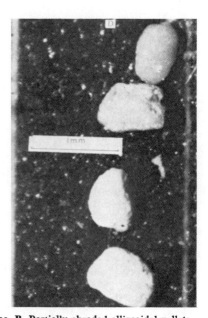

FIG. 4.—**A.** Peneroplid Foraminifera being altered by blue-green algae. **B.** Partially abraded ellipsoidal pellets. **C.** Spherical pellets. **D.** Irregular pellets. **E.** Quartz with carbonate jackets.

and the low specific gravity of 2.724 (Sollas, 1921, p. 196). However, it does not explain why the test shows low polarization colors and does not exclude the possibility that the low specific gravity may be caused by the presence of organic matter (Sollas, 1921, p. 196). If the voids were filled with air, the test would be black.

In thin sections, tests show that alteration proceeds from discrete patches that become enlarged and gradually obliterate the original texture. This process may be accompanied by the infilling of the chambers by microcrystalline aragonite (Figs. 8, 9). The altered areas are characteristically microgranular and lack the brown body color shown by fresh tests. Under crossed nicols the altered areas show high polarization colors, but do not extinguish. The size of the crystals is of the order of 1.5 μ (Fig. 8, C). Feigel's solution reacted only patchily

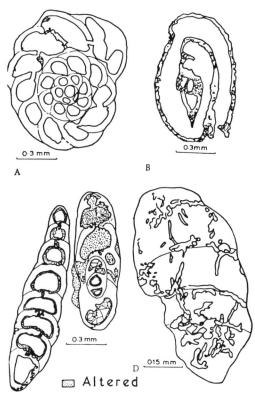

Altered

FIG. 6.—Sketch of foraminifers of Figure 5 under plane polarized light.

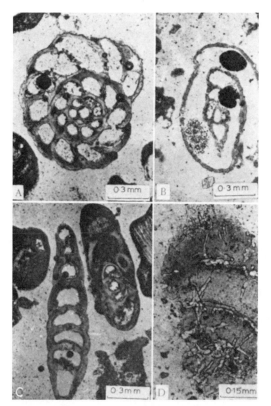

FIG. 5.—**A.** Modern peneroplid in plane polarized light. Unaffected by algae. **B.** Modern *Quinqueloculina* under plane polarized light. Partly bored by algae. **C.** Modern peneroplids under plane polarized light; aragonite rims chambers. **D.** Modern peneroplid under plane polarized light. Bored by algae.

on completely altered foraminifers, but X-ray analysis showed that those tests are almost entirely aragonite. The failure of Feigel's test may be the result of films of organic matter covering the crystals.

Hand-picked peneroplids from the medium-sand fraction of one of the samples were separated into five groups on the basis of the extent of external alteration. X-ray analyses showed that the five groups progressively changed from high-magnesian calcite to aragonite.

The algae growing on some of the peneroplids also were found by John Twyman (personal commun., 1964) on the oölites from Abu Dhabi. They were identified by Dr. Stewart of Westfield College, who found them to be *Entophysalis deusta* (Menegh), Drouet and Daily. Newell *et al.* (1960) found this genus common in the Bahamas.

Algae collected in Abu Dhabi survived 3 years in storage jars. Upon exposure to sunlight they started growing vigorously. The algae of several samples were separated from the associated carbonate sand by a solution of weak

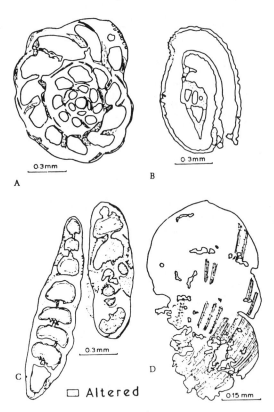

□ Altered

FIG. 7.—Sketch of foraminifers of Figure 5 under crossed nicols.

HCl. They consisted of several different types of globular cells and tangles of filaments in a transparent mucilaginous matrix, believed to be mainly a secretion of algae (Fritsch, 1952).

The algal cells and filaments were found to lie just beneath the surface of the altered tests, as shown by (1) progressively dissolving the carbonate of the grains in 5 percent acetic acid or 0.2 percent HCl, and (2) rendering the surface transparent with 40 percent hydrofluoric acid. Algae could be made even more apparent by staining with malachite green, which stains algal cells and the mucilage with different intensities.

If the unaltered tests are completely decalcified, they leave behind a thin, diaphanous, soft elastic membrane that retains the original form of the Foraminifera. This is the "tectin" of Hyman (1940) and is believed to be the original organic material of the test which, in life, occurred in conjunction with the calcite.

Decalcified altered Foraminifera leave behind a much more translucent material, which also molds the test and fills its chambers, but is markedly different in appearance from the "tectin." In all ways it resembles the mucilage that is present in association with the blue-green algae. It contains algal cells and small quantities of minute mineral grains which probably adhered to the mucilage as the test rolled on the sea floor.

If Foraminifera are treated with 40 percent hydrofluoric acid for more than 10 minutes, all the calcium carbonate is replaced, molecule by molecule, by transparent calcium fluoride (Grayson, 1956). The mucilage envelope and algae resist solution (Fischer, 1897) and consequently stand out as translucent and green in altered areas. Treatment with 40 percent hydrofluoric acid for 5 minutes clarifies the surfaces of the Foraminifera and reveals the network of radiating filaments and globular algae. Fresh Foraminifera become completely transparent. The best optical results with the calcium fluoride specimens are obtained by viewing them under water.

Partially altered Foraminifera examined in thin section show the algal filaments penetrating both the unaltered and altered areas (Figs. 5–7). The filament "bores" generally are filled with microcrystalline aragonite, which preserves the filament shapes. Thus the unaltered test may be penetrated by tubes of altered material. The boundaries separating the unaffected high-magnesian calcite areas of the test from the aragonitic altered areas commonly resemble the boring of algal filaments (Bathurst, 1966). Under plane polarized light parts of the test adjacent to the bore shapes may show faint granulation, which is believed to be the first sign of alteration.

Staining thin sections of the altered material with aqueous malachite green results in a patchy effect. Deeper colors are present at the outside edges and in the chambers devoid of carbonate. Etching thin sections with weak acid solutions of malachite green dissolves the calcium carbonate, leaving the stained insoluble algae and mucilage. The deeper stain is usually at the edge of the former test and in some chambers, showing algal cells and filaments both at the surface and penetrating the interior. Parts of the chambers not filled with algal cells generally are stained less intensely, an indication that they are filled with mucilage. Solution and staining of the altered walls of the Foraminifera leave only a trace of mucilage that stains lightly, possibly because the microcrystalline aragonite crystals displace the mucilage as they form. The presence of this finely dispersed

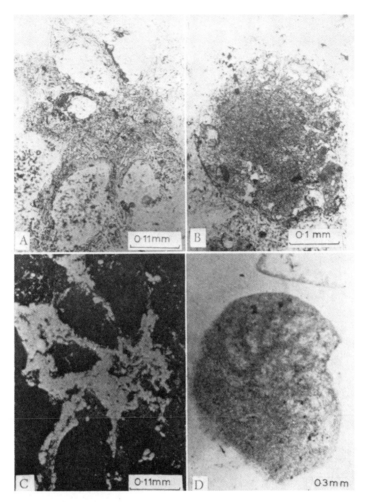

Fig. 8.—**A.** Modern peneroplid under plane polarized light. Well bored by algae. **B.** Modern peneroplid under plane polarized light totally altered. Chambers are unfilled. **C.** Modern peneroplid (as above) under crossed nicols. Algal bores filled by microcrystalline aragonite. **D.** Modern peneroplid under plane polarized light. Totally altered. Chambers are filled by acicular aragonite which also is being altered.

mucilage is confirmed by gently dissolving the whole tests. If the dissolution is too vigorous, the mucilage is ruptured and removed.

The intimate relation of blue-green algae with alteration can be demonstrated. All plants photosynthesize and respire. Dalrymple (1965) observed the ability of blue-green algae to precipitate calcium carbonate. Parks and Curl (1965) showed in the laboratory that a culture of blue-green algae in sea water produces measurable change in the conductivity of the water between day and night. They inferred that, in light, photosynthesis transforms bicarbonate ions to carbonate, and in the dark respiration causes the opposite effect.

In the peneroplids of the Persian Gulf, the mucilaginous envelope generally creates a microenvironment within each test. Carbon dioxide given off during respiration would promote solution of calcium carbonate in this restricted environment. Conversely, the carbon dioxide utilized during photosynthesis would cause precipitation. Although high-magnesium calcite is dissolved, it is aragonite that is precipitated because it is apparently the stable form of calcium carbonate under such conditions (Kinsman, 1964b). In this way the high-magnesium calcite of the Foraminifera test is progressively dissolved and replaced by aragonite on a piecemeal basis. Once a foraminifer is altered, solu-

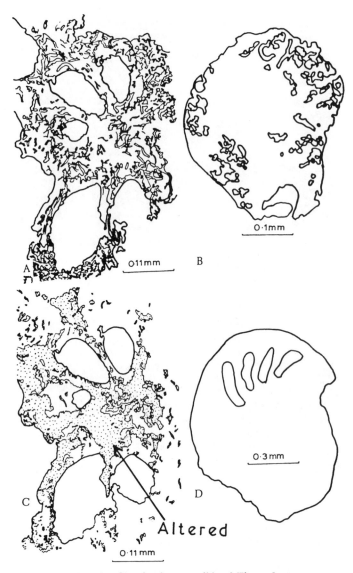

Fig. 9.—Sketch of peneroplids of Figure 8.

tion and reprecipitation of aragonite continue (Fig. 8), ultimately destroying all evidence of its origin.

This process alone could account for alteration. Bathurst (1964) suggested that alteration is accomplished by algae boring and reboring the calcite test, and that solution of the test occurs only where it is in direct contact with the algal filaments. However, the fact that the algae are restricted to the area near the surface of the test and do not necessarily extend into all the affected parts argues against alteration as the result of boring alone. Also, the mucilagi-

nous envelope secreted by the algae confines the carbon dioxide, which thus can react at any point within the envelope.

The alteration of carbonate grains could be controlled by several factors—the length and frequency of exposure of the grains to direct sunlight, the dimensions of the grains, and the size of their component crystals. For example, if the grain is buried quickly, it may not have time to alter or will only alter if it has a thin, delicate shell and small component crystals.

The shoal areas of the Khor al Bazam from which the altered Foraminifera were collected

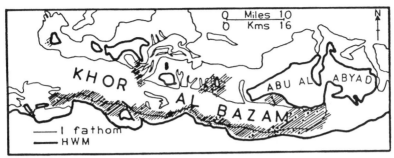

FIG. 10.—Distribution of algally altered peneroplids in Khor al Bazam Persian Gulf. After Murray (1966).

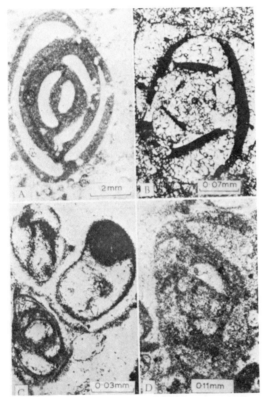

FIG. 11.—**A.** Tertiary miliolid under plane polarized light, exhibiting algally produced bores. Rest of test, though unaffected by algal alteration, has lost original body color. **B.** Tertiary miliolid under plane polarized light, exhibiting effects of algal alteration. Dark unaffected parts of test have lost distinctive body color and are finely granular. Affected test was probably preserved in remains of algal mucilage, which was dispersed when aragonite went into solution during first phases of cementation. **C.** Quaternary Foraminifera under plane polarized light, exhibiting results of algally induced alteration. Upper two are preserved by algal mucilage but lower one is only partly affected by algae and still exhibits its body color. **D.** Tertiary miliolid under plane polarized light almost unrecog-

(Fig. 10) are rippled. If the ripples migrate slowly, the Foraminifera being altered will be exposed only occasionally to direct sunlight. Consequently the alteration process may be slow. Alteration is most effective on the weed-covered shoals of the coastal terrace and parts of the offshore bank. Skeletal grains from those areas have abundant algally-filled pits. The prolific plant growth must upset the carbon dioxide balance of the sea water and alteration therefore may be more rapid.

Wolf and Conolly (1965, p. 108) pointed out that algae are generally good indicators of shallow water because photosynthesis is depth controlled. Thus, alteration would not be expected to take place below the photic zone. Off Qatar, bioclastic material is rounded and the surface character obliterated in water less than 20 m deep (Houbolt, 1957).

In rocks from similar ancient carbonate environments, the microcrystalline aragonite of the altered material generally is replaced by a mosaic of microcrystalline calcite in which algal alteration is still recognizable (Figs. 11–14; Shearman and Skipwith, 1965). The altered material commonly forms Bathurst's (1966) micritic envelope.

Blackening of Grains

In the central and western parts of the Khor al Bazam, shells and other carbonate particles are commonly blackened. Houbolt (1957) noted blackening in sediments off the Qatar peninsula. Studies in the Khor al Bazam suggest that discoloration of organic matter is the result of chemical reactions a few inches below

nizable because of algal alteration. Algally affected areas are preserved by coarser calcite than original test which was darker and of finer grain.

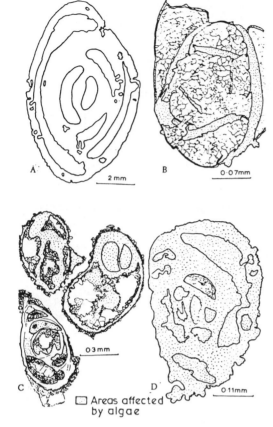

FIG. 12.—Sketch of miliolids of Figure 11.

Calcium carbonate grains
1. Mollusk shells and fragments
2. Ostracod valves
3. Foraminifera tests
4. Calcareous algae
5. Coral fragments
6. Bryozoan skeletal structures
7. Fecal pellets
8. Oölites
9. Unidentified particles

Grains other than calcium carbonate
10. Quartz
11. Heavy-mineral grains
12. Unidentifiable sugary brown grains
13. Evaporite minerals
14. Flocculent material

Mixed noncarbonate and carbonate grains
15. Aggregates

The following minor components were iden-

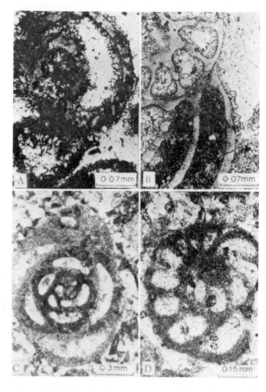

FIG. 13.—**A.** Quaternary *Quinqueloculina* under plane polarized light. Partially altered by algae. Small calcite crystals rim test and fill mucilage. **B.** Quaternary peneroplid under plane polarized light. Partially altered by algae. Small calcite crystals rim test and infill mucilage caught in chambers. **C.** Tertiary miliolid under plane polarized light. Outer part of test shows evidence of algal alteration preserved in calcite. Central darker area has lost original body color. **D.** Tertiary miliolid under plane polarized light. Totally altered by algae and preserved in calcite.

the sediment surface in the reducing zone. All stages of blackening can be traced to organic matter containing living blue-green algal cells. The blackest grains generally are in the center of the lagoon. In ratio of components, the blackened sand and the unblackened sand are the same (Murray, 1966); contemporaneous origin therefore is suggested (Fig. 15). The fact that benzene bleaches the black grains indicates the organic nature of the black color. At a few sample localities, particularly on rises on the lagoon floor, the grains are brown instead of being blackened. Tests made on all types of discolored grains for phosphate and iron were negative.

Components

The following major components were identified in the sediments from the study area. A major component constitutes more than 1 percent of the sieve grade.

tified. All form less than 1 percent of sieve fractions.

1. Serpulid tubes
2. Sponge spicules
3. Echinoid fragments
4. Fish bones
5. Crustacean fragments

The sediment components are similar to those of the Bahamas (Illing, 1954, p. 17, 22) and Florida Bay (Ginsburg, 1956, p. 2421). There are exceptions. Bryozoan skeletal structures are present in greater amounts than in Florida and the Bahamas, and detrital grains of quartz and other noncarbonates are not present in Florida or the Bahamas, nor are gypsum or anhydrite crystals. In the following description each sediment component is related to the major lithofacies units of the Khor al Bazam (Fig. 16) which are separated on the basis of the component content and texture. They can be equated with the geomorphic units as follows.

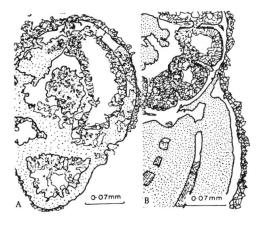

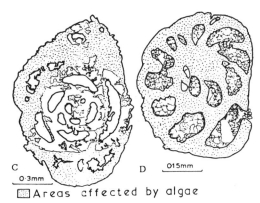

Areas affected by algae

FIG. 14.—Sketch of foraminifers of Figure 13.

Molluscan sand ....Lagoon and open sea
Oölitic sand ....Coastal terrace
Pellet aggregate and pellet sand ..Coastal terrace and south edge of off-shore bank
Coral and coralline-algae sand ..Seaward shoals and outer edge of off-shore bank
Fecal-pellet sand ....Southern shoal and channel area
Carbonate mud and pellets .....Protected lagoon

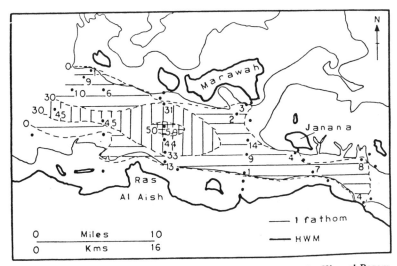

FIG. 15.—Ratio of blackened to white and black peneroplids in the Khor al Bazam, Persian Gulf. After Murray (1966).

229

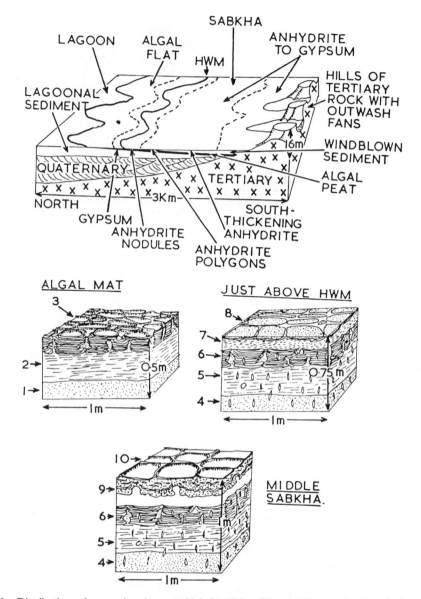

Fig. 16.—Distribution of evaporites in supratidal flat lining Khor al Bazam, Persian Gulf. **1.** Lagoonal carbonate sands and/or muds. **2.** Poorly laminated carbonate rich algal peat. **3.** Algal mat formed into polygons. **4.** Lagoonal sediments with gypsum crystals. Possible early dolomite present. **5.** Algal peat with large gypsum crystals. **6.** Algal peat with large gypsum crystals. **7.** Mush of gypsum and carbonate. **8.** Anhydrite polygons and windblown carbonate and quartz. **9.** Anhydrite layer replacing gypsum mush and forming diapiric sructures capped by anhydrite layers and nodules in windblown carbonate and quartz. **10.** Halite crust formed into compressional polygons.

	and mangrove swamp
Algal mat	.................Intertidal
Evaporites	.................Supratidal

Each of the lithofacies is related to the hydrodynamics and water chemistry of the environment. Flora and fauna, though affected by the same physical agents, also control lithofacies.

Mollusk shells and fragments.—The following mollusks collected in the Khor al Bazam were identified by J. Biggs of the British Museum.

Bivalvia

Angulus (Fabulina) rhomboides Quoy and Gaimard
Angulus (Fabulina) immaculata Philippi
Arca plicata (Chemn) Dillwyn
Arca lacerata Linné
Arca tortuosa Linné
Arcopagia robusta Hanley
Brachidontes variabilis Krauss
Cardium sueziensis Issel
Chlamys ruschenbergeri Tryon
Circe corrugata (Dillwyn)
Circe scripta (Linné)
Codakia fischeriana Issel
Corbula acutangula Issel
Corbula subquadrata Melvill
Crenella adamisiana Melvill and Standen
Diplodonta ravayensis Sturany
Diplodonta sp.
Dosinia histrio Gmelin
Dosinia sp.
Grastrochaena cuneiformis Spengler
Glycymeris hoylei Melvill
Glycymeris pectunculus (Linné)
Glycymeris spurcus Reeve
Lima tenuis A. Adams
Lima (Limatula) leptocarya Melvill
Lioconcha picta Lamarck
Laevicardium papyraceum (Bruguière)
Lucinae dentula (Linné)
Mactra cf. *olorina* Philippi
Macoma jeanae Dance and Eames
Malleus regula Forskal
Meretrix sp.
Motirus irus (Linné)
Nuculana confusa (Hinds)
Phacoides semperianus (Issel)
Pinna sp.
Pinctada radiata (Leach)
Pitaria spp. (incl. *hagenhowi* Dunker)
Quadrans pristis (Lamarck)
Solenocurtus strigullatus (Linné)
Spondylus exilis Sowerby
Sunetta effosa Hanley
Tapes undulata Born
Tellina pygmaea Loven
Tellydora pellyana A. Adams
Timoclea sp.
Trachycardium lacunosum Reeve
Trachycardium maculosum (Wood)
Venus (Callanaitis) calophylla Hanley

Gastropoda

Atys cylindrica Jan
Ancilla cinnamomea (Sowerby)
Ancilla eburnea Deshayes
Bullaria ampulla (Linné)
Calyptraea pellucida Reeve
Cerithidea cingulatus Gmelin
Cerithium morus Lamarck
Cerithium scabridum Philippi
Cerithium petrosum Wood
Clava (Clava) fasciata (Bruguière)
Columbella sp.
Cypraea caurica Linné
Cytherea sp.
Diodora funiculata (Reeve)
Drupa margariticola (Broderip)
Euchelus bicinctus Philippi
Finella pupoides A. Adams
Finella scabra A. Adams

Laemodonta (Laemodonta) bicolor (Pfr.)
Minolia gradata Sowerby
Minolia holdsworthiana (Nevill)
Monilea obscura (Wood)
Murex (Chicoreus) anguliferus Lamarck
Mitrella (Mitrella) blanda (Sowerby)
Perrinia stellata (A. Adams)
Persicula sp.
Phasianella nivosa Reeve
Phasianella sp.
Pterigia sp.
Pyrene (Seminella) phaula (Melvill)
Retusa omanensis Melvill
Ringicula propinquans Hinds
Rissoina sismondiana Issel
Rissoina distans (Anton)
Rissoina savignyi Jousseaume
Scaliola arenosa A. Adams
Scaliola elata Semper
Strombus sp.
Tricolia foridlana Pilsby
Trochus (Infundibulops) erythraeus Brocchi
Turbo radiatus Gmelin
Turbo coronatus Gemelin
Turitella aurocincta v. Martens
Umbonium vestiarium Linné
Xenophora caperata Philippi

The mollusks are so distributed that the heavy, robust forms, mainly gastropods, are on the coastal terrace and bank, but thin-shelled lamellibranchs, commonly still articulated, dominate the lagoon sediments. The shells are commonly uncolored or lightly colored, as are those of the interior of the Bahama Banks (Illing, 1954, p. 18).

It was found that many of the shells are being destroyed by boring and crushing organisms and by waves. Boring organisms include small mollusks, worms, sponges, and blue-green algae. Crushing organisms include the Crustacea which, though varied in species, roam the whole of the Khor al Bazam and offshore bank. Crabs are particularly active on rock platforms covered by green algae. At low tide the sound of shells being crunched is common. Breakage caused by waves is confined largely to the edges of the offshore bank and the coastal terrace.

Mollusk shells normally form a high percentage of the sediments on, and directly in the lee of, the offshore bank, in the center of the lagoon, and on the coastal terrace. On the bank and terrace the coarse shells and fragments are probably formed *in situ* and are not being carried into the lagoon. Similarly, the thin-shelled mollusks of the lagoon are probably *in situ*. Articulated shells and lack of abrasion support this belief. The most clearly defined of the mollusk environments is lee of the offshore bank, where molluscan development coincides with the higher percentages of carbonate mud. Near the head of the eastern Khor al Bazam and beyond the offshore bank where currents are active, the shells are better sorted and free of flocculent material. There short transport may take place.

From Khusaifa westward the same general pattern continues. It is affected by changes in relief of the lagoon floor; blue-green algae are most active on areas of high relief in the midlagoon, and the apparent percentage composition of mollusks decreases over those areas.

The highest concentration of shells is found in certain beaches, where they form nearly 100 percent of the components in the coarser grades. Common to most beaches are cuttlefish skeletons. Blackened shells are common in the central parts of the west Khor al Bazam.

Ostracods.—Ostracods generally are articulated in the lagoon samples, but in shelf samples they are disarticulated and bored. The ostracods appear to be living in the central lagoon and in the muddy areas in the lee of the bank.

Foraminifera.—The Foraminifera collected in the Khor al Bazam were identified by Murray (1966). He found very few living individuals in the samples. Representatives of the superfamily Miliolacea make up most of the populations, the dominant forms being species of *Quinqueloculina, Triloculina,* and *Peneroplidae.* He showed their areal distribution and suggested they are mostly untransported death assemblages. He concluded that the Foraminifera must be living on the seaweed, which had not been sampled.

The Foraminifera generally are spread evenly over the lagoon. However, just to the lee of the bank Foraminifera transported from the bank cause a small rise in percentage composition. In the intertidal areas of the coastal terrace and certain beaches, the Foraminifera percentages are high, but they decrease landward over algal flats. In the shoal areas around the margins of the Khor al Bazam, Foraminifera tests may be partially or wholly altered by the activities of blue-green algae (Fig. 10).

The following representatives of the superfamily Miliolacea are present on Khor al Bazam, Trucial Oman Coast (after Murray, 1966).

Quinqueloculina sp.	20–30% (lagoon) and algal mat
Triloculina sp.	20–30% (6-ft depth) in lagoon
Peneroplis pertusus	10–29% intertidal at head of Khor al Bazam
Peneroplis planatus	10–32% lagoon subfacies
Spirolina aciculavia	10–23% (shelf)
Spirolina arietna	10%

Articulina sp.	10% (lagoon)
Parrina bradyi	1–2%
Miliolinella sp.	1–2%
Spiroloculina sp.	
Ammonia beccani	1–10% (shelf)
Elphidium	

Murray (1966) also discussed the blackening of Foraminifera. His map of the distribution of this phenomenon is shown in Figure 15.

Calcareous algae.—Extensive encrustations of robust red calcareous algae including *Litho-thamnium* sp. and *Archaeolithothamnium* sp. are living along the reef front of the offshore bank and parts of the edge of the coastal terrace. They undergo rapid alteration, becoming indistinguishable from the microcrystalline grains a short distance from their source areas. Illing (personal commun., 1966) reported that the calcareous algae *Goniolithon* are common in reefs in Qatar. It is probable that these algae are present widely in the reefs of the west Khor al Bazam. The delicate red calcareous alga, *Jania,* is a more important component, particularly in finer grades of sand. Over much of the coastal terrace it grows in small colonies that break up to form rodlike grains. The influence of the algal detritus begins at the headland west of Mirfa (Fig. 1) and continues to the terrace just south of Bazam al Garbi, a distance of 40 km. The percentage of algal fragments decreases abruptly lagoonward and increases shoreward.

Like other red algae, rods of *Jania* are subjected to blue-green algal alteration, but remain recognizable because of their distinctive shape.

Coral fragments.—It was very difficult to identify coral fragments with the stereoscopic microscope, partly because of rapid alteration of the grains. Where alteration has not occurred, the grains abrade readily and thus their characteristic surface texture is destroyed.

Many of the components which were classed as unidentifiable on the offshore bank and coastal terrace may be of coralline origin. Offshore-bank samples have high percentages of coral when traced across the bank from the reefs.

Bryozoan skeletal structures.—Mrs. Patricia Cook of the British Museum identified the following specimens of Bryozoa.

Cheilostomata, Anasca: *Thalamoporella gothica* var. *indica* Hincks
Ctenostomata, Carnosa: *Sundanella sibogae* Harmer

Thalamoporella gothica is very common, and grows as encrustations on the stems of the "bladder wrack" seaweed. It forms a very light-weight cellular structure that accumulates as detritus in the areas of maximum development of bladder wrack, particularly on the terrace and southern part of the west Khor al Bazam between Mirfa and Khusaifa. When the bladder wrack breaks free from the sea floor, it drifts as "Sargassum-like" rafts which eventually wash up on the shore. The strandline of the beaches is lined by accumulations of this weed, which becomes incorporated in beach ridges as they build seaward. The weed rots in beach ridges and leaves *Thalamoporella gothica* to form lenses 20 cm or more thick.

Pellets.—Three types of pellets are forming in the Khor al Bazam. They range in size from 0.05 to 1 mm and may be cylindrical and ellipsoid (Fig. 4B), spherical (Fig. 4C), or irregular (Fig. 4D).

Solution of all pellets with HCl leaves a mucilaginous sludge which commonly contains cells of blue-green algae. The process by which the pellets are hardened may be related to the organic sludge, particularly if it includes blue-green algae.

Cemented ellipsoid and cylindrical pellets of aragonite mud are rare in most Khor al Bazam samples, but are numerous just seaward of algal mats. Pellets have been found in the guts of some worms. This suggests that they are of fecal origin and are formed by the myriads of worms that burrow in the sediment. Alternatively, some are excreted by mollusks. Kornicker and Purdy (1957) and Folk and Robles (1964) found that the gastropod *Battilaria minima,* a species of cerithium, produced pellets similar in shape to the ones observed in the Khor al Bazam.

The pellets are initially very soft. If they are subjected to violent wave or current movement, they are easily disrupted, but if conditions are quiet, the pellets are cemented and harden. In the Bahamas, Illing showed this hardening to be very rapid. The fact that only a few soft pellets are present in the Khor al Bazam suggests that the hydrodynamic environment is too violent to allow many to be preserved. The surfaces of some of the grains suggest that they were abraded during cementation (Fig. 4B). A similar occurrence of pellets was recorded in the "crab flats" near Abu Dhabi by Kinsman (1964b).

Spherical pellets are distinguished from the fecal pellets by their high sphericity, and are more common than fecal pellets in the Khor al Bazam. They normally are confined to shoal areas such as the Quala embayment and the broad shelf west of Rodain. They have four probable origins: they either (1) represent material altered by blue-green algae (Dalrymple, 1965), (2) are derived from broken fecal-pellet fragments, (3) are comparable to friable aggregates (Illing, 1954), or (4) are quiet-water oölites (Freeman, 1962).

More irregular and consequently less readily identified pellets also are present in the lagoon (Fig. 4D). In the quieter waters of the coastal terrace these grains commonly are surrounded by flocculent material.

Oölites.—Oölites were recognized in thin sections from one or two localities on the shelf. At Rodain they were in the form of perfect spheres of the fine- to medium-grained sand. They probably formed on the swash bars of the area. Although they contain a series of concentric mucilaginous layers which contain algal cells, few of these oölites are bored by blue-green algae, as are the oölites in the lee of the Abu Dhabi tidal deltas (John Twyman, pers. commun., 1964). This lack of boring is the result of either the turbulence of the environment or its water chemistry.

Unidentifiable grains.—There are several unidentifiable grains of extremely variable appearance and probably diverse origins. Most of them are products of alteration of other carbonate grains. An exception is detritus in the fine sand and silt grades which may represent disaggregated or poorly formed pellets. Unidentifiable material is commonly blackened, particularly in the western part of the Khor al Bazam.

Unidentifiable components constitute a high percentage of the sediment throughout the lagoon and decrease in abundance on the coastal terrace. In coarse sand the unidentifiable material is concentrated on the edges of the shoals and in areas of high relief in the lagoon. There are exceptions, in areas where local conditions encourage algal alteration, as indicated by intense pitting of the grain surface.

Quartz grains.—Though some quartz grains may be of eolian origin from the south, they are derived largely from outcrops of Tertiary and Quaternary rocks. Most of the identified source outcrops lie above low-water mark on the coastal strip and on the offshore bank. At Jebel Mirfa, sand commonly contains as much as 50 percent quartz. Similarly, at the peninsula of Quala adjacent beaches and intertidal flats are composed of nearly 100 percent quartz sand which drowns any carbonate material. Bayward the percentage of quartz decreases and carbonate-coated grains increase (Fig. 4E). Many of the jackets show one or more projecting hemispheroids of carbonate. In the Khor al Bazam, quartz percentage decreases markedly away from shoal areas, particularly in coarser grades. Finer grades have an even spread and hence may be transported more easily.

Quartz grains are generally in the fine sand grade and are subround in shape. Surfaces of many quartz grains have a partial carbonate coating similar to that of the quiet-water oölites (Freeman, 1962; Fig. 4E). Sand grains in a carbonate province accrete until an optimum size is reached (Illing, 1954). This process is controlled by the hydrodynamics of the environment. The Khor al Bazam grains generally have thicker carbonate jackets in shoal areas and thinner ones in the lagoonal areas.

Heavy minerals.—Heavy minerals, like the quartz and saccharoidal brown grains, are derived largely from the Quaternary and Tertiary rocks.

Unidentified sugary brown grains.—Saccharoidal brown grains composed of stained quartz, calcite, and feldspar are widespread but not abundant. They are unimportant except that they parallel quartz in distribution and hence probably have the same source.

Evaporite minerals.—Unlike previously described grains, evaporite minerals are forming in intertidal and sabkha (supratidal) sediments after deposition. There, progressive landward increase in the salinity of groundwater, as a result of evaporation (Butler, 1965; Kinsman, 1964), produces four parallel gradational belts of distinct mineral assemblages (Fig. 16).

1. Upper intertidal: gypsum and celestite crystals (Evans and Shearman, 1964) and dolomitized calcium carbonate within the capillary zone.
2. High water: calcium sulfate hemihydrate (Skipwith, 1966), anhydrite nodules, and dolomite accompanying solution of gypsum in the capillary zone; dolomite and large "sand crystals" of gypsum below the water table; halite precipitated from capillary water, and by the evaporation of stranded tidal waters, at the air-sediment interface.
3. Above high water: anhydrite polygons (which are festoon in cross section) and diapirlike structures within the capillary zone; gypsum and dolomite below the water table.
4. Adjacent to outwash fans: anhydrite converted to gypsum by influx of less saline groundwater.

Occurrence of these mineralogic belts is variable and one or more is commonly absent.

Traced landward, gypsum normally appears in the capillary zone of the algal flats as a mush of lenticular crystals (Fig. 16). Each crystal is about 0.5 mm in diameter and is flattened perpendicular to its c-axis (Masson, 1955). Inland from the intertidal zone the layer of crystals thickens to 20 cm. The gypsum generally is covered by a thin algal mat. At the landward edge of this mat small blebs of anhydrite are included in the surface layer (Fig. 17A). Some of the gypsum crystals in this zone show signs of solution, and calcium sulfate hemihydrate ($CaSO_4.0.02H_2O$) may be present (Skipwith, 1966). Just seaward of the anhydrite are buried algal and lagoonal sediments which contain larger flattened gypsum crystals up to 15 cm in diameter. Their size is presumably the result of slow growth at depth within the water table (Masson, 1955). Where this gypsum is in sandy lagoonal sediments, it contains many inclusions, but in algal sediments it characteristically contains few impurities.

At some distance from the sea, large gypsum crystals commonly protrude through the sabkha surface. There the water table has fallen and the surface sediments, no longer bound together by capillary water, are removed by deflation. The gypsum crystals are fragmented on exposure; Shearman (pers. commun., 1965) explains this as the result of diurnal thermal expansion and contraction assisted by crystallization of small halite crystals along cleavage cracks.

Anhydrite first forms as small nodules about 0.5–1 mm in diameter. They lie within the sediment surface and are of a soft, white cheese-like texture. They are thixotropic. Like the gypsum mush, the anhydrite forms in the capillary zone. Traced into areas of higher salinity, the nodules become more abundant and larger, in places as much as 4–6 cm across (Fig. 17B). They form in eolian and storm-washover sediments that begin to accumulate on the upper algal flats and extend back onto parts of the sabkha. Landward of the nodules a surface layer of anhydrite develops into a series of interlocking saucer-shaped structures which are polygonal in outline (Fig. 17C; Butler, 1965). Inland, as the sediments overlying the algal mat thicken further, the anhydrite also thickens and forms layers at depth, near, and at the surface. The lower entrolithic layers commonly are contorted, and they form small crenulations and tight folds similar to those of ptygmatic quartz veins (Fig. 17D). In the Khor al Bazam, as in the sabkha inland from Abu

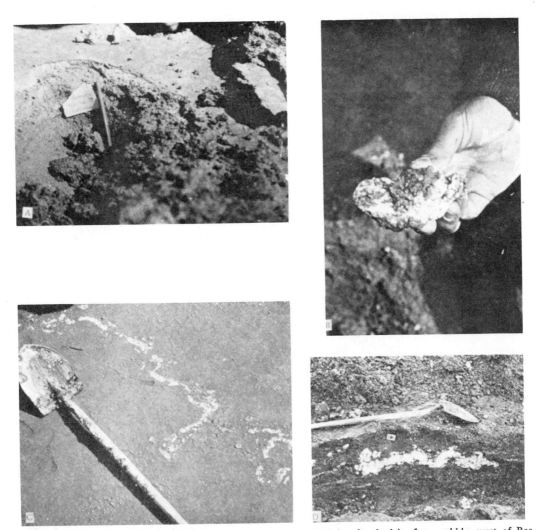

FIG. 17.—**A.** Small anhydrite blebs in sabkha surface. **B.** Nodule of anhydrite from sabkha west of Ras al Aish. **C.** Abu Dhabi sabkha. Edge of anhydrite polygons exposed by erosion. **D.** Contorted anhydrite passing laterally into gypsum after anhydrite.

Dhabi, some of the entrolithic anhydrite layers thicken to form layers more than 20 cm thick. West of Tarif are antiform structures more than 20 m in diameter (Fig. 18).

As Butler (1965) first reported near Abu Dhabi, the sequence of anhydrite development in the Khor al Bazam ends with the influx of groundwater into the sabkha via outwash fans. Anhydrite is hydrated to gypsum (Fig. 17D), which forms white, coarse, elongate, toothlike monoclinic crystals about 2–3 cm long and 0.5 cm in diameter. The hydration process disturbs the surface of the sabkha so much that it resembles a plowed field (Fig. 19). The blistered

surface contains large quantities of halite.

Elongate wispy crystals of rock salt are common in surface sediments of the sabkha at the

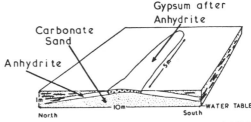

FIG. 18.—Diapir of anhydrite, located west of Mirfa on the south coast of the Khor al Bazam, Persian Gulf.

235

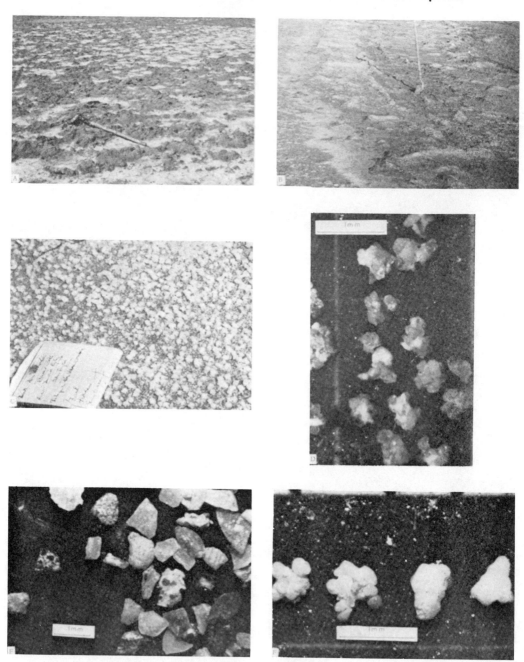

FIG. 19.—A. Khusaifa sabkha. Contorted surface where gypsum has formed after anhydrite. B. Compressional ridges in salt crust overlying algal flat. C. Khusaifa sabkha. Halite in form of hopper crystals at surface. D. Grapestones. E. Skeletal material and aggregates showing blackening. F. Sequence of grapestones to botryoidal grains.

top of the capillary zone. After flash floods and marine incursions, surface waters evaporate and leave a thin white crust of halite or, alternatively, sandy salt blisters (Fig. 19B). The salt is removed by wind and may be replaced by evaporation of capillary water. Where depressions in the sabkha retain pools of flood water for any length of time, "hopper" crystals develop (Fig. 19C).

X-ray analysis, by J. Lucia of Shell Oil

Company, of a series of core samples from algal flats and sabkha west of Ras al Aish indicates the presence of dolomite, which forms with the gypsum and anhydrite (Figs. 20, 21; Table I). The percentage of dolomite appears to be a function of grain size. The proportion of dolomite in calcareous mud is generally greater than that in calcareous sand. One sabkha mud from west of Ras al Aish contains more than 80 percent dolomite.

Flocculent matter.—Flocculent matter is a mixture of carbonate mud and a high percentage of organic material. It is very common in the lagoon samples, but forms a low percentage of all except the most sheltered intertidal sediments. One sheltered area was the Quala embayment. Like pellets, flocculent matter is a good indication of an environment protected from vigorous water movement.

Aggregates.—In the Bahamas, Illing (1954) recognized the following sand-size accretionary grains, all of which are present in the Khor al Bazam.

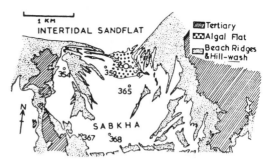

FIG. 20.—Map of sabkha west of Ras al Aish, on Khor al Bazam, Persian Gulf.

Grain Type	Composition	Locality
Friable aggregates	Discrete silt-size particles bound by organic fibers	Lagoonal grass areas
Lumps		
Grapestones	Discrete sand-size particles bound by, and protruding from, an aragonite cement	Lagoon floor, coastal terrace, and offshore bank
Botryoidal lumps	Similar to grapestones, but with polished aragonite envelope which disguises components	Coastal terrace
Encrusted lumps	Similar to grapestones, but with surface so bored by blue-green algae and altered as to be unidentifiable except in thin section	Lagoon floor, coastal terrace, and offshore bank
Worm tubes[4]	Discrete sand-size particles bound by aragonite cement to form tubes	Coastal terrace and offshore bank
Shell infillings[4]	Discrete sand-size particles bound by aragonite cement and enclosed or partially enclosed by mollusk shells	Lagoon, coastal terrace, and offshore bank

The friable aggregates, grapestones, botryoidal grains, encrusted lumps, and shell infillings are all represented as aggregates on the map (Fig. 22). These aggregates form on the lagoon

[4] Also recognized in the Khor al Bazam but not listed by Illing for the Bahamas.

floor and on some of the adjacent shoal areas. Material from the shoal areas is not transported far into the lagoon.

Samples taken across the algal mats show that, although aggregates are forming in front of them, the aggregates are not transported across them. The percentage of aggregates decreases sharply landward, but the old buried frontal algal mats contain a higher proportion.

Friable aggregates are composed of discrete silt-size particles. They are generally lightly bound together by threads of seaweed, algae, or some other organic tissue (Illing, 1954). Component particles of aggregates in the Khor al Bazam consist of nearly equal proportions of quartz and carbonate. In contrast, those described by Illing are exclusively carbonate. The complete sequence illustrated by Illing, of progressive cementation of these grains to form ovoid grains, was not found; however, the two end members were found. The highest percentage of uncemented friable aggregates is present in sea-grass areas of the lagoon floor, and the ovoid grains are found both in sea grass and on the coastal terrace.

Grapestones are aggregates composed of sand-size particles bound together by an aragonite cement. Component grains protrude, giving the appearance of a bunch of grapes. The components are predominantly aragonite, with up to 20 percent quartz (Fig. 20D). Illing's (1954, p. 30) description of the Bahamian examples fits the Khor al Bazam grapestone.

The cement that joins the grains is finely divided aragonite of varying texture. It forms first around the points of contact, and is friable and chalky white, in contrast to the greasy or matt-textured grains. It can easily be scraped off with a pin point, and is composed of aggregated particles of mud dimensions. From its occurrence, it is clear that it is being precipitated from sea-water, yet the particles show no recognizable crys-

FIG. 21.—Cores from sabkha west of Ras al Aish (Jerry Lucia of Shell, photograph).

talline shape, and are similar to the material that forms the matrix of the grains themselves.

Externally the chalky white cement is restricted to the crevices between the protruding grains. As cementation proceeds, the cement beneath this surface layer becomes firmer and matt-textured. Further grains or small lumps may be joined on by the same sequence of stages. However, the forces of mechanical disintegration prevent unlimited growth, and finally all traces of chalky white texture are lost.

In the Khor al Bazam, grapestones form on the grassy lagoon floor and shoal areas. In the lagoon during periods of minimum wave activ-

ity, surface sediment is bound initially by an organic slime forming a thin crust (Nesteroff, 1956). On intertidal shoals the initial binding agent is either organic mucilage or incipient aragonite beach-rock cement; the latter is precipitated at low tide during hot summer months (Ginsburg, 1953). In both areas, if the cementation process continues long enough, photosynthesis by blue-green algae precipitates aragonite. It cements and fuses the carbonate grains and is indistinguishable from them. The

Table I. Estimated Percentage of Crystalline Components by X-Ray Diffraction
(after J. Lucia, Shell Research)

Sample	Halite NaCl	Dolomite CaMg(CO₃)₂	Calcite CaCO₃	Aragonite CaCO₃	Feldspar	Quartz SiO₂	Gypsum CaSO₄·2H₂O	Gamma Calcium Sulfate γ-CaSO₄
354-1	15	10			20	45	10	
354-2	5	10			5	80		
359-1	10				20	15	50	5
359-2	P		P	P			P	
360-1	20			20		5	55	
360-2	10		15	65		10		
361-1	35		20	20		10	15	
361-2	55		15	15		10	5	
361-3	95		3			2		
361-3A	95		3			2		
365-1	10		15			5	70	
365-2	15		5	15		10	55	
367-1	5		5			75	15	
367-2			25			15	60	
367-3		15	50		15	20		
367-4	5	85				10		
367-5	5	85				10		
368-1	10		5		5	80		
368-2	15	5	5	20	5	50		
368-3	15		10	20		55		
368-4	20	5	15		10	50		
368-5	5	5	10	40	5	35		
368-6	5	5	10	20	10	50		

P = Present but percentage not estimated.

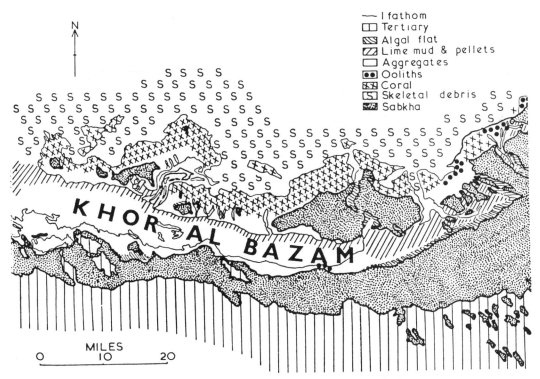

FIG. 22.—Lithofacies of Khor al Bazam, Persian Gulf.

acicular crystals of beach-rock cement are altered and the whole grain ultimately becomes homogeneous. The longer the surface is undisturbed by wave action, the better developed the binding.

Once formed, the surface crust may be broken into small lumps by wave action. The size of the lumps depends on the strength of the cement and the force of the waves. With a heavy sea and lightest of bindings, the crust may be separated into its constituent grains, but with gentle action and strong cementation, aggregates are produced. Thus, aggregates in the protected grass areas of the lagoon on the average are larger and more irregular than those of the exposed shoal regions.

In the intertidal region of Quala bay, aggregates are composed of 20 percent quartz, 60 percent aragonite pellets, and 20 percent aragonite cement. The sediment in which they are present contains a ratio of aggregate to quartz to pellet of approximately 3:3:1. If the sediment were cemented unselectively, the aggregates would be expected to contain more quartz than they do. The low percentage of quartz in aggregates may be explained in three ways.

1. The percentage of quartz in the sediment is not the usual one, but is abnormally high because the samples were collected only a short time after a strong on-shore gale. The excess quartz was carried in from the headlands.

2. When the partly cemented surface crust is broken by waves, it generally fractures along the boundaries of the quartz grains rather than along the fused boundaries of carbonate grains. The separated quartz grains may retain patches of the cement. This could account for the fact that many of the quartz grains have carbonate jackets, some even with one or more hemispheroidal aragonite shapes protruding from the surface. The latter are fused aragonite pellets (Fig. 4E).

3. Another possible process of aggregation involves flocculent blebs of blue-green algae and their investing mucilage. They float free in sediment, and act as centers of aragonite precipitation to form pellets. In their early stages these pellets have "sticky" surfaces which adhere easily to other grains. Thus aggregates, quartz grains, and pellets quickly acquire aragonite pellets. The loosely bound aggregates become cemented by algally precipitated aragonite. Pellets formed by this process are unlikely to be free because of their adhesive qualities. Although algae and their mucilage are present on all carbonate grains, there is no evidence that algae colonize noncarbonate grains. If they do not, or only do so poorly, this ecologic factor also may influence the ratio of quartz to carbonate in the final aggregate.

In thin section, it was found that the initial cement binding grapestones is of darker color and is apparently coarser grained than the final cement. As cementation proceeds, the textures of the cement and the component grains become indistinguishable. Purdy (1963) observed a similar occurrence in the Bahamas.

In the Khor al Bazam the presence of acicular aragonite crystals as a cement in the voids of grapestones is rare. A few similar developments of crystals are found below the intertidal level, and Kinsman (1964, pers. commun.) found them in the base of corals from the Abu Dhabi area.

Some of the component carbonate particles commonly are blackened, but are bound by white aragonite cement (Fig. 18E). In places the cement itself is blackened. As with the sediment components heretofore described, blackening is largely confined to the lagoon samples and follows the pattern shown by Murray (1966, Fig. 15).

Botryoidal lumps are like grapestones in that they are composed of discrete sand-size particles. However, they differ in being covered by a sheath of polished aragonite. As with grapestones, the mammillated surface of these grains is grapelike. In the Khor al Bazam, few grow larger than 0.539 mm in diameter. The smaller the grain, the more complete the cover and the better the polish. Botryoidal lumps represent grapestones that, by a process similar to oölite formation, acquire successive coatings of aragonite, which eventually disguise the component grains and fill the cracks between them (Fig. 19F). The lumps form in shallow and intertidal waters, the highest energy environments of the Khor al Bazam.

Encrusted lumps are grapestones whose surfaces have been modified by intense algal activity. The process of alteration so changes the surfaces that it is impossible to distinguish the lumps from other amorphous grains (Illing, 1954). They are most abundant in the areas of greatest development of bladder wrack seaweed.

Worm tubes are the cemented walls of worm burrows. When fresh, they are easily distinguished by their shape, but if they are broken and abraded, it becomes increasingly difficult to separate them from grapestones.

Normally the wall of the worm tube is no more than three layers of sand grains in thickness. Other types of tubes may have walls more than 2 cm thick. They probably are formed by lamellibranchs or arthropods. Tubes broken free from reworked intertidal sand commonly litter the beach.

Shell infillings are cemented aggregates that

accumulate in the protection of shells. These, like the worm tubes, can be an important source of grapestones. When fresh, they are easy to identify, but if broken, abraded, or algally altered, it is difficult to establish their origin.

Serpulid structures.—Serpulid structures form accretionary nodules up to 30 cm or more in diameter on intertidal flats and shoals. The nodules are colonized by gastropods, lamellibranchs, and barnacles.

Sponge spicules.—Sponges grow in abundance in parts of the bladder wrack weed zone. Spicules released by decay of sponges are present in the finer grades of sand and silt.

Echinoid fragments.—Common echinoids are the irregular, flattened sand dollars in the lagoons and the regular ones in the reefs. Their fragmented skeletons are dispersed so quickly that they do not form an important constituent of the sand.

Fish bones and teeth.—In spite of the high salinity of the Khor al Bazam, there is an extensive necron, and notable in it are mackerel, rays, sharks, sea snakes, turtles, dolphins, and manatees, all of which may be found in the lagoon or offshore. The coral banks support a reef necron which includes groupers and parrot fish. Their geologic importance is discussed by Cloud (1959) and Kaulback *et al.* (1962). A few fish bones and teeth were found in one or two samples.

Crustacea.—Though living crustacea are common, their skeletal remains are quickly fragmented and dispersed, and they form a minor constituent of the sediments. Barnacles colonize boulders of serpulid tubes but otherwise are rare. They include *Chthalmulus* cf. Hoelc and *Banalus amphitrite* Darwin.

FLORA

Blue-green algal mats (Kendall and Skipwith, 1968), mangroves (*Avicenna marina*), and the bushy halophyte *Arthroconeum glaucum* are found in protected areas (Fig. 23). Bladder wrack grows on a hard substrate in waters where the bottom is stirred by currents and waves. In quieter waters are large areas of *Halodule (Diplanthera) uninervis*, a grass like weed which is extremely effective in binding sediment and in trapping silt- and mud-sized particles. Its role is equivalent to that of the *Thalassia* of the Bahamas (Illing, 1954; Purdy, 1963). In the most protected waters of the lagoon where the sediment is not affected by current action, a weed having small, rounded "cresslike" leaves grows; in the lee of the offshore bank where muddy sediments accumulate, a weed having elongate leaves reminiscent of "mint" grows.

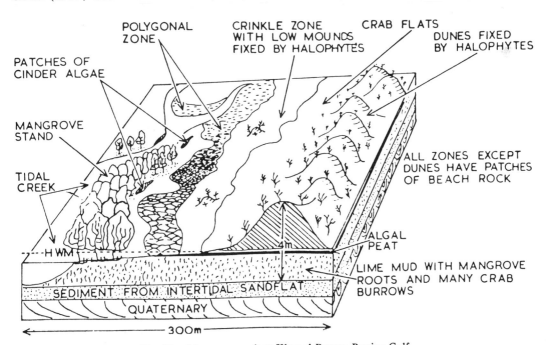

FIG. 23.—Mangrove creek at Khor al Bazam, Persian Gulf.

Conclusions

A hypothesis on the hydrodynamic sequence shown in the Trucial Coast by calcium carbonate particles of physicochemical origin was developed from the foregoing sections. It has been found that calcium carbonate mud and a range of accretionary grains are present as components of the carbonate sediments which are forming now in the shallow waters of lagoons and shelves off the Trucial Coast (Figs. 2, 24). Broadly, these environments may be classed as (1) protected, (2) moderately exposed, and (3) exposed.

Protected Environments

Protected environments include enclosed lagoons and lee areas of banks and islands. The sediments range from aragonite mud to soft and cemented aragonite pellets. The aragonite mud is confined mainly to sheltered lagoonal areas of eastern Abu Dhabi, and the lee of banks and islands of the Khor al Bazam area. The soft and cemented aragonitic pellets are found in sheltered lagoons and shallow shelves of moderate exposure to wind and waves.

In areas of low hydrodynamic energy, aragonite mud accumulates and is seldom disturbed. It is uncemented, probably because the static pore waters remain in equilibrium with the sediment. Associated with the mud are blebs of flocculent organic mucilage, containing blue-green algae. The blebs are common to most of the environments along the coast. In this protected environment the gentlest movement of water causes the mucilage blebs and recently excreted fecal pellets of low density to saltate. Surfaces of mucilage blebs and pellets are sticky and they possibly accrete detrital aragonite. During saltation the surfaces of accreting grains are exposed continuously to sunlight and to fresh solutions of seawater, probably supersaturated with respect to calcium carbonate. Thus, by the dual process of photosynthesis and respiration, aragonite is precipitated within the mucilage and the grains are cemented and harden to form almost indistinguishable pellets. The longer the pellets saltate, the harder they are cemented. The balance between soft and hard pellets is determined by the relative intensity of the turbulence of the water and the length of saltation. If one pellet adheres to another, the effective weight is of the two together; they cease to leave the sea floor because water movement in pellet areas is too gentle to lift them and algal growth binds them to other

grains. Grains lying on the sea floor become buried beneath other grains. Cementation ceases because the algal cells within the grains are unable to photosynthesize when shielded from sunlight and, in addition, the sediment is in chemical equilibrium with the surrounding pore waters.

Moderately Exposed Environments

Moderately exposed environments are lagoon shores and offshore banks which are protected only partly from heavy seas and intense hydrodynamic activity. The sediments formed there are characterized by aggregates of cemented aragonite pellets.

The aggregates probably form when pellets are at rest long enough to adhere and form a thin crust, the adhesion being induced either by mucilage or beach-rock cementation. In this environment heavy wave action is common enough and intense enough to break the thin surface crust, but not strong enough to split the grains into their original separate form. Aggregates develop and are joined to other grains and acquire cement both externally and internally. If washed into the exposed environment, they develop polished oölitic coats.

Exposed Environments

The exposed environments mark the seaward edge of the complex of banks and lagoons of the Abu Dhabi coastline and the headland of Rodain in the Khor al Bazam. The sediments include oölites and vast lateral accumulations of coral.

In this environment of great turbulence the adhesion of one carbonate grain to another does not occur. The grains are never at rest long enough for a strong cement to develop which will not be broken by wave action. As in the other environments, the grains grow by the dual process of mechanical and chemical accretion. However, instead of being ellipsoid, they develop as spheres because of their physical environment. Moreover, the momentum acquired by these grains, it is suggested, leads on impact to a flattening of the elongate aragonite crystals tangentially to the surface at the point of impact.

The sequence from mud to oölite follows the progessive increase in the turbulence of the waters in which they form, and is similar to the succession suggested by Illing (1954) and Purdy (1963) for carbonate grains of physicochemical origin in the Bahamas. This apparent order in the development of carbonate grains

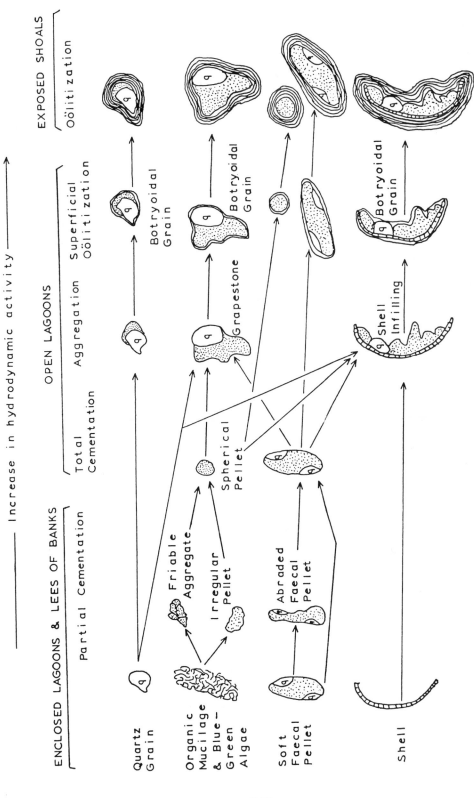

Fig. 24.—Relation of carbonate grains to hydrodynamic environment. Blue-green algae may modify initial texture of all carbonate grains.

may be related also to salinity variation. Variation probably does not change the chemical processes by which the particles acquire aragonite, though it probably affects the speed of the processes. For instance, if the water body in which the particles are being formed is enclosed, its size will control the concentration of salts by evaporation. Thus there may be chemical reasons for the absence of oölites at the mouths of the most easterly and smallest lagoons of Abu Dhabi (Evans, pers. commun., 1965).

The presence of living coral, large patches of living marine weed, and living coralline algae also may modify the CO_2 balance so that precipitation or crystallization of aragonite by chemical means is inhibited.

The sediments of Khor al Bazam can be separated into facies that are distinctly different from each other in the content and the surface texture of the constituent components. These facies are found in distinct geomorphologic environments and can be related to variations in the hydrodynamic conditions of the environment of deposition. Thus, in deeper waters below the action of tidal currents there is skeletal sand with a fairly high content of interstitial mud. The original surface texture of skeletal components is destroyed and most grains have a saccharoidal texture. Where tidal currents sweep the bottom sediments, the interstitial mud generally is swept away and the component grains become joined together to form "lumps." Within the influence of the wave base, sand grains may become rounded and coated. There is a sequence of sediment types related to increasing turbulence of the water at the sediment-water interface within the lagoon. Outside the barrier complex, waves and currents follow a different pattern. There is an abrupt change from the skeletal sand of the open gulf floor to the coral and coralline algae of the reef banks. This change is related also to the bottom topography, because there is an abrupt break in slope between the gulf floor and the banks, in contrast to the gentle gradations found within the lagoon of the eastern Khor al Bazam.

Each sediment type now is forming in a small area but, because of the rapid coastal accretion along the Trucial Coast, there is a good chance that these facies will be preserved. Vertical sections in the supratidal zone show lagoonal sediments at the base capped by intertidal sediments and overlain by windblown and storm-washover sediments. It is the addition of

quartz sand and evaporites which distinguishes the sabkha from the intertidal sediment. These facies are modified further by diagenetic alteration of aragonite to dolomite.

REFERENCES CITED

Banner, F. T., and G. V. Wood, 1964, Recrystallization in microfossiliferous limestones: Geol. Jour. (Liverpool), v. 4, p. 21–34.

Bathurst, R. G. C., 1964, The replacement of aragonite by calcite in the molluscan shell wall, *in* Approaches to paleoecology: New York, John Wiley and Sons, p. 357–376.

——— 1966, Boring algae, micrite envelopes, and lithification of molluscan biosparites: Geol. Jour. (Liverpool), v. 5, p. 15–32.

Butler, G. P., 1965, Early diagenesis in the recent sediments of the Trucial Coast of the Persian Gulf: Ph.D. thesis, London Univ., 251 p.

Cloud, P. E., Jr., 1959, Geology of Saipan, Marianna Islands—Part 4, Submarine topography and shoal water ecology: U.S. Geol. Survey Prof. Paper 280 K, p. 361–445.

——— et al., 1962, Environment of calcium carbonate deposition west of Andros Island, Bahamas: U.S. Geol. Survey Prof. Paper 350, 138 p.

Dalrymple, D. W., 1965, Calcium carbonate deposition associated with blue-green algal mats, Baffin Bay, Texas: Inst. Marine Sci. Pub., v. 10, p. 187–200.

Emrich, G. H., and F. J. Wobber, 1963, A rapid visual method for estimating sedimentary parameters: Jour. Sed. Petrology, v. 33, p. 831–843.

Evans, G., 1966, The recent sedimentary facies of the Persian Gulf region: Royal Soc. London Philos. Trans., ser. A, v. 259, p. 291–298.

——— and D. J. Shearman, 1964, Recent celestine from the sediments of the Trucial Coast of the Persian Gulf: Nature, v. 202, p. 385–386.

Fischer, A., 1897, Untersuchungen über der Ban der Cyanophycean und Bacterien; Jena, 136 p.

Folk, R. L., 1955, Note on the significance of "turbid" feldspars: Am. Mineralogist, v. 40, p. 356–357.

——— 1962, Spectral subdivision of limestone types, *in* Classification of carbonate rocks—a symposium: Am. Assoc. Petroleum Geologists Mem. 1, p. 62–84.

——— 1965, Some aspects of recrystallization in ancient limestones, *in* Dolomitization and limestone diagenesis, a symposium: Soc. Econ. Paleontologists and Mineralogists Spec. Pub. 13, p. 14–48.

——— and R. Robles, 1964, Carbonate sands of Isla Perez, Alacran reef complex, Yucatan: Jour. Geology, v. 72, p. 255–293.

——— and C. E. Weaver, 1952, A study of the texture and composition of chert: Am. Jour. Sci., v. 250, p. 498–510.

Freeman, T., 1962, Quiet water oölites from Laguna Madre, Texas: Jour. Sed. Petrology, v. 32, p. 475–483.

Fritsch, F. E., 1952, Structure and reproduction of the algae, v. 2: Cambridge, England, Cambridge Univ. Press, 939 p.

Frondel, C., W. H. Newhouse, and R. F. Jarrell, 1942, Spatial distribution of minor elements in single crystals: Am. Mineralogist, v. 27, p. 726–745.

Ginsburg, R. N., 1953, Beachrock in south Florida: Jour. Sed. Petrology, v. 23, p. 85–92.

——— 1956, Environmental relationships of grain size and constituent particles in some south Florida carbonate sediments: Am. Assoc. Petroleum Geologists Bull., v. 40, p. 2384–2427.

Grayson, J. F., 1956, The conversion of calcite to fluorite: Micropaleontology, v. 2, p. 71–78.

Hay, W. W., K. M. Towe, and R. C. Wright, 1963, Ultramicrostructure of some selected foraminiferal tests: Micropaleontology, v. 9, p. 171–195.

Houbolt, J. J. H. C., 1957, Surface sediments of the Persian Gulf near the Qatar peninsula: The Hague, Mouton and Co., 113 p.

Hyman, L. H., 1940, The invertebrates, vol. 1, Protozoa through Ctenophora: New York, McGraw Hill Book Co., 726 p.

Illing, L. V., 1954, Bahaman calcareous sands: Am. Assoc. Petroleum Geologists Bull., v. 38, p. 1–95.

Kaulback, J. A., C. G. St. C. Kendall, and P. A. d'E. Skipwirth, 1962, Cyclothems on the islands of Kharg and Khargu, Persian Gulf: Unpub. rept., Iranian Oil Exploration and Producing Companies, 19 p.

Kendall, C. G. St. C., and P. A. d'E. Skipwith, 1968, Recent algal stromatolites of the Khor al Bazam, southwest Persian Gulf [abs.]: Geol. Soc. America Spec. Paper 101, p. 108.

—— et al., 1966, On mechanical role of organic matter in diagenesis of limestone [abs.]: London Geol. Assoc. Circ. 681, p. 1–2.

Kinsman, D. J. J., 1964a, The recent carbonate sediments near Halat el Bahrani, Trucial Coast, Persian Gulf, in Deltaic and shallow marine deposits—developments in sedimentology, v. 1: Amsterdam, Elsevier Pub. Co., p. 185–192.

—— 1964b, Recent carbonate sedimentation near Abu Dhabi, Trucial Coast, Persian Gulf: Ph.D. thesis, London Univ., 315 p.

Kornicker, L. S., and E. G. Purdy, 1957, A Bahamian faecal-pellet sediment: Jour. Sed. Petrology, v. 27, p. 126–128.

Masson, P. H., 1955, An occurrence of gypsum in southwest Texas: Jour. Sed. Petrology, v. 25, p. 72–77.

Murray, J. W., 1966, The Foraminiferida of the Persian Gulf, Part 4, Khor al Bazam: Palaeogeography, Palaeoclimatology, and Palaeoecology, v. 2, p. 153–169.

Nesteroff, W. D., 1956, La substratum organique dans les dépôts, calcaires sa signification: Soc. Géol. France Bull., sér. 6, v. 6, p. 381–390.

Newell, N. D., E. G. Purdy, and J. Imbrie, 1960, Bahamian oölitic sand: Jour. Geology, v. 68, p. 481–497.

Orme, G. R., and W. W. M. Brown, 1963, Diagenetic fabrics in the Avonian limestones of Derbyshire and North Wales: Yorkshire Geol. Soc. Proc., v. 34, pt. 1, no. 3, p. 51–66.

Parks, N., and H. C. Curl, Jr., 1965, Effects of photosynthesis and respiration on electrical conductivity of seawater: Nature, v. 205, p. 274–275.

Purdy, E. G., 1963, Recent calcium carbonate facies of the Great Bahama Bank: Jour. Geology, v. 71, p. 334–355, 472–497.

—— 1965, Diagenesis of recent marine carbonate sediments [abs.], in Dolomitization and limestone diagenesis: Soc. Econ. Paleontologists and Mineralogists Spec. Pub. 13, p. 169.

Shearman, D. J., and P. A. s'E. Skipwith, 1965, Organic matter in recent and ancient limestones and its role in their diagenesis: Nature, v. 208, p. 1310–1311.

Skipwith, P. A. d'E., 1966, Recent carbonate sediments of the eastern Khor al Bazam, Abu Dhabi, Trucial Coast: Ph.D. thesis, London Univ. 407 p.

Sollas, W. J., 1921, On Saccammina carteri Brady and the minute structure of the Foraminifera shell: Geol. Soc. London Quart. Jour., v. 77, p. 193–212.

Wolf, K. H., 1965a, "Grain diminution" of algal colonies to micrite: Jour. Sed. Petrology, v. 35, p. 420–427.

—— 1965b, Petrogenesis and palaeoenvironment of Devonian algal limestones of New South Wales: Sedimentology, v. 4, p. 113–178.

—— and J. R. Conolly, 1965, Petrogenesis and palaeoenvironment of limestone lenses in Upper Devonian red beds of New South Wales: Palaeogeography, Palaeoclimatology, and Palaeoecology, v. 1, p. 69–111.

Wood, A., 1948, The structure of the wall of the test in the Foraminifera—its value in classification: Geol. Soc. London Quart. Jour., v. 104, p. 229–256.

Editors' Comments
on Paper 9

9   LOGAN and CEBULSKI
Excerpts from *Sedimentary Environments of Shark Bay,
Western Australia*

SHARK BAY, AUSTRALIA

Shark Bay, like the Persian Gulf, represents carbonate deposition
in a very arid climate. In Paper 9 Logan and Cebulski have given a very
lucid account of the carbonate tidal flats in Shark Bay that approximate the dimensions of a small epicontinental sea having a span of
roughly 12,000 km² and water depth of 10 m. The work covers the
detailed bathymetric features of the area and hydrologic structures of
the water mass, and subdivides each of them into three distinct types.
There is a salinity gradient from normal marine to hypersaline with a
transition of metahaline from north to south. Thus bathymetric and
salinity gradient coupled with wave and tidal activity control the biotic
community and sedimentary facies in each of the subenvironments.
Paper 9 provides a detailed description of each of the subenvironments and demonstrates how a modern sedimentary environment
should be looked at and studied.

9

Reprinted from pages 1-2, 6-8, 11-16, and 21-37 of Am. Assoc. Petroleum Geologists Mem. 13, 1970, 37p.

Sedimentary Environments of Shark Bay, Western Australia[1]

BRIAN W. LOGAN[2] **and DONALD E. CEBULSKI**[2,3]

Nedlands, Western Australia 6009

Abstract Shark Bay is a large embayment at lat. 25°S on the western coast of Australia. Studies of the environment have been carried out in conjunction with investigations of carbonate sedimentation and diagenesis in the region.

In dimension, Shark Bay approaches a small epicontinental sea; the area is approximately 5,000 sq mi and average depth is 30 ft. The water mass is cut off from the Indian Ocean by a ridge of calcareous eolianite and is subdivided internally into numerous inlets, gulfs, and basins by dune ridges and submerged banks (sills). Influx of oceanic water is only through openings in the northern part of the outer barrier.

The embayment is adjacent to a low-relief, arid to semi-arid hinterland. Runoff influx is negligible and evaporation greatly exceeds precipitation. These factors, combined with the hydrologic structure of the water mass and restriction imposed by banks and sills, result in increasing gradients of salinity into the closed southern parts of the embayment. The salinity is from 36 ‰ in the north to as high as 65 ‰ in the south. The water mass has a layered structure with nearly vertical attitude. Major clines subdivide the water body into three major types: oceanic (36–40 ‰), metahaline (40–56 ‰), and hypersaline (56–70 ‰).

The water types are limiting on the biota. There are three biotic zones, the distribution of which is essentially similar to the distribution of water types. Within the broad environmental zones, numerous local environments are limited by such factors as wave action, tides and tidal currents, and depth. Depth is of major importance because most other parameters are linked to it.

INTRODUCTION

This paper is an account of the sedimentary environments in Shark Bay, a marine embayment on the western coast of Australia (Fig. 1). Since the first studies of carbonate sedimentation in Shark Bay in 1964, several investiga-

[1] Manuscript received, June 6, 1969; revised, October 21, 1969.

[2] Department of Geology, The University of Western Australia.

[3] Deceased, July 1965.

Early work on the hydrologic system of Shark Bay was part of a Ph.D. project at The University of Western Australia carried out by Donald E. Cebulski, who died in an automobile accident at Shark Bay.

Research reported in this paper is part of an investigation of carbonate sedimentation and diagenesis in Shark Bay which has been supported by the American Petroleum Institute (as Research Project 71-B), the

tions on various aspects of sedimentation have been conducted. The projects have been concerned with the relations of environment, organisms, and sediments mainly in local areas. However, local environments are but components of a larger system which has developed as a result of regional processes and geologic events during the Quaternary. The purposes of this paper are to outline the broad environmental framework of the embayment and to provide a background to which studies of more localized processes can be related.

Shark Bay is a large shallow embayment which was formed by marine transgression into a terrain composed mainly of Pleistocene dunes. The flooding created a series of broad gulfs and narrow inlets (Fig. 2), which are partly cut off from the Indian Ocean. The embayment is adjacent to a hinterland of low relief in an arid to semi-arid climate. Two rivers

Australian Research Grants Committee, and grants from The University of Western Australia; this funding is gratefully acknowledged.

The Western Australian Department of Fisheries and Fauna (through the Western Fisheries Research Committee) has given valuable assistance in the way of ship time with the research vessels *Peron, Lancelin,* and *Flinders* and the patrol vessel *Vlaming.* Appreciation is expressed to A. J. Fraser, former Director of Fisheries and Fauna, B. Bowen, present Director of Fisheries and Fauna, and R. J. Slack-Smith for the cooperative spirit in which this help was given. Field sampling would not have been possible without able assistance from officers and crews of the vessels; special thanks are due W. White, J. Seabrooke, H. Petersen, E. Campbell, A. McKenzie, and D. Wright.

Data were collected on a series of cruises carried out from February 1964 to February 1968. Several members of the Department of Geology, The University of Western Australia, took part in these operations. The writers are indebted to R. G. Brown, G. R. Davies, J. F. Read, W. M. O'Beirne, and F. Billing for their assistance. Sandra Gray and Kathleen Taber gave valuable technical assistance in analysis of samples. The help given by F. Billing and T. Cooke, in preparation for cruises, and by C. Hughes and C. Bell, in photography and illustration for the report, also is appreciated. The manuscript was read critically by R. G. Brown and J. F. Read.

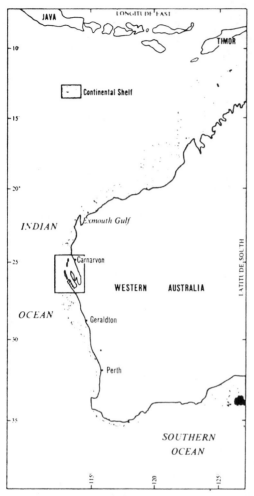

Fig. 1.—Location of Shark Bay on coast of Western Australia. Shaded area is continental shelf.

drain into Shark Bay, but their flow is intermittent and runoff influx is small. Climatic factors combined with low runoff and restricted circulation (due to shoaling and hydrodynamic factors) result in a general increase of salinity values from normal oceanic in the northern part of the bay to hypersaline in the southern extremities. The salinity gradient exerts a broad control on the distribution of environments throughout the length of the embayment. Increasing restriction on benthonic organisms is reflected in the type and distribution of carbonate sediments being deposited.

BATHYMETRY

Shark Bay is a shallow marine embayment of approximately 5,000 sq mi total area. The average depth in this vast body of water is about 30 ft, but about a quarter of the total area is occupied by shoals at depths of a few feet below tide level (Fig. 2). Depths are greatest in the northern part where the bottom slopes from 60 to 120 ft (Fig. 2). The water mass is partly cut off from the Indian Ocean by a barrier ridge composed of eolianite dunes. The embayment within the barrier is broken into a series of gulfs, inlets, and basins by north-trending ridges composed of Pliocene-Pleistocene dune rocks. Table 1 summarizes the dimensions of major bathymetric features. An account of physiographic subdivisions is given by Logan et al. (this volume).

Bathymetric Zones

Bathymetric features in the embayment contain a zonal morphology (Fig. 2) of three parts: (1) intertidal-supratidal platform, (2) sublittoral platform, and (3) embayment plain. The sublittoral platform borders the embayment like a miniature continental shelf. It slopes seaward from the intertidal zone at a gentle gradient; the outer margin is marked by a steepening in slope toward the embayment plain. Configuration of the platform is varied, being related to weather aspect, frequency of wave action, and sedimentary processes in local areas. Platforms facing north and east are narrow (100 yd–½ mi), having a slope break at 5–15-ft depth and a steep descent to the embayment plain. Platforms of southwest and south aspect are generally broad (1–5 mi) and slope gently to the embayment plain from about 15–20-ft depth. The flat and featureless embayment plain is at 25–30-ft depth in the southern part of the embayment and slopes down to about 120-ft depth at Geographe Passage. The intertidal-supratidal platform is of varied width, ranging from a few feet on rocky headlands to several miles in tidal-supratidal flat developments. The surface slopes up from LWS (low-water springs) level to about 6 ft above HWS (high-water springs) level.

[*Editors' Note:* Material has been omitted at this point.]

REGIONAL FACTORS

Climate, tide, and winds are the major external factors shaping the environment of Shark Bay.

Climate

Table 2 is a summary of meteorologic data from weather stations at Carnarvon and Hamelin Pool. Meteorologic data available prior to 1942 are summarized by the Royal Australian Air Force (1943). The climate is semi-arid; average annual rainfall is 8–9 in. (20–22 cm) and annual evaporation is 80–90 in. (200–220 cm). Most precipitation is in winter (May, June, July), but summer averages are boosted by occasional heavy cyclonic rains which may occur in December-March. Relative humidity is high (70–80 percent) in oceanic areas but is considerably lower onshore; Carnarvon has humidity of about 40 percent, but it varies on a seasonal and diurnal basis. Hamelin Pool, which is farther from oceanic influences, has lower values of about 30 percent and is characterized by small seasonal and diurnal fluctuation. Relative humidity in the afternoon commonly exceeds that of the morning because of oceanic air influx with afternoon sea breezes (southwest wind).

Winds

Figure 3 is a summary of wind data taken at Carnarvon during 1961–1965, inclusive, available by courtesy of the Commonwealth Bureau of Meteorology. The area is in the belt of southeast tradewinds, and prevailing winds are from the south. Tradewind influence locally is altered and reinforced by a strong sea-breeze system during summer, and by winds associated with a series of depressions occurring in the Southern Ocean during winter. Strong southerly winds (Fig. 3) have average velocity ranging from 10 to 15 knots in summer (October to April, inclusive), and average maximum velocity ranging up to 25 knots in January and February. Southerly "gales" may be sustained for 3–5 days, the usual pattern being strong southeast to south-southeast winds in the morning and stronger south-southwest to southwest winds in the afternoon and night. Southerly winds also prevail during winter (May to September, inclusive), but average velocity is lower, in a range of 5–8 knots, and periods of calm are common. Strong northerly winds

Table 2. Summary of Meteorologic Data for Carnarvon[1] and Hamelin Pool[2]

Station	Month	Av. Max. Temp. (°F)	Av. Min. Temp. (°F)	Rainfall (in.)	Evaporation (in.)	Min. Mean Relative Humidity (1,500 hr)
	Jan.	87.7	71.7	0.41	11	48
	Feb.	88.3	72.3	0.70	9	52
	Mar.	87.5	71.1	0.66	10	46
	Apr.	91.5	64.4	0.64	7	48
	May	81.7	56.2	1.49	4	48
Carnarvon	June	74.8	49.9	2.40	3	44
	July	71.4	51.2	1.56	4	38
	Aug.	73.0	53.1	0.68	5	40
	Sept.	75.5	57.0	0.23	6	45
	Oct.	77.6	60.7	0.12	7	49
	Nov.	81.4	65.5	0.03	10	44
	Dec.	84.6	69.0	0.16	10	54
		Mean 81.2	Mean 61.8	Total 9.08	Total 86	
	Jan.	98.0	68.4	0.26	11	29
	Feb.	97.3	69.3	0.50	9	28
	Mar.	93.9	67.3	0.57	10	28
	Apr.	86.9	62.3	0.39	7	32
	May	77.0	55.3	1.24	4	42
Hamelin Pool	June	70.4	50.3	1.95	3	39
	July	68.9	48.0	1.51	4	48
	Aug.	71.6	48.7	0.76	5	44
	Sept.	77.5	51.5	0.35	6	32
	Oct.	82.0	54.7	0.14	7	29
	Nov.	89.2	60.2	0.14	10	24
	Dec.	94.7	64.8	0.09	10	29
		Mean 83.9	Mean 58.4	Total 7.90	Total 86	

[1] Average for 44 years of observation.
[2] Average for 45 years of observation (from Royal Australian Air Force, 1943).

occur occasionally in winter, depending on the intensity of depressions in the Southern Ocean.

The northwest coast of Australia is subject to cyclonic disturbances (hurricanes) which originate in the Timor Sea and follow a southwest course along the coast before turning inland and diminishing. Most cyclones cross the coast north of Exmouth Gulf (Fig. 1) and bring heavy rains to the Gascoyne and Wooramel river basins; flooding and transportation of terrigenous detritus into Shark Bay then may occur. Cyclones also may pass inland at points between Exmouth Gulf and Geraldton (Fig. 1). In such instances Shark Bay comes under the direct influence of the depression, and heavy rainfall and high winds occur. Wind duration is commonly up to 12 hours and winds of 40–60 knots are usual, gusting to 100 knots. Wind direction varies according to location, but a cyclone passing across western Shark Bay is likely to produce gale-force east to southeast winds followed by gale-force northeast winds.

Wind Drift

Littoral drift propelled by prevailing southerly winds is directed north-northwest to northwest along coasts of westerly aspect; these include the western margins of the outer barrier, Peron Peninsula, the coast from Hamelin Pool to Carnarvon, and other west-facing coasts in smaller inlets and gulfs. In contrast, on coasts of easterly aspect, water level is lowered by wind set-up toward the northwest, which results in a general strengthening of incoming flood-tide currents. These coasts include the eastern margins of Peron Peninsula, Edel Land, and Dirk Hartog Island.

The interaction of tidal currents with wind drift is an important factor in developing dominance of either ebb or flood currents and thereby creating a regional circulation pattern. On west-facing shores the overall effect of the prevailing northerly drift is to reinforce ebb currents and to retard flood-tide currents. Thus ebb-current dominance holds on these coasts.

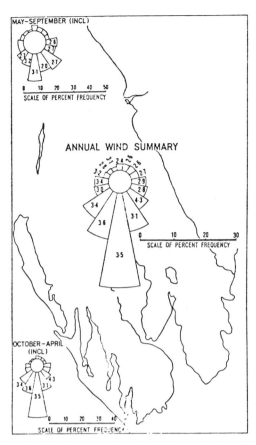

FIG. 3.—Wind summary for Shark Bay region based on
data from Carnarvon.

On east-facing shores there is flood-tide dominance.

The asymmetry of flood dominance on east-
facing coasts and ebb dominance on west-facing coasts leads to a circulation in which net
movement is from west to southeast and then
east, and finally northwest. This movement is
probably along the density (salinity) trend
lines (Fig. 6).

[*Editors' Note:* Material has been omitted at this point.]

Tides

Tide is the major factor causing water movement in the embayment. The pattern of tides in northern Shark Bay is known from continuous recording taken over a 3-year period at Carnarvon; the pattern is mixed diurnal. Neap tides have a diurnal oscillation with one high water and one low water during each 24-hour period; spring tides are mixed, having LLW, HLW, HHW, and LHW peaks. Maximum spring tides coincide roughly with first quarter and full moon phases. The tide range at Carnarvon is about 5.5 ft (1.70 m) on maximum spring tides and about 2 ft (0.61 m) on neap tides. The pattern in Hopeless Reach is similar to that at Carnarvon. Recordings during 1966 showed a time lag of 2–4 hours behind Carnarvon, and amplitude of tide was about 0.5 ft (15 cm) less. The Australian National Tide Tables (1968) give a 6-hour lag for HWS peak in Hamelin Pool. Tide records from Freycinet Reach show an extremely complex pattern, but mixed tides predominate.

Current flow on flood tide.—Current-direction data for flood-tide conditions are shown in Figure 5. The general direction of flood currents throughout Shark Bay is south-southeast, but there are numerous local departures from this direction. Naturaliste and Geographe Passages are roughly cotidal. Flow paths in Geographe Passage are south-southeast into Uranie Strait, and the same direction is maintained on the eastern margin of Hopeless Reach. A southerly flow from Geographe Passage along the east side of Bernier Island is deflected eastward in response to some northward flow from Naturaliste Passage. Inflow at Naturaliste Passage initially is directed south and south-southeast into Denham Sound, but current direction gradually swings into a westerly path toward Cape Peron with progress of the flood-tide phase. Waters entering through Blind Strait (Fig. 2) initially are directed north, but are deflected to east and south-southeast across the Bar Flats Sill into Freycinet Reach (Fig. 5).

Tidal currents in Denham Sound show departures from the general south-southeast direction. Water in southern parts of the sound is driven southeast into the Bar Flats Sill, and there is flow through channels to the south-southeast. However, much of the water impinging on the sill and the northwest margin of Peron Peninsula is deflected to give a northeast flow around Cape Peron (Fig. 5). This flow, which probably is reinforced by a strong littoral drift during periods of southerly winds, is directed into northern Hopeless Reach in a direction normal to the general south-southeast flow from Uranie Strait. A station on the west side of Cape Peron showed a strong flood dominance with peak velocities of 1.6–1.8 ft/sec (49–55 cm/sec); these velocities contrast with 0.6–1.0-ft/sec (18–30-cm/sec) peak velocities for flood tide in Uranie Strait and Hopeless Reach. Because of the velocity contrast, water crossing Cape Peron Bank tends to retain an identity as a current stream which flows northeast and east in Hopeless Reach parallel with the trend of isohalines in the area (Fig. 5). The stream, which consists of mixtures of oceanic and metahaline water (37–43 ‰), may be identified with the Cape Peron salinocline (37–41 ‰), discussed hereafter. A station in north-central Hopeless Reach showed an east and southeast current flow at 0.5–0.6 ft/sec (15–18 cm/sec) at peak flood.

Current flow on ebb tide.—Current-direction data for ebb-tide conditions are shown in Figure 5. The general direction of currents during this tidal condition is north-northwest, but there are local departures. Current directions in Hopeless Reach and Uranie Strait are north-northwest toward Geographe Passage, and current directions in Denham Sound are in a similar sense toward Naturaliste Passage. Data from a station on the west side of Cape Peron indicated a short southwest flow from Hopeless Reach into Denham Sound. In open-reach and sound areas, ebb-current velocities range from 0.3 to 1.0 ft/sec (9 to 30 cm/sec), and velocities in restricted channels may reach about 2 ft/sec (60 cm/sec).

HYDROLOGIC ENVIRONMENT

Studies of the hydrologic environment were carried out over a period of years. They began

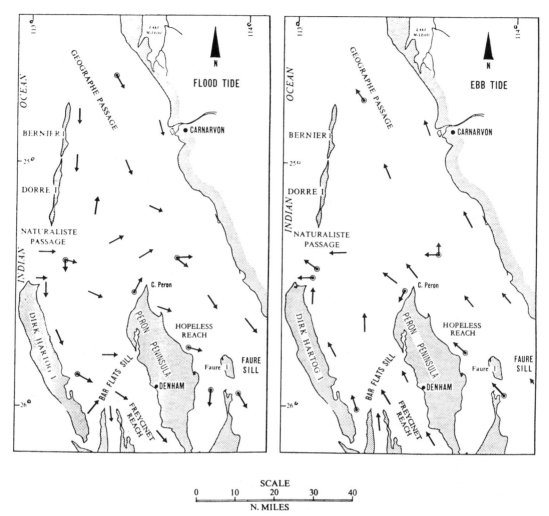

SCALE

| 0 | 10 | 20 | 30 | 40 |

N. MILES

FIG. 5.—Tidal-current flow directions for flood and ebb conditions; note reverse flow with flood tide in vicinity of Cape Peron. Directions were determined by observation at stations shown by circles; other directions from Admiralty Chart 518.

with preliminary surveys in 1956–1958 and continued with more systematic coverage in February 1964, November-December 1964, February 1965, July-August 1965, August 1966, February 1967, and July 1967. Environmental parameters measured were salinity, temperature, and oxygen content. Work also has been done on distribution of the nutrient element phosphorus in the water and on nitrogen in bottom sediments.

Salinity and Hydrologic Structure

Negative gradients.—The distribution of salinity in the embayment is similar to that of a "negative" lagoon ("negative" used in sense of Emery *et al.,* 1957), where there is a gradient from oceanic salinity at the seaward opening to hypersalinity in bay heads (Fig. 6A, B). The gradient in Shark Bay is from oceanic salinity of 35–38 ‰ in the north to high values of 60–65 ‰ in Hamelin Pool and Lharidon Bight and 46–53 ‰ in Freycinet Basin. Small inlets in Edel Land also have negative salinity gradients. Salinity in southern extremities of Useless, Boat Haven, and Depuch Inlets is 53–58 ‰, whereas salinity at the mouths of these inlets generally is 40–46 ‰.

During the period of observation, waters in

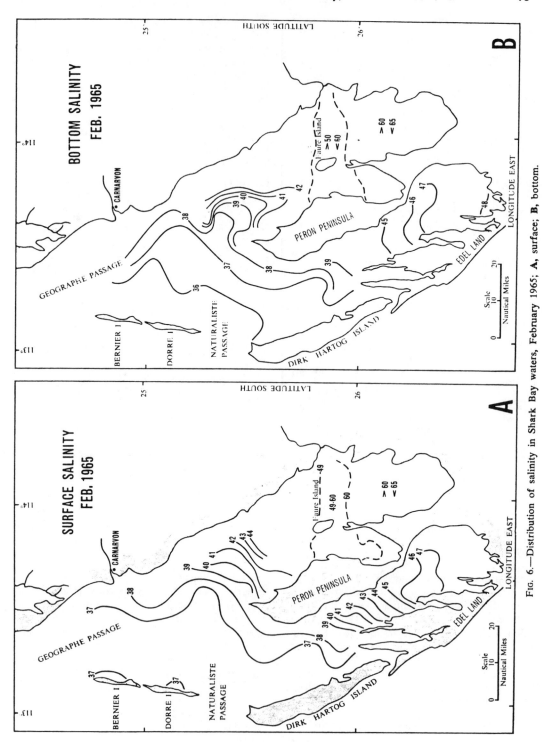

Fig. 6.—Distribution of salinity in Shark Bay waters, February 1965; A, surface; B, bottom.

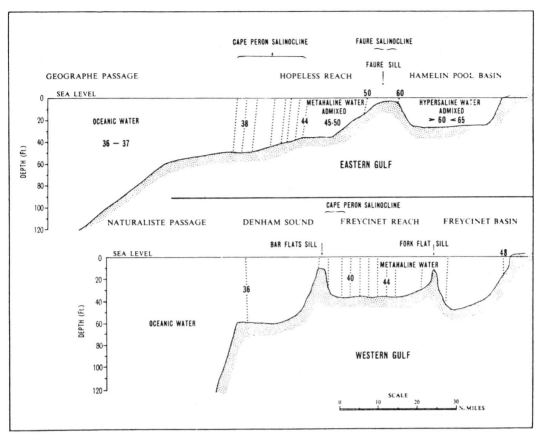

Fig. 7.—Diagrammatic cross sections illustrating hydrologic structure in Shark Bay. Water mass is layered; layers are nearly vertical.

the embayment were isosaline, surface to bottom. Observations of stratification are limited to local areas (and certain times); for example, a high-salinity tongue was observed in Geographe Passage during cruises in February and November 1964. Other stratification was associated with momentary overflow phenomena occurring in channels at about slack-tide periods. The isosaline water column set against pronounced horizontal gradients suggests that turbulent admixing, surface to bottom, by currents and wave turbulence is intense. This interpretation is supported by data concerning oxygen, which show high levels of O_2 saturation in bottom water, and also by temperature data, which indicate rapid diurnal fluctuation in bottom-water temperature in response to atmospheric fluctuation.

Hydrologic structure.—The water body has a layered structure, each layer having essentially uniform salinity and density characteristics; the layers are arranged in nearly vertical attitude (Fig. 7). The validity of this structure is supported by salinity gradients, isosaline water column, and observations of full-column homogeneous flow in surface-to-bottom current profiles. The spacing of the isohalines down the length of the embayment is irregular. Isohalines are crowded in salinoclines or zones of steep gradient which are localized either by shoaling (sills) or by hydrodynamic effects. Salinoclines localized by shoaling are permanent features, whereas clines formed by hydrodynamic interaction are subject to decay on a seasonal or other basis. In broad, open embayment areas gradients are commonly low or ill defined, and water there may be characterized broadly as cells or masses in which salinity and density values are between defined limits.

Major salinoclines are the basis for subdivision of embayment waters into three major types:

1. Oceanic, characterized by salinity of 35–40 ‰; in the northern embayment.

2. Metahaline, characterized by salinity of 40 to about 56 ‰; in Hopeless Reach, Denham Sound, and Freycinet Basin.

3. Hypersaline, 56–70 ‰ salinity; in Hamelin Pool and Lharidon Bight. Small bodies of hypersaline water also are present in the southern parts of inlets in Edel Land.

Oceanic water in northern Shark Bay is separated from metahaline waters in midbay areas by the Cape Peron salinocline, which extends

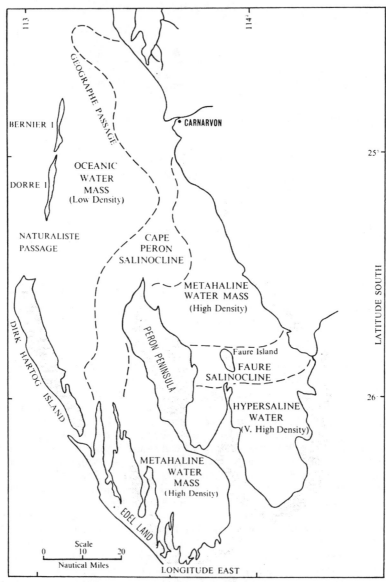

Fig. 8.—Distribution of major salinoclines in Shark Bay waters; salinoclines are viewed as interfaces which partition water body into three major types: oceanic, metahaline, and hypersaline. Compare with Figure 6.

in a narrow zone along the northern margin of the Bar Flats Sill to Cape Peron and across the north of Hopeless Reach to the vicinity of Carnarvon (Fig. 8). This interface is a hydrodynamic barrier which changes in response to external factors. Hypersaline waters in Hamelin Pool and Lharidon Bight are separated from metahaline waters by the Faure salinocline, which is located on the Faure Sill. Other salinoclines of small magnitude are present in several parts of the embayment.

[*Editors' Note:* Material has been omitted at this point.]

Temperature

Semisynoptic data for hydrologic parameters were obtained during February and August 1965 when several vessels were operated simultaneously. The patterns of temperature distribution during these sampling periods are summarized in Figure 12.

In shallow marine waters, temperature is a parameter which fluctuates with changes in atmospheric temperature. Diurnal and seasonal effects are strongly marked in shallow water over the sublittoral platform and other shoal areas in the southern embayment. Diurnal fluctuation generally is about 1°C, but may be extended in shallow areas to as much as 5°C. Winter minimums in the southern embayment are 15–18°C and summer maximums are in the range 26–30°C. Temperature is a more conservative property in deeper waters of

northern Shark Bay, where it is influenced largely by temperature of oceanic waters entering from the Dirk Hartog Shelf. Temperatures in northern bay waters fluctuate on a seasonal basis under the influence of seasonal changes in the waters on the shelf. Temperature data for shelf waters first were summarized by Rochford (1951c, 1953) and Rochford and Spencer (1957). Minimum temperatures are recorded in August, when the 21°C isotherm is in the vicinity of Carnarvon; maximum temperatures occur in March, April, and May (24–25°C), and thereafter temperature declines to the August minimum. From August to December, temperatures remain relatively constant in the 21–22°C range.

Oxygen

Bottom waters within Shark Bay generally are saturated with oxygen. Large diurnal fluctuations in O_2 content occur in proximity to seagrass stands.

Organic Matter

Shark Bay is inhabited by a rich fauna which depends on a supply of nutrient elements for existence. In many estuaries and lagoons, rivers are an important external source of nutrients. However, river influx into Shark Bay is negligible, and there is little possibility of abundant food from this source. The ocean is another potential source of nutrients, but Rochford (1951b) has shown that waters of the Dirk Hartog Shelf have very low concentrations of nutrient elements—nitrogen and phosphorus. Values of dissolved P_2O_5 in waters from northern Shark Bay were in the range 0.1–1.0 ppm, and in many samples no P_2O_5 was detected. The small quantities of nutrient elements entering Shark Bay with oceanic influx probably are retained and distributed through the system by biologic agents.

The development of nutrient-rich waters and substrates in the embayment is probably dependent on the prolific growth of plants; in oceanic and metahaline areas, dense growths of seagrasses contribute large quantities of organic material—rhizomes, fronds, and finely macerated debris. Hypersaline basins have rich floras of blue-green and green algae, and sediments in these areas commonly contain algal mats, filaments, sporangia, and fine gelatinous material.

The most satisfactory method of estimating the relative amounts of organic matter in Shark

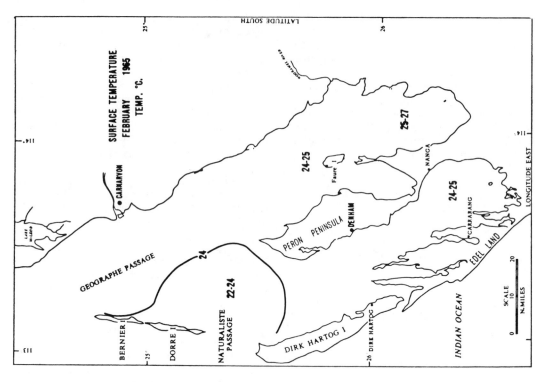

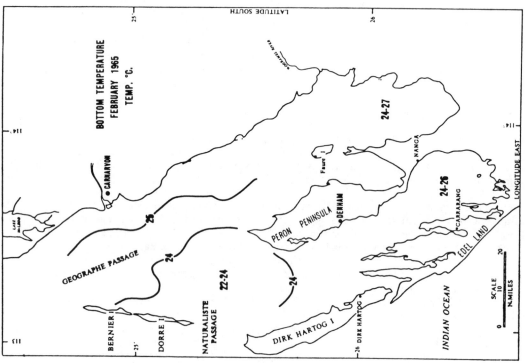

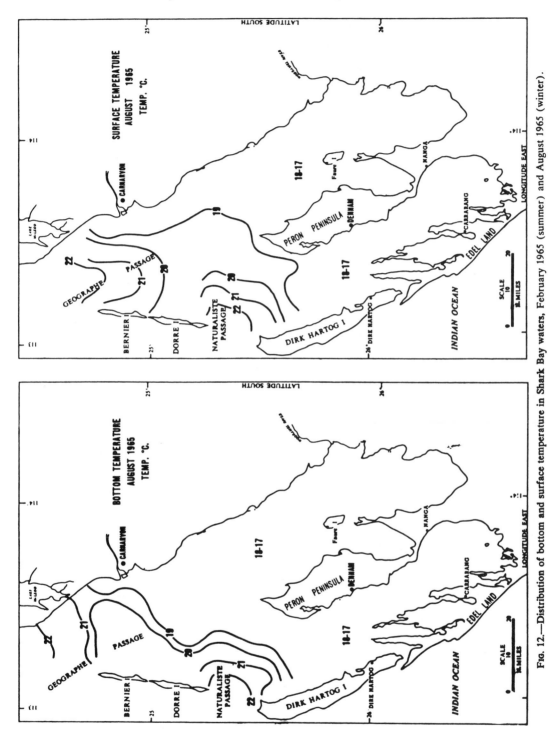

Fig. 12.—Distribution of bottom and surface temperature in Shark Bay waters, February 1965 (summer) and August 1965 (winter).

Table 5. Concentration of Total Nitrogen
($^o/_{oo}$) in Shark Bay Environments

Environment	Oceanic	Metahaline	Hyper-saline
Sublittoral platform	0.08–0.2	0.08–0.2	0.8
Seagrass meadow	0.40–1.0	0.40–1.0	—
Open strait or reach	0.15–0.7	0.30–0.8	—
Restricted channel	0.18	0.30	—
Channel fan or delta	0.70	1.10–1.7	2.0
Basin	—	0.80	1.4
Mean total nitrogen	0.4	0.75	1.3

Bay sediments has been the determination of total nitrogen content. Concentrations of this element can be used as a rough index of organic matter and relative nutrient level of sediments. Table 5 is a summary of total nitrogen values in sediments from various environments.

Because much organic matter is in fine, particulate form, it is essentially similar in depositional behavior to other silt- and clay-size particles. Trask (1932) found a general relation between grain size and organic content of sediments, *i.e.*, an increase of organic matter with decreasing grain size. This relation also can be demonstrated in Shark Bay, where fine, silty sediments generally contain higher concentrations of nitrogen than sandy sediments. The relation of sediment size to organic-matter content reflects a balance between deposition and/or erosion of fine particles. Organic-rich sediments are located in deeper basin areas, seagrass meadows, channel deltas (commonly stabilized and protected by seagrass), and intertidal mud-flat environments. Sediments on the sublittoral platform, which are subject to continual wave agitation, and sediments in channel floors and open reaches, where tidal currents are strong, generally have low values of total nitrogen. There is also a marked relation between water type and nitrogen content of sediments. Mean total nitrogen levels in sediments are 0.4 ‰ in oceanic areas, 0.75 ‰ in metahaline areas, and 1.3 ‰ in hypersaline areas.

BIOLOGIC FACTORS

Shark Bay is inhabited by a rich and diverse assemblage of organisms. Many organisms and organic communities are important in carbonate sedimentation. The shell-secreting benthos is a major source of carbonate particles; the contributors, in general order of abundance, are coralline algae, mollusks, foraminifers, echinoids, serpulids, and bryozoans. Organisms, mainly plants, also influence sedimentation by modifying the environment. Seagrass communities stabilize the sea floor, function as sediment traps and baffles, and provide a substrate for habitation by a rich epibiota which contributes skeletal debris; seagrasses also may be a basic link in food chains. Algal mats function as sediment-binding and trapping agents in intertidal locations, and mangroves considerably alter intertidal environments in limited areas. Organisms also are important in early diagenesis. Activities of burrowing and scavenging forms, such as echinoids, mollusks, crustaceans, and worms, destroy primary depositional structures and textures, and generally admix sediments into homogeneous strata. Scavenging, burrowing, and boring species are also important in fragmentation of skeletal elements and alteration of carbonate grains; common boring forms include blue-green algae, sponges, and pelecypods.

Little information has been published concerning the biota; the best known invertebrate groups are foraminifers (Logan, 1959) and mollusks. Clark (1946) records various species of echinoderms, but virtually nothing has been published on coelenterates, bryozoans, crustaceans, and worms, all of which are abundant. There has been intensive study of penaeid crustaceans, which form the basis of a commercial fishery (Slack-Smith, 1967). Shark Bay is the center of an important fishing industry based chiefly on populations of mullet (*Mergil* sp.), whiting (*Sillago* sp.), and snapper (*Chrysophrys* sp.)

Biogeography

Coastal waters of Western Australia are part of the Indo–West Pacific biogeographic province (Eckman, 1953), and the floras and faunas have marked affinities with those described from other areas within the province. Shark Bay is near the northern margin of a subtropical biogeographic region (Flindersian province), which extends from the southern coast of Australia along the west coast to the vicinity of the 24th parallel. North of this latitude there is a transition from the subtropical assemblages of southern waters to the tropical assemblages of the northwestern shelf. A few species in Shark Bay have northern tropical affinity, but tropical species are mostly absent.

The subtropical biota is a reflection of low water temperatures during winter months, when the range is 15–18°C, similar to that of southern Australian waters.

Major Biotic Subdivisions

The biota is recruited from marine stocks which can tolerate high salinity and environmental fluctuations. There is a large number of biotopes, each inhabited by characteristic communities of organisms. Limits of these biotopes appear to be related to depth, current action, wave action, substrate type, and other factors. Salinity controls distribution of organisms on a broad scale and causes a threefold biotic zonation: (1) oceanic (salinity 35–40 ‰), (2) metahaline (salinity 40–56 ‰), and (3) hypersaline (salinity 56–70 ‰).

Limits of biotic zones roughly coincide with limits of water types described on a preceding page (Fig. 13). Thus the Cape Peron salinocline marks the transition from oceanic faunas to those of metahaline aspect; similarly, there is a transition from metahaline to hypersaline biotas across the Faure salinocline. The overall trend in distribution of benthonic organisms through the salinity gradients is of increasing restriction at all taxonomic levels. For example, benthonic foraminiferal populations in the oceanic zone contain 93 species of 22 families; in the hypersaline zone, foraminiferal population is composed of 16 species belonging to 5 families. Similar restrictions are imposed on other benthonic groups.

Metahaline and oceanic biota.—Metahaline and oceanic biotic zones occupy middle-bay areas and northern Shark Bay (Fig. 13). Detailed work on foraminifers in these areas has been reported (Logan, 1959). Davies (1970a) gives an account of communities along the eastern margin of Shark Bay. Oceanic and metahaline zones have many species in common; distinctions are based on presence or absence of stenohaline components, which are limited to areas with salinity of less than 39–40 ‰. Stenohaline components of the oceanic biota include several foraminifers, most corals, and a few mollusks.

The most important plants are seagrasses which extend from the lower intertidal zone to a depth of about 50 ft. Common species are *Cymodocea antarctica* (Labill.) Endl. and *Posidonia australis* Hook. F.; other seagrass species are *Halophila spinulosa* (R.Br.) Aschers, *Hal-*

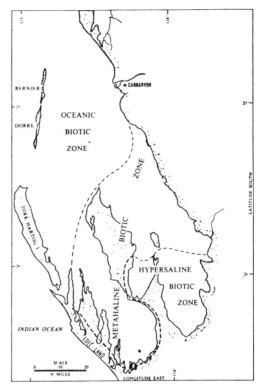

FIG. 13.—Biotic zones in Shark Bay based principally on foraminiferal populations. Compare with Figures 6 and 8.

ophila ovalis (R.Br.) Hook. F., *Cymodocea isoetifolia, Zostera muelleri,* and *Diplanthera* sp. Seagrass stands support abundant epibionts including coralline algae, foraminifers, gastropods, bryozoans, and calcareous worms. Organic-rich substrates between the plants support a benthonic community of foraminifers, mollusks, echinoderms, numerous crustaceans, and fish. Mangrove thickets are developed along the eastern coast and are particularly dense on the Boodalia subdelta, south of the Gascoyne River (Fig. 2). The range of mangroves apparently is limited by salinity; they extend south along the eastern coast, decreasing in size and density of growth. South of lat. 25°30′S, mangrove communities give way to algal-mat communities (Davies, 1970a). Algae are abundant in oceanic and metahaline zones. Continuous mats of blue-green and green algae are developed in some intertidal locations, behind spits and beach ridges. Algae of various types are common elements of seagrass communities (Davies, 1970a).

Foraminiferal faunas of oceanic and metahaline areas are known from earlier work (Logan, 1959). Many species are common to both biotic zones. Representatives of the Miliolidae, Peneroplidae, and Nonionidae are dominant. Species of the families Camerinidae, Heterohelicidae, Amphisteginidae, Cymbaloporidae, Globigerinidae, and Globorotalidae are restricted to oceanic waters. These families are apparently stenohaline marine stocks which can penetrate lagoons only where conditions are similar to those of the open sea. The mollusks are one of the most important organic groups contributing skeletal material to carbonate sediments. Knowledge of molluscan faunas has been accumulated during the course of studies of sedimentation. The fauna is diverse, having many common species; community associations are described hereafter. Knowledge of the occurrence of other organic groups is scanty. Echinoderms are apparently abundant, as numerous plates are found in bottom sediments. Bryozoans and calcareous worms commonly are observed in the epibiota on seagrass fronds. Commercial fishing is carried out in the metahaline and oceanic zones, and species of noncommerical fish and crustaceans are numerous.

Hypersaline biota.—The hypersaline biota consists of a small, distinctive assemblage of organisms which tolerates salinity of 56–70 ‰; transitions from metahaline assemblages occur in areas where salinity is normally in the range of 53–56 ‰. Species in the hypersaline assemblage also are present in metahaline and oceanic assemblages but are not common. Restrictions imposed on the biota in hypersaline areas probably result in lower competition and predation; thus species able to tolerate high salinity may become abundant.

Filamentous and unicellular blue-green algae form widespread mats on intertidal surfaces. The mats are responsible for formation of stromatolitic sediments (Logan, 1961; Logan et al., 1964; Davies, 1970b). Similar mats are present in hypersaline parts of Laguna Madre, Texas (Rusnak, 1960), and Laguna Ojo de Leibre, Mexico (Phleger and Ewing, 1962). A thin, incoherent layer of blue-green algae also may develop on the sublittoral platform, but is observed only at certain times and does not have a sediment-binding function. Species of green algae (*Acetabularia calyculus* and *A. peniculus*) are abundant and make notable contributions of skeletal material to sediments. *Penicillus nodulosus* (Lamx.) Blainville is common on muddy substrates in basin areas.

Living foraminiferal populations in the hypersaline zone have been studied (Logan, 1959). Biotic restriction is well illustrated by the foraminifers: the live population is limited to three common species, *Miliolinella circularis* var. *cribostoma* (Heron-Allen and Earland), *Peneroplis planatus* (Fichtel and Moll), and *Spirolina hamelini* (Logan, 1959), and the less common *Quinqueloculina laevigata* d'Orbigny, *Q. neostriatula* (Cushman), *Q. seminulum* (Linne), and *Spirolina acicularis* (Batsch). These species are members of the families Miliolidae and Peneroplidae.

The molluscan contribution to the assemblage is limited to a few species of pelecypod and a few small gastropods. The small pelecypod *Fragum hamelini* Iredale is present in vast numbers. This species, which secretes a shell 1 × 1 cm, is the dominant organism on the sublittoral platform in Hamelin Pool and Lharidon Bight. The *Fragum* population appears in vast numbers only at certain times. Living *F. hamelini* were abundant in 1956 and 1957. However, during February 1965 no living individuals were found, and coquina surfaces were covered with a semicoherent organic slime. In February 1966 large numbers of living *F. hamelini* were observed: as many as 70 individuals occupied 0.5 sq ft (4.5 sq cm) of the sea floor, and individuals in this population were all about the same size and about half the size of mature individuals. The size distribution suggests that the population represents a single generation and that repopulation possibly is from reproductive populations living outside the hypersaline zone.

The death assemblage in the hypersaline zone contains organisms which populated the area during periods of lower salinity; some foraminiferal species also may have been transported from the adjacent metahaline zone. The foraminiferal species *Discorbis vesicularis* var. *dimidiata* (Parr), *Elphidium crispum* Linne, and *Marginopora vertebralis* Blainville, which are present in trace quantities in the hypersaline zone, are common in metahaline communities. A large component of the death assemblage consists of forms reworked from Pleistocene strata which underlie postglacial sediments. This *remanié* component, which consists chiefly of robust forms of pelecypods, gastropods. and a few anthozoans, is most evident in sublittoral-platform and intertidal lithotopes where Pleistocene marine units have been eroded deeply.

MARINE ENVIRONMENTS OF SHARK BAY

Depth and salinity exert the principal control on distribution of environments within Shark Bay. Depth-controlled environments extend across oceanic, metahaline, and hypersaline subdivisions imposed by the distribution of salinity. There are three main depth zones: (1) embayment plain, (2) sublittoral platform, and (3) intertidal-supratidal platform. The zones contain several localized environments limited by such variables as current and wave action, substrate, biota, turbidity, and tide.

Embayment-Plain Environments

The embayment plain includes all areas in which depths are greater than 25–30 ft. Its dis-

tinctive environments, which are limited by local bathymetry, salinity, and other factors, are (1) oceanic strait, (2) metahaline reach, (3) metahaline basin, and (4) hypersaline basin. Environmental parameters for these environments are summarized in Table 6. The common characteristic in the environments is that the sea floor is at depths where prevailing wave currents do not move sedimentary particles. Wave generation is fetch limited. In the landlocked southern areas, "wave base" is at depths of 10–15 ft; in the wide northern embayment, "wave base" may be at depths of 30–35 ft (Fig. 4). It is likely that most embayment-plain sediments are stirred by wave currents during cyclones. Embayment-plain environments are gradational into sublittoral-platform environments. The embayment plain is set apart from other marine environments by a sparser development of plants (Fig. 14-3). Seagrasses grow in patchy developments, becoming more abundant in shallow marginal areas.

Table 6. Summary of Parameters, Embayment-Plain Environments

Parameter	Oceanic Strait		Metahaline Reach	Metahaline Basin	Hypersaline Basin
	Uranie Strait, Denham Sound		Hopeless Reach, Freycinet Reach	Freycinet Basin ←Lharidon Bight→ Hamelin Pool Basin	
Depth (ft)	30–120		40–60	30–50	25–30
Salinity Range (°/∞)	35–38; diurnal 1–2		38–48; diurnal 2–4	45–48; diurnal 1–3	50–65; diurnal 5
Temp. (°C) Summer Winter	24–25 22–19		24–25 18–17	24–25 18–17	25–27 18–17
Oxygen Content	Saturated 4–5 ml/l; diurnal fluct. 2–6 ml/l near seagrass stands			As in oceanic strait	As in oceanic strait
Turbidity	Moderate	High		Moderate, fine particulate matter in suspension	Low
	Fine particulate matter in suspension				
Substrate Nitrogen(°/∞)	0.15–0.77		0.3–0.8	0.8	1.4
Sediments	Variable, coarse- to medium-grained skeletal-fragment sands, silty in patches			Silty, fine-grained skeletal-fragment sands	Silty, coarse- to medium-grained microcoquinas
Wave Action	Slight; areas near Naturaliste and Geographe Passages may come under influence of oceanic swell		Slight under prevailing conditions	Slight under prevailing conditions	Slight under prevailing conditions
Tidal Currents	0.3–1.0 ft/sec (9–30 cm/sec)		0.5–1.5 ft/sec (15–45 cm/sec)	No data available	Zero to 0.1 ft/sec (0–3 cm/sec)
Biologic Factors	Flora sparse, seagrasses in scattered stands in shallow marginal areas; shell-secreting benthos generally sparse, mainly foraminifers, pelecypods. and irregular echinoids. Major assemblage is *Corbula* community. Sediments are extensively burrowed and skeletal grains are fragmented.			Flora sparse, *Cymodocea* in scattered stands in shallow marginal areas; shell-secreting benthos sparse, mainly foraminifers, pelecypods, and echinoids. Sediments heavily burrowed; skeletal grains fragmented.	Flora sparse, mainly green algae. Fauna restricted to a few species not prolific in basin areas; sediments not burrowed; skeletal grains intact.

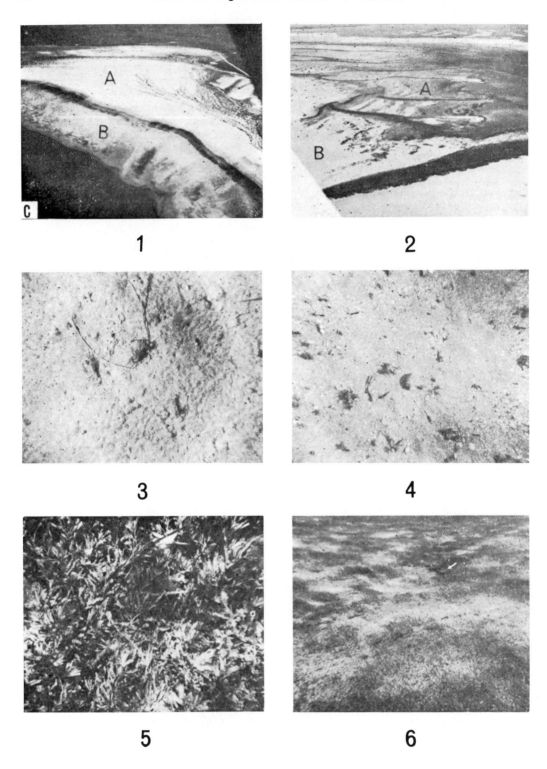

The density of shelly benthos in most embayment-plain habitats is low, although motile benthonic forms, especially crustaceans and fish, may be abundant at times. A notable feature of pelecypod faunas is the large number of small or juvenile individuals.

Oceanic-strait and metahaline-reach biota. —The main plants inhabiting these environments are seagrasses, which grow in scattered stands; the main species is *Posidonia australis*. The seagrasses support a rich epibiota of calcareous algae, foraminifers, bryozoans, gastropods, and worms. Variations in density of plant growth are reflected in the composition of bottom sediment and benthonic communities. The foraminiferal faunas are variable but typically are composed of *Amphistegina lessonii* d'Orbigny, *Cibicides refulgens* Montfort, *Discorbis vesicularis* var. *dimidiata*, *Elphidium crispum*, *E. craticulatum* (Fichtel and Moll), *Marginopora vertebralis*, *Peneroplis planatus*, *Planispirina exigua* H. B. Brady, *Quinqueloculina anguina* Terquem var. *arenata* Said, *Q. laevigata* d'Orbigny, *Q. vulgaris* d'Orbigny, *Rotalia audouini* d'Orbigny, *Textularia agglutinans* d'Orbigny, and *Triloculina tricarinata* d'Orbigny. *A. lessonii* and *C. refulgens* are stenohaline forms limited to oceanic areas and absent from the metahaline-reach habitat. The molluscan fauna is imperfectly known, but a general account can be given from data obtained in dredging and trawling. The most abundant mollusk is a small pelecypod, *Corbula* sp., which is an important contributor of skeletal debris to bottom sediments. The *Corbula* community extends throughout deeper areas of the oceanic strait and metahaline reach (Tuppin. 1969). Commercial populations of the saucer scallop, *Amusium pleuronectes* Linne, occur in part of Denham Sound, and individuals of *Chlamys asperrima* Lamarck are numerous. *Katelysia strigosa* Lamarck is common in dredgings. and other pelecypod species are present in smaller numbers. Very few gastropods are found in dredge samples. Few data are available on the abundance or distribution of other organisms. Bottom photographs of embayment-plain habitats show numerous mounds, but very few animals are present on the surface (Fig. 14-3). Irregular echinoids are abundant in many trawls, and it is likely that these organisms form the mounds. Large populations of penaeids are present at certain times, and crabs (*Portunus* sp.) are also abundant.

Metahaline-basin biota.—Plants are scarce in deeper parts of the metahaline basin, but seagrass stands grow on shallower fringes. The sparse fauna is poorly documented. The foraminiferal death assemblages consist of *Quinqueloculina laevigata*, *Peneroplis planatus*, *Discorbis vesicularis* var. *dimidiata*, *Elphidium crispum*, *Triloculina rotunda* d'Orbigny, *Elphidium simplex* Cushman, *Quinqueloculina neostriatula*, and *Triloculina trigonula* (Lamarck).

Hypersaline-basin biota.—The biota in the hypersaline basin is restricted to a few salt-tolerant species which are present in small numbers. *Penicillus* and *Acetabularia* grow in shallow marginal areas, and the fauna consists of a few forams, small mollusks, serpulids, and juvenile crustaceans.

Sublittoral-Platform Environments

The sublittoral platform is a shallow, submerged zone which ranges in width from a few hundred yards to several miles (Fig. 14-1). The inner margin of the platform is gradational from the lower intertidal zone at LWS level, and the surface generally extends in a gentle, even slope to depths between 5 and 25 ft. From there the slope steepens and descends to the embayment plain (Fig. 2). The major physical parameter operative in the environment is wave-current action. There is a marked relation between wave action and configuration of the platform. Platforms of north and north-

← ≪≪

FIG. 14.—Environments of Shark Bay. **1.** Aerial view of coastal zone at southern end of Hamelin Pool, showing relation of intertidal-supratidal platform (*A*), sublittoral platform (*B*), and embayment plain (*C*). **2.** Aerial view of Faure Sill showing seagrass meadows (*A*), sand flats (*B*), and blind channels in sill. **3.** Underwater scene of embayment plain in Hopeless Reach. Sediments are burrowed extensively by organisms. Mottles are 2–4 cm. **4.** Underwater scene of sublittoral sand-flat environment, metahaline zone. Vertical photograph. Coarser shell debris is 2–3 cm. **5.** Underwater scene of dense seagrass meadow (*Cymodocea*). Leaves are 1–2 cm. **6.** Underwater scene of sublittoral sand-flat environment, hypersaline zone. Sediment is *Fragum* coquina with population of green algae (*Acetabularia*). Hammer shows scale.

east aspect are leeward locations protected from prevailing wave trains. Exposed platforms of south and southwest aspect are subject to frequent wave action; they typically are broad and have gentle slopes.

The environment of the sublittoral platform is rigorous. Oscillatory wave currents are capable of moving sand-size debris, and coarser debris may be moved during storms. Parts of the platform also are swept by tidal currents ranging up to about 25 cm/sec that can move sand-size particles. There is considerable diurnal fluctuation in temperature and commonly in salinity. Oxygen content of the waters is usually high (saturated), but there may be considerable fluctuation between saturation and undersaturation near dense seagrass stands. Seagrasses, mainly *Cymodocea*, colonize the platform in oceanic and metahaline areas. The plants form dense, meadowlike stands which considerably modify the environment (Fig. 14-5). The roots and rhizomes stabilize the substrate, and the fronds form a baffle which inhibits working of bottom sediments by current action.

There are many sublittoral-platform environments, limited by local environmental factors; however, three subdivisions characterized by somewhat uniform sediments and organic assemblages can be described: (1) sand-flat environment in oceanic and metahaline zones; (2) sand-flat environment in hypersaline areas; and (3) seagrass-meadow environment, limited to metahaline and oceanic areas. Parameters for these environments are summarized in Table 7.

Biota of sand-flat environment, oceanic and metahaline zones.—The sand-flat environment extends from the intertidal zone to a depth of about 15 ft. Broad sand flats are developed in many areas, but in other areas sand-flat conditions hold in bare patches of varied size in seagrass meadows (Fig. 14-2). The sand-flat environment is rigorous; mobile substrates and considerable fluctuations in hydrologic parameters are characteristic (Fig. 14-4). The sand-flat biota is generally sparse. The flora is limited to sparse growths of seagrass in protected or deeper areas. Foraminifers are limited and the fauna is restricted to *Peneroplis planatus*, *Miliolinella circularis* var. *cribostoma* (in protected areas), and *Spirolina* sp., along with a few miliolids and rotalids. Mollusks are the most abundant elements of the biota. The faunas are dominated by sand-burrowing pelecypods; gastropods become more abundant near seagrass stands. There is considerable variation in composition of the fauna, but several fairly consistent assemblages are evident: *Pinna-Pinctada*, *Fragum-Hemicardium*, *Costacallista-Anomalocardia*, and *Peronella*-Mollusca.

The *Pinna-Pinctada* assemblage, which is associated with sparse seagrass growth in shallow subtidal sand flats in the oceanic and low-salinity areas of the metahaline zone, is described by Davies (1970a). The razor clam *Pinna bicolor* Gmelin is dominant and *Pinctada carchariarium* Jameson is subdominant. Specimens of *Pinna* generally are embedded in the sea floor, whereas those of *Pinctada* are attached to seagrass fronds and larger shell debris by a strong byssus. These species support an epibiota consisting of *Chama* sp., serpulids, foraminifers, and bryozoans. Vagile and sessile forms include the pelecypods *Cardita incrassata* Sowerby, *Fragum unedo* Linne, *Costacallista impar* Lamarck, *Placamen* sp., and *Anomalocardia squammosa* Linne, and numerous other less abundant species. Gastropods are not common, but *Cerithium aluco* Linne, *Euplica bidentata* Menke, and several other species are present. Other organisms associated with the assemblage are bryozoans, irregular echinoids, asteroids, and the green alga *Penicillus*.

The *Fragum-Hemicardium* community inhabits shallow areas which are exposed during spring tides. The pelecypods *Fragum unedo* and *Hemicardium hemicardium* are characteristic; other bivalves include *Divaricella ornata* Reeve, *Placamen* sp., *Gafrarium intermedia* Reeve, *Circe scripta* Linne, *Pitarina citrina* Lamarck, and *Costacallista impar*. Gastropods, which are not common, include *Diodora (Austroglyphis) rugosa* Theile, *Calliostoma marginata* Woods, *Prothalotia* sp., *Phasianella solida* Born, *Tricolia* sp., *Cerithium aluco*, *Rhinoclavis vertagus* Linne, and *Euplica bidentata*. The scaphopod *Laevidentalium* is common.

The *Costacallista-Anomalocardia* association inhabits protected sandy areas in southern (high-salinity) parts of the metahaline zone (salinity 46–56 ‰). The association is dominated by pelecypod species *Costacallista impar*, *Bassina berryi* Gray, and *Anomalocardia squammosa*; other pelecypods common in the assemblage are *Circe scripta*, *Gafrarium intermedia*, *Hemicardium hemicardium*, *Fragum unedo*, *Fragum hamelini*, *Divaricella ornata*, *Tellina virgata* (Linne), *Tapes literata*

Table 7. Major Environments of Sublittoral Platform

	Sand Flats (Oceanic and Metahaline Zones)	Sand Flats (Hypersaline Zone)	Seagrass Meadows (Oceanic and Metahaline Zones)
Area	Widespread margins of Peron Peninsula, Faure Sill, Edel Peninsula, Freycinet Reach and Basin, inlets of Edel Land, east side of Bernier Is.	Margins of Hamelin Pool, Lharidon Bight	Widespread in oceanic and metahaline zones, Uranie Strait, Uranie Bank, Cape Peron Bank, Bar Flats Sill, Freycinet Reach and Basin, Faure Sill
Limits	Transitional to intertidal zone in shallower parts and into seagrass meadows and embayment-plain environments		Transitional into intertidal zone and sand-flat environments at shallow limits and into embayment-plain environments at depth
Depth	Upper part exposed at spring tides and during southerly gales; av. depth 5 ft, extends to depth of approx. 15 ft		LWS level to about 40 ft, densest grass development in metahaline areas of southerly aspect
Salinity	35–56 °/oo extends through regional gradient; diurnal fluctuation variable depending on local gradient and movement of saline front; 1–3 °/oo in open reaches, 5–10 °/oo in salino-cline zones	56–70 °/oo; diurnal 1–10 °/oo	35–56 °/oo, diurnal fluctuation variable depending on local gradient and movement of saline front; 1–3 °/oo in open reaches; up to 10 °/oo in sill areas
Temp. (°C) Summer Winter Diurnal Rge.	22–30 15–18 1–2	22–30 15–18 1–2	22–30 15–18 1–2
Oxygen Content	Saturated, 4.0–5.5 ml/l	Saturated, 4.0–6.0 ml/l	Large diurnal fluctuations, 2.0–6.0 ml/l
Turbidity	Low	Low	Often high because of milky suspension of fine particulate matter
Substrate Nitrogen (°/oo)	0.08–0.2	0.8	0.4–1.5
Sediments	Medium-grained quartz sands, and lithoclastic sands; ripple marks and cross-bedding common	Coquinas, ooid sands, quartz sands, skeletal-fragment sands; ripple marks and cross-bedding common; sediments commonly lithified	Variable depending on density of grasses; silty skeletal-fragment sands (dense growth), clean skeletal-fragment sands (sparse growth); sediments commonly burrowed and skeletal grains fragmented
Wave Action	Dependent on weather aspect; areas of southerly aspect subject to frequent attack; oscillatory currents 20–70 cm/sec		Dependent on weather aspect and depth, as in sand-flat environments
Tidal Currents	Dependent on location, usually moderate to strong up to about 1.5 ft/sec (45 cm/sec)		Dependent on location, up to 2.0 ft/sec (60 cm/sec)

(Linne), and mytilids; *Pinna bicolor* and *Pinctada carchariarium* are conspicuously absent. Gastropods in the assemblage are *Rhinoclavis vertagus*, *Cerithium aluco*, and numerous small cerithiids. The scaphopod *Laevidentalium* is common.

The *Peronella*-Mollusca assemblage is limited to local areas and patches in seagrass meadows, generally in depths of 10–20 ft. The assemblage is characterized by an abundance of the sand dollar *Peronella lesueri* Agassiz, as well as large irregular echinoids and abundant asteroids. The associated molluscan fauna is essentially similar to the *Fragum-Hemicardium* and *Costacallista-Anomalocardia* assemblages.

Biota of sand-flat environment, hypersaline zone.—The main floral elements in the environment are blue-green and green algae. At certain times an incoherent layer of blue-green algal material coats all surfaces, but at other times it is absent. Common green algal species are *Acetabularia peniculus* and *Acetabularia calyculus*, which grow attached to shells and rocks. The fauna is typical of the hypersaline biota. The main foraminiferal species are *Peneroplis planatus*, *Miliolinella circularis* var. *cribostoma*, and *Spirolina*. The richest foraminiferal population is developed as an "infauna," inhabiting the interstitial spaces in pelecypod coquinas. The most abundant organism is the pelecypod *Fragum hamelini,* which inhabits more protected parts of the environment. The pelecypod contributes large quantities of disarticulate valves to the sea floor, and one of the most typical sediments on the platform is a coquina composed almost entirely of valves of this species (Fig. 14-6).

Biota of seagrass-meadow environment.—Davies (1970a) gives a detailed account of seagrass communities and habitats on the eastern shore of Shark Bay. Modified versions of ecologic units proposed by Davies are observed elsewhere in the embayment. Three basic communities are recognized: (1) *Cymodocea* seagrass community, (2) *Posidonia* seagrass community, and (3) *Posidonia*-algal community.

The *Cymodocea* community is composed primarily of plants of *Cymodocea antarctica* and some *Posidonia australis, Halophila spinulosa,* and *Syringodium isoetifolium* (Aschers) Dandy; the brown alga *Cystophyllum muricatum* (Turn.) J. Ag. also is present (Fig. 14-5). *Cymodocea* commonly supports a large epibiota at the extremities of the leaves. The epibionts include the foraminifers *Peneroplis planatus, Vertebralina striata,* encrusting nubeculariids, discorbids, and *Marginopora vertebralis.* Articulate coralline algae, *Metagoniolithon* and *Amphiroa,* are found on the leaf axes, as well as common encrusting *Melobesia* and bryozoans. The sheltered benthos includes the gastropods *Calliostoma marginata, Prothalotia, Phasianella solida, Tricolia, Notocochlis, Euplica bidentata,* and small cerithiids. The bivalves *Pinctada carchariarium, Pinna bicolor,* and *Cardita incrassata* are uncommon. The *Cymodocea* community extends from the lower intertidal zone across the sublittoral platform to a depth of about 40 ft. *Cymodocea* has greater salinity tolerance than *Posidonia* and replaces that species on channel floors in high-salinity areas such as the Faure Sill.

The *Posidonia* community is characterized by large individuals of the seagrass *Posidonia australis; Cymodocea antarctica* is also present. An abundant epibiota lives on the plants; it includes the foraminifers *Peneroplis planatus, Marginopora vertebralis, Vertebralina striata,* encrusting nubeculariids, *Spirillina, Patellina, Discorbis vesicularis* var. *dimidiata,* and several other rotaline species. The most common coralline alga on the leaves is an encrusting species of *Melobesia;* encrusting bryozoans are also common. The sheltered benthos of the *Posido-*

nia community includes gastropods and codeacean algae. The gastropods include *Thalotia* and *Notocochlis* and numerous small cerithiids. Algae are scarce, but *Penicillus nodulosus* and *Udotea argentea* (Zanardini) are present. The *Posidonia* community is developed mainly in tidal channels, in depths from low-water level to about 30 ft.

The *Posidonia*-algal community is characterized by an association of small plants of *Posidonia australis* and numerous individuals of coralline and codeacean algae. *Cymodocea antarctica* and other seagrasses are also present. The algal elements of the community are the coralline algae *Lithophyllum* and *Goniolithon* and the codeacean algae *Penicillus nodulosus, Udotea argentea,* and *Halimeda.* Members of the sheltered benthos in the community include the gastropods *Calliostoma marginata, Prothalotia* sp., *Phasianella solida. Tricolia, Cerithium aluco, Rhinoclavis vertagus, Notocochlis, Euplica bidentata,* and small cerithiids. Pelecypod species include *Pinctada carchariarium, Pinna bicolor,* and *Cardita incrassata.*

Intertidal and Supratidal Environments

Intertidal and supratidal zones have marginal environments; they are subject to periods of marine inundation, but these periods are separated by intervals during which the substrate is exposed. Emery *et al.* (1957) have suggested a general classification of intertidal environments into three gradational categories —rocky headland and rocky shore, sandy beach, and tidal flat. These categories are valid for Shark Bay and serve as the framework for the discussion of environments. The main factors which determine the nature of intertidal environments are character of tides, frequency of exposure to waves, and salinity of influxing waters.

Conditions in the supratidal and intertidal zones vary with elevation. Areas low in the intertidal zone are submerged more frequently than areas in higher parts, whereas the supratidal zone is inundated by marine waters only during abnormal high tides and storms. The surface of the supratidal zone is commonly damp because of proximity of the groundwater table; groundwaters may be saline or brackish and seawater and brackish water commonly mix in this zone. The general intertidal environment exists in a series of zones in which conditions are determined by frequency of tidal influx.

The zonation is most marked in flats where gradual surface slope combined with tidal levels produces broad areas; zonation is condensed and less defined in localities where the slope is steeper. Variation of energy conditions with exposure to wave action is similar to that on the adjacent sublittoral platform. Intertidal areas of open southerly aspect are more frequently subject to wave action than areas of northerly aspect.

Rocky Intertidal Environments

Rocky intertidal environments exist on exposed shores where Pleistocene dune rocks or Cretaceous limestone have been eroded by marine processes. The form of the substrate is influenced greatly by lithology; there are three basic forms: (1) intertidal platform–notch–bench configuration, (2) seacliff-intertidal talus, and (3) rocky intertidal flat.

Calcareous eolianite crops out along inlets in the western part of Shark Bay. The eolianite units are markedly indurated and retain a rock platform–notch–bench morphology similar to that described from other eolianite coasts by Fairbridge (1950). The usual form is a smooth rocky platform tens to hundreds of feet wide that slopes gently up from LWS level to a notch at about HWS level. The platform is in rock which is covered intermittently by lithoclastic sand. The notch is formed in hardened eolianite which has a "honeycombed" or scalloped appearance due to chemical corrosion. The platform-notch feature extends across the regional salinity gradient in the western area, and salinity has a controlling influence on the biota. A community with prolific rock oysters (*Crassostrea*) lives on rocks in the oceanic zone and parts of the metahaline zone. As salinity increases, the oysters decrease in abundance and are replaced by a sparse community of mytilids, barnacles, and blue-green algae.

In the Peron Province, truncation of major dune trends by marine erosion has produced a series of bluffs. They rise from sea level to a calcrete rim at elevations up to 150 ft. Dune sandstone is weakly cemented and generally breaks down to quartz sand. However, at the bases of some cliffs there is a talus of sandstone and calcrete boulders broken from the rim; part of the talus lies in the intertidal zone and forms a rocky substrate. In oceanic areas and parts of the metahaline zone, the rocks are thickly encrusted with oysters (*Crassostrea*). The oysters

gradually diminish in number through the metahaline zone to a point where the rocks support only a sparse population of mytilids, barnacles, and blue-green algae.

The eastern coast of Hamelin Pool is composed of flat-lying Cretaceous and Pleistocene limestone beds which have been eroded into a series of platforms. The typical morphology is a smooth, rocky surface sloping gently up to heights of 1–2 ft above HWS level. The platforms range in width from tens of feet to several hundred yards. They lie within the hypersaline zone and are inhabited by a very restricted assemblage of organisms. Major elements in the community are blue-green algae which form mats and leathery crusts over much of the surface from below LWS level to HWS level. The algal mats function as sediment-binding agents and are responsible for formation of reeflike clusters of stromatolite structures. The stromatolites are headlike masses in the lower intertidal zone (SH structures of Logan *et al.*, 1964). The height of structures decreases shoreward toward high-water level, and higher parts of the platform are thinly veneered by sheets, platy fragments, and pebbles of indurated stromatolitic sediments.

Intertidal-Beach Environments

Beach environments are present throughout the embayment. The beaches are composed of material carried shoreward from the sublittoral platform. In oceanic and metahaline zones the inner parts of this platform are inhabited by sparse communities of shell-secreting organisms, and the rate of accumulation of skeletal debris is usually low. Most of the sediment on the inner parts of the platform consists of particles eroded from underlying strata. In Edel Land the beach sediments are composed mainly of lithoclasts and grains derived from eolianite; in the Peron Province the beach sediments are mainly of quartz grains derived from underlying sandstone. The main communities found in the beach environments of oceanic and metahaline areas are the *Fragum-Hemicardium* and *Costacallista-Anomalocardia* assemblages. The sublittoral platform in the hypersaline zone is a site of rapid sedimentation. The sediment consists of pelecypod coquinas (*Fragum hamelini*), foraminiferal-pelecypod coquinas, and ooid sands. This material is carried shoreward and deposited in beach ridges. Algal mats cover the surface in many parts of the hypersaline zone and bind much debris into stromatolitic structures and sediments. Where depositional rate exceeds the capacity of mats to bind sediment, beaches of coquina and ooid sand are developed.

Tidal-Supratidal Flats

Extensive intertidal-supratidal flats are present in southern parts of the embayment and along the eastern coast between the Gascoyne and Wooramel Rivers. There are few tidal flats in the central part, where coastal areas are characterized by rocky headlands and sandy beaches. Deltaic tidal-supratidal flats are developed on the subdeltas of the rivers. The river flats are underlain by red-brown silty sediments and are crossed by distributary and tidal channels (Davies, 1970a). The flats of the Gascoyne subdeltas merge with wide tidal flats which border the eastern coast of Shark Bay; these flats, characterized by carbonate sediments. have been developed during Pleistocene marine phases and by outbuilding of the "Wooramel" seagrass bank during the postglacial interval (Davies, 1970a). Tidal-supratidal flats occupy low, embayed areas in the terrain of Cretaceous and Pleistocene outcrops on the eastern margin of Hamelin Pool. These flats are subcircular and up to several miles wide; they are underlain by thin postglacial sequences over Pleistocene units or bedrock of Cretaceous age (limestone and dolomite). Elongate tidal-supratidal flats are developed in the interdune depressions of Edel Land. Outcrops of eolianite and fossiliferous limestone of Pleistocene age fringe these flats.

Tidal-supratidal flats may be subdivided into several gradational zones related to tide levels and other factors such as depth of the groundwater table. The mean tidal range in the northern part of Shark Bay is about 3 ft and is 5.5 ft on spring tides; mean tidal range in southern parts of the embayment is 1.5–2 ft and up to 3 ft on spring tides. The water level is greatly affected by wind set-up. During prevailing southerly winds in summer, vast areas of the flats are exposed for periods of up to a week. The gradient of the flats ranges from subhorizontal to about 2 ft/mi. Gradient largely determines the areal extent of any zone. The zones are (1) upper supratidal, (2) lower supratidal, (3) upper intertidal (inundated by spring tides), (4) middle intertidal, and (5) lower intertidal (exposed by spring tides).

Upper supratidal zone.—This zone is at heights of 5–8 ft above HWS level. It is a remnant of a Pleistocene marine phase which probably returned briefly to a supratidal location during postglacial higher sea levels (5–8 ft). The area now is never inundated by marine waters, but the surface is commonly damp because of capillary rise of groundwater. Soil formation is occurring, and in most places a low heathlike vegetation of salt-tolerant plants, *Salicornia* and *Arthrocnemum*, is present. The surface is undulate in a series of subcircular pans and low swales. The pans collect water after rains but are dry during summer. The swales are vegetated but the pans are generally devoid of plant growth. The pans are subject to deflation by winds which leave a lag concentrate of fossils derived from underlying Pleistocene units; in flats along the eastern margin of Hamelin Pool, Cretaceous bedrock commonly is exposed at the surface. The boundary between the upper supratidal zone and the lower supratidal zone is marked by stranded beach ridges which probably were formed during a 5–8-ft high sea level during postglacial time.

Lower supratidal zone.—This zone, which extends from HWS level to heights of about 5 ft above mean sea level, occasionally is flooded by marine waters; frequency of flooding is probably about once in 2–3 years. Desiccation is the normal condition, but sediments at depths of a few inches are commonly moist because of the capillary rise of groundwaters; salts (gypsum, halite, and ?dolomite) are precipitated in surface sediments. The area is dissected by shallow channels which may carry tidal waters during spring tides. These channels are lined by indurated pelmicrite and generally are floored by muddy carbonate sediments which contain pelecypod valves and coarse debris eroded from underlying Pleistocene and postglacial units. The surface of the lower supratidal zone is commonly formed by a layer of indurated aragonitic pelmicrite which is developed in continuous crusts and pavements of tabular, platy fragments ranging from pebble to boulder size. Sheets and crusts have a texture of unsupported voids (fenestrae), and show considerable evidence of solution and reprecipitation of carbonate minerals and contain abundate root casts. Crust pavements are covered by thin films of blue-green algae and a heathlike vegetation of *Salicornia* and *Arthrocnemum*.

Upper intertidal zone.—This zone is transitional between the intertidal zone, which is flooded daily or twice daily by tidal waters, and the supratidal area, which is flooded only during abnormal tides. Spring tides of 4–5 days duration recur every 10 days; during spring tides the zone is flooded and in the intervening 10 days it is desiccated. In dry periods the water table is only inches below the surface and sediments generally are kept moist by capillary rise of water. The upper intertidal zone has a development of pustular algal mats on the surface, and active mats are interspersed with areas of indurated aragonitic sediment. The lithified sediments are developed as continuous sheets, and crusts and fragments of cobble to pebble size also are common. The surface is dissected by tidal channels which carry tidal waters during neap-tide periods.

Middle intertidal zone.—This zone is between mean high-water level and mean low-water level. The surface is flooded by tidal waters daily in a 6- or 12-hour cycle, depending on tides. In hypersaline areas this zone is surfaced by smooth mats of blue-green algae which bind sediment. In oceanic areas the intertidal zone is covered by mangrove thickets.

Lower intertidal zone.—This zone is below mean low-water level; it is exposed only during spring-tide periods on a cycle similar to that of the upper intertidal zone (*i.e.*, 4–5 days of exposure on a 6- or 12-hour cycle, followed by about 10 days of submergence). The zone is transitional, and most of its characteristics reflect the predominance of marine over intertidal conditions. Marine floras and faunas are commonly diverse and similar to those described from sublittoral-platform environments.

SUMMARY OF SHARK BAY ENVIRONMENT

The fundamental factor in determining the environmental system is the configuration of the sea floor and the surrounding land—distribution of depth, length and width of the embayment, and particularly the presence of barriers. A major barrier separates the embayment from the Indian Ocean and internal barriers partition the inclosed water body into inlets, gulfs, and basins. Influx of oceanic water is only through the passages in the western barrier, and circulation is further inhibited by internal barriers and shoals, and by localization of salinoclines. The north-south elongation of the gulfs is also important in determining internal energy conditions because prevailing southerly winds generate heavy seas. Also, tidal currents in long, constricted gulfs may attain high velocities.

The main external factors are climate, wind, and tide. The climate is semi-arid to arid as evaporation exceeds precipitation; the result is a general lack of runoff and a tendency for salinity to increase in shallow, confined waters. However, evaporation alone would not be effective in producing the salinity gradients if tidal influx and exchange of oceanic water were not inhibited by shoaling and by factors related to wind and tide. Tides provide the main drive for water movement. Flood-tide currents are directed generally south-southeast and ebb currents north-northwest and northwest. Important local deviations in direction occur where currents impinge on shoals and where tidal waters are confined to narrow channels. Current velocity may reach 60–90 cm/sec in restricted channels and 15–30 cm/sec in open reaches and inlets. Tidal currents bring large quantities of oceanic water into the embayment, but most of it is removed on ebb phases. Comparison of tidal volume with evaporation loss suggests that tidal influx should be ample to maintain oceanic salinity *throughout* if free exchange were a feature of the system. The region is influenced by strong prevailing southerly winds, which generate turbulent admixing of waters, surface to bottom. Intense local admixing creates a homogeneous water column and the consequent vertical attitude of density layers characteristic of the water mass. The vertical attitude of density layers inhibits free circulation and exchange.

The interaction of tidal currents and wind-generated littoral drift promotes regional circulation. Littoral drift propelled by prevailing southerly winds is directed north along coasts of westerly aspect. The overall effect of wind drift has set up a prevailing ebb-current dominance on west-facing coasts. In contrast, on coasts of easterly aspect, lowering of water levels by wind set-up toward the north results in flood-tide dominance. The asymmetry of tide dominance leads to circulation involving net movement from west to southeast and then to east and finally northwest. The circulation helps to explain (in empirical terms) the observed characteristics of the hydrologic system: (1) patterns of salinity and density distribution with trend lines oriented generally west to east, changing to northeast and finally to northwest, (2) development of the Cape Peron salinocline and other hydrodynamic interfaces, and (3) steady-state conditions with slow rates of exchange. It follows that maintenance of the hydrologic regime is very much dependent on the prevailing winds and that long-term cessation of southerly winds is likely to enhance circulation and cause lowering of salinity.

The hydrologic system is summarized best in terms of salinity. The water body is partitioned into layers separated by nearly vertical interfaces. There is a regional gradient from oceanic water (35 ‰) in the northern bay to hypersaline waters (56–70 ‰) in the southern extremities (Fig. 6). The three main water types are (1) oceanic (salinity 35–40 ‰), in the northern embayment; (2) metahaline (salinity 40 to about 56 ‰), in Hopeless Reach, Denham Sound, Freycinet Basin, and inlets of Edel Land; and (3) hypersaline (salinity 56–70 ‰), in Hamelin Pool, Lharidon Bight, and inlets of Edel Land.

Boundaries of the main water masses are located at salinoclines. The Faure salinocline separates hypersaline waters in Hamelin Pool and Lharidon Bight from metahaline waters in Hopeless Reach. This cline is localized by the Faure Sill. The Cape Peron salinocline separates oceanic water in northern Shark Bay from metahaline waters in Denham Sound and Hopeless Reach. The cline is partly localized by the Bar Flats Sill, but it is mainly a hydrodynamic barrier maintained by tidal flow and littoral drift. The nature of salinoclines is variable depending on the interaction of the system with other factors.

The salinity gradients and other characteristics of the hydrologic system exert a control on distribution of environments at a regional level. The tripartite subdivision of the water mass is expressed in three broad subdivisions—oceanic, metahaline, and hypersaline—which are inhabited by distinctive assemblages of organisms and which contain distinctive suites of sediments. The distribution of organisms through the regional gradient is a pattern of increasing restriction of diversity in floras and faunas. Salt-tolerant forms, however, may be prolific in high-salinity conditions, probably because of limitations on predation and competition.

Other factors are limiting within the broad zones imposed by the hydrologic system. The main factors are waves, tidal currents, turbid-ity, and substrate. Plants are the main biologic factors. All of these factors interact with depth to produce a variety of local environments. Depth is the chief connecting factor and it is convenient to summarize environments and parameters in terms of three subdivisions: embayment plain, sublittoral platform, and intertidal-supratidal platform.

The embayment-plain environments are in depths from 30 to 120 ft and are mainly beyond the limit of wave action under prevailing conditions. Storm waves occasionally may scour the sea floor. Moderate to strong tidal currents sweep the embayment plain in open-reach and strait areas, but in enclosed basins tidal flow may be negligible. There is a relative constancy in oxygen content, temperature, and to some extent salinity. One of the important characteristics of embayment-plain environments is the general, sparse growth of plants. Sediments on the sea floor therefore are winnowed by tidal currents. The benthonic communities generally are composed of burrowing and vagile species of pelecypods, crustaceans, and echinoids. Standing crops are commonly low, and assemblages typically are composed of small species and juvenile forms.

The sublittoral-platform environment is subject to frequent wave action, depending on depth and weather aspect; substrates are usually sandy. Strong tidal currents may sweep the platform, thus limiting accumulation of fine sediment and nutrient materials. Oxygen levels are normally high and waters remain supersaturated; diurnal temperature fluctuation is rapid and salinity also may fluctuate greatly, depending on local movement of the regional saline front and on evaporation. Shallow sand flats have relatively sparse communities of benthos with sand-burrowing and robust vagile forms. Seagrasses are of major importance on the sublittoral platform because these plants greatly modify the environment by inhibiting wave and current action and stabilizing the substrate. The grasses also trap fine sediment, and nutrient levels in bottom sediments are enhanced so that a more numerous sheltered benthos can exist. The plants function as substrates for a prolific epibiota of foraminifers, coralline algae, bryozoans, worms, and mollusks that are attached to the stems and leaves.

There is a great variety of intertidal and supratidal environments, ranging from rocky headlands to vast tidal-supratidal flats. The factors involved in the environmental variation are substrate (morphology and type), weather aspect, tidal range, and the nature of the groundwater table. The single outstanding characteristic of tidal and supratidal environments is the range of fluctuation in conditions.

REFERENCES CITED

Australian National Tide Tables, 1968: Australia Dept. Natl. Development Div. Natl. Mapping.

Baker, B. B.. Jr., W. R. Deebel, and R. D. Geisenderfer, 1966, Glossary of oceanographic terms, 2d ed.: U.S. Naval Oceanog. Office Spec. Pub. SP-35, 204 p.

Bigelow, H. B., and W. T. Edmondson, 1947, Wind waves at sea, breakers and surf: U.S. Naval Oceanog. Office Pub. 602, 177 p.

Clark, H. L., 1946, The echinoderm fauna of Australia: its composition and its origin: Carnegie Inst. Washington Pub. 565 (4), 576 p.

Davies, G. R., 1970a, Carbonate bank sedimentation, eastern Shark Bay, Western Australia: this volume.

———— 1970b. Algal-laminated sediments, Gladstone embayment, Shark Bay, Western Australia: this volume.

Eckman, Sven, 1953, Zoogeography of the sea: London, Sidgwick and Jackson, 417 p.

Emery, K. O., R. E. Stevenson, and J. W. Hedgpeth, 1957, Estuaries and lagoons: Geol. Soc. America Mem. 67, v. 1, p. 673–750.

Fairbridge, R. W., 1950, The geology and geomorphology of Point Peron, Western Australia: Royal Soc. Western Australia Jour., v. 34, no. 3, p. 35–72.

Hjulstrom, F., 1939, Transportation of detritus by moving water, in P. D. Trask, ed., Recent marine sediments: Am. Assoc. Petroleum Geologists, p. 5–31.

Inman, D. L., and N. Nasu, 1956, Orbital velocity associated with wave action near the breaker zone: U.S. Army Corps Engineers Beach Erosion Board Tech. Memo. 79, 43 p.

Johnstone, D.. M. H. Condon, and P. E. Playford, 1958, Stratigraphy of the lower Murchison River area and Yaringa North Station, Western Australia: Royal Soc. Western Australia Jour., v. 41, no. 1, p. 13–16.

Logan, B. W., 1959, Environments, foraminiferal facies and sediments of Shark Bay, Western Australia: Unpub. Ph.D. thesis, Univ. Western Australia, 287 p.

———— 1961, Cryptozoon and associated stromatolites from the Recent, Shark Bay, Western Australia: Jour. Geology, v. 69, no. 5, p. 517–533.

———— J. F. Read, and G. R. Davies, 1970, History of carbonate sedimentation, Quaternary Epoch, Shark Bay, Western Australia: this volume.

———— R. Rezak, and R. N. Ginsburg, 1964, Classification and environmental significance of algal stromatolites: Jour. Geology, v. 72, no. 1, p. 68–83.

Menard, H. W., 1950, Sediment movement in relation to current velocity: Jour. Sed. Petrology, v. 20, no. 3, p. 148–160.

Munk, W. H., 1949, The solitary wave theory and its application to surf problems: New York Acad. Sci. Annals, v. 51, art. 3, p. 376–424.

Neumann, Gerhard, 1963, On ocean wave spectra and a new method of forecasting wind-generated sea: U.S. Army Corps Engineers Beach Erosion Board Tech. Memo. 43, p. 1–42.

Phleger, F. B., and G. C. Ewing, 1962, Sedimentology and oceanography of coastal lagoons in Baja California, Mexico: Geol. Soc. America Bull., v. 73, p. 145–182.

Pierson, W. J., Jr., Gerhard Neumann, and R. W. James, 1955 Practical methods for observing and forecasting ocean waves by means of wave spectra and statistics: U.S. Naval Oceanog. Office Pub. 603, 284 p.

Rochford, D. J., 1951a, Studies in Australian estuarine hydrology. I. Introduction and comparative features: Australian Jour. Marine Freshwater Resources, v. 2, no. 1, p. 1–116.

———— 1951b, Hydrological and planktological observations by F. R. V. Warreen in south-western Australian waters, 1947-50. Oceanographical station list: C.S.I.R.O. Div. of Fisheries, v. 3.

———— 1951c, A comparison of the hydrological conditions off the eastern and western coasts of Australia: Indo-Pacific Fisheries Council Proc., Sec. II, p. 1–8.

———— 1953, Onshore hydrological investigations in eastern and south-western Australia, 1951. Oceanographical station list: C.S.I.R.O. Div. of Fisheries, v. 14.

———— and R. Spencer, 1957, Onshore and oceanic hydrological investigations in eastern Australia and south-western Australia, 1955. Oceanographical station list: C.S.I.R.O. Div. of Fisheries and Oceanography, v. 27.

Royal Australian Air Force, 1943, Weather on the Australian station; local information. Pt. 11. The eastern part of the Indian Ocean: RAAF Pub. 252, v. 11, p. 1–49.

Rusnak, G. A., 1960, Sediments of Laguna Madre, Texas, in F. P. Shepard et al., eds., Recent sediments, northwest Gulf of Mexico: Am. Assoc. Petroleum Geologists, p. 153–196.

Slack-Smith, R., 1967, The prawn fishery of Shark Bay, Western Australia: F. A. O. World Sci. Conf. on Biology and Culture of Shrimps and Prawns, p. 1–18.

Stokes, G. G., 1847, On the theory of oscillatory waves: Cambridge Philos. Soc. Trans., v. 8, p. 441–455.

Sundborg, A., 1956, The river Klaräven. A study of fluvial processes: Geog. Annaler, v. 38, no. 2–3.

———— 1967, Some aspects of fluvial sediments and fluvial morphology. I. General views and graphic methods: Geog. Annaler, v. 49A, no. 2–4, p. 333–343.

Trask, P. D., assisted by H. E. Hammar and C. S. Wu, 1932, Origin and environment of source sediments of petroleum: Houston, Gulf Publishing Co.

Tuppin, N. K., 1969, Carbonate sediments and sedimentation, Hopeless Reach, Shark Bay, Western Australia: Honors thesis, Univ. Western Australia, 89 p.

U.S. Beach Erosion Board, 1954, Shore protection, planning and design: U.S. Army Corps Engineers Tech. Rept. 4.

Vaughan, T. W., 1940, Ecology of modern marine organisms with reference to paleogeography: Geol. Soc. America Bull., v. 51, p. 433–468.

Part II

CARBONATE PRODUCTION

Editors' Comments
on Papers 10 Through 13

10 FRIEDMAN
On the Origin of Aragonite in the Dead Sea

11 MATTHEWS
Excerpts from *Genesis of Recent Lime Mud in Southern British Honduras*

12 STOCKMAN, GINSBURG, and SHINN
The Production of Lime Mud by Algae in South Florida

13 CHAVE, SMITH, and ROY
Carbonate Production by Coral Reefs

Until recently the origin of carbonate mud, which occupies a volumetrically important part in the stratigraphic record, had been a constant source of headache to carbonate sedimentologists. From the study of modern lime muds in thoroughly studied areas, such as Andros Island on the Great Bahama Bank, Smith (1940), Cloud and Barnes (1948), and Cloud (1962) concluded that carbonate mud is the result of physico-chemical precipitation. Wood (1941), however, from a study of Carboniferous limestone, postulated an algal origin of carbonate mud ("algal dust"). But the message of algal origin of carbonate mud appears to have been lost until 1955, when it was first pointed out by Lowenstam (1955) that carbonate mud originates in spontaneous postmortem disintegration of common green algae (Lowenstam and Epstein, 1957).

Presently it appears that carbonate mud forms in a spectrum of environments by a variety of mechanisms. Papers 10, 11 and 12 represent a cross-section of viewpoints in regard to the genesis of lime mud.

Friedman, from his studies of Dead Sea sediments, suggests that lime mud there has been produced through nonbiological inorganic processes (Paper 10). He suggests that intense evaporation is the triggering mechanism for precipitation and strengthens his claims through mineralogical and isotopic studies of the sediment. He highlights the role of sulfate-reducing bacteria *(Desulfovibrio)* in generating calcite layers in Dead Sea sediment.

Matthews (Paper 11) adopted a direct approach to study the genesis of lime mud in southern British Honduras. He cautioned carbonate sedimentologists that not all carbonate muds are composed of aragonite needles derived through postmortem disintegration of green algae. Matthews suggests other possible modes of formation of carbonate muds including physical breakage and abrasion of skeletal particles and biological processes of particle-size reduction through boring, mastication, ingestion, and removal of organic binding from shells.

Stockman, Ginsburg, and Shinn (Paper 12) demonstrate, on the other hand, that disintegration and breakdown of fragile and semiresistant skeletons of different calcium-carbonate secreting algae can well take care of the total lime mud accumulation in modern southern Florida; they infer from their studies that ancient micritic limestones may have had a similar origin.

Biologists and geologists have studied both modern and ancient coral reefs for a better insight into organic communities, their ecology, and the resulting sedimentary facies. But few works exist relating to quantitative aspects of carbonate sediment production in a coral reef. Chave, Smith, and Roy (Paper 13) estimate in two steps: (1) they erect a hypothetical reef and determine its potential, gross, and net carbonate production, and (2) they compare the carbonate production from this hypothetical reef with that of actual reefs. From these two approaches they conclude that a close similarity exists between the two with a certain amount of qualification. In their calculation of net production, they have not considered gains through chemical diagenesis because of paucity of data. Following this model of comparable potential carbonate production rates for most reef producers, Smith (1973) suggested that it is the physical-chemical setting rather than biologic composition that is the most important factor controlling calcification rates in marine communities. (An additional paper concerning reefs, Paper 5, is included in Part I.)

REFERENCES

Cloud, P. E., 1962, Environment of Calcium Carbonate Deposition West of Andros Island, Bahamas, *U.S. Geological Survey Prof. Paper 350,* 138p.

Cloud, P. E., and V. E. Barnes, 1948, Paleoecology of the Early Ordovician Sea in Central Texas, in *Report of the Committee on a Treatise in Marine Ecology and Paleoecology, No. 8,* Division of Geology and Geography, National Research Council, Washington, D.C., pp. 29–83.

Lowenstam, H. A., 1955, Aragonite Needles Secreted by Algae and some Sedimentary Implications, *Jour. Sed. Petrology* **26:**270–272.

Lowenstam, H. A., and S. Epstein, 1957, On the Origin of Sedimentary Aragonite Needles of the Great Bahama Bank, *Jour. Geology* **65:**364-375.

Smith, C. L., 1940, The Great Bahama Bank, I. General Hydrographic and Chemical Factors; II. Calcium Carbonate Precipitation, *Jour. Marine Research* **3:**147-189.

Smith, S. V., 1973, Carbon Dioxide Dynamics: A Record of Organic Carbon Production, Respiration, and Calcification in the Eniwetok Windward Reef Flat Community, *Limnology and Oceanography* **18:**106-120.

Wood, A., 1941, "Algal dust" and the Finer-Grained Varieties of Carboniferous Limestone, *Geol. Mag.* **78:**192-200.

10

Reprinted from Israel Jour. Earth-Sci. **14**:79-85 (1965)

ON THE ORIGIN OF ARAGONITE IN THE DEAD SEA

GERALD M. FRIEDMAN*

Department of Geology, The Hebrew University of Jerusalem

ABSTRACT

The aragonite-calcite ratio in Dead Sea carbonate sediments can be correlated with C^{13}/C^{12} and O^{18}/O^{16} ratios. Increasing aragonite content conforms to an increase in the heavier carbon and oxygen isotopes. The enrichment of the heavier isotopes is explained by strong evaporation during the formation of aragonite when the lighter isotopes are preferentially removed as part of CO_2. The layers enriched in the lighter isotopes are predominantly composed of calcite. Calcite was formed by the degradation of gypsum through the activity of bacteria which preferentially use lighter isotopes from their carbon source.

INTRODUCTION

The formation of aragonite during periods of whitening has been observed in several bodies of water in diverse parts of the world. Whitenings were noted as early as 1836 in the Bahamas, where they were reported by the British nautical survey. The cause of whitenings and the formation of aragonite have been a subject of controversy for many years. Four theories for the origin of aragonite have been proposed: 1. a bacterial origin, 2. a derived origin, 3. a physico-chemical origin, and 4. a skeletal disintegration origin. Purdy (1936, p. 485-492) reviewed these theories and attempted to reconcile the views of Lowenstam and Epstein (1957), and Cloud (1962) postulating that the sedimentary aragonite needles in the Bahamas are of algal origin and that physico-chemical precipitation forms overgrowths on suspended clay-sized carbonate particles, most of which consist of algal aragonite needles. Cloud (1962, p. 19-22) studied three periods of whitenings in the Bahamas and noted (p. 21) that characteristics common to all areas of whitenings were their elongate form and their tendency to drift with the wind and tidal current. The geometry of these whitenings led some observers to the suggestion that they were caused by schools of fish (Ginsburg 1956, p. 2398), but Cloud found no evidence of schooling fish. Cloud (p. 22) instead called upon photosynthesis as playing an important role in initiating and maintaining a high rate of aragonite precipitation. Accelerated uptake of photosynthetic CO_2 might serve as a triggering device to induce widespread aragonite precipitation in the water mass. Revelle and Fairbridge (1957, p. 258) suggested that

---

* Fulbright Visiting Professor.
Permanent address : Department of Geology, Rensselaer Polytechnic Institute, Troy, New York
Received June 12, 1964
Revised MS July 26, 1966

temporary morning clouding of still waters in the lagoon of Houtman's Abrolhos (southwestern Australia) might be due to photosynthetic reduction of CO_2 pressure.

In the Persian Gulf large-scale aragonite precipitation in the form of whitenings were observed in many places by Wells and Illing (1964). The whitenings appear as isolated patches of milky water and occur with monotonous regularity throughout the years. These whitenings were thought to be initiated by CO_2 consumption by phytoplanktonic organisms, which in turn triggers off aragonite precipitation. Thus the conclusions of Wells and Illing for the Persian Gulf whitenings appear to agree with those of Cloud and Revelle and Fairbridge for other parts of the world.

Whitenings in the Dead Sea, with formation of aragonite, were described by Bloch et al. (1944), Shalem (1949), Rasumny (1962), and Neev (1963, 1964). The first three authors discussed the August 1943 whitening, whereas Neev observed and interpreted a whitening in August 1959. According to Neev (1963, p. 153) aragonite is precipitated during whitenings when the water temperatures rise to their annual maximum. In contrast to the Bahamas and the Persian Gulf, where whitenings occur in isolated patches or elongate streaks, aragonite precipitation appears to be uniform all over the Dead Sea. Moreover, according to Neev, whitenings in the Dead Sea occur only once in a few years, whereas in the Bahamas and the Persian Gulf they are very common and occur in different local areas several times a year.

Sediments of the Dead Sea include gypsum, aragonite and calcite. Calcite and aragonite layers can be visually distinguished. Calcite layers are dark whereas aragonite layers are white. They are rhythmically interbedded and resemble varves in appearance.

The purpose of the present contribution is to provide independent evidence from isotope data and mineralogical analyses for the origin of the aragonite which is formed during whitenings in the Dead Sea.

PROCEDURE

Samples were collected with a grab sampler along the shore and by boat in near-shore areas at a depth ranging from 10 cm to 4 m below water level. In addition, bottom sediments were collected from greater depths (5 to 325 m) by D. Neev* in both basins of the Dead Sea. At two onshore stations particular care was taken to sample separately black and white layers of the carbonate sediments.

In the laboratory, the samples were pulverized, treated with commercial Clorox** and washed with distilled water. Mineralogical analyses were made by X-ray diffraction methods. The percentage of aragonite in each sample was obtained from working curves by plotting known amounts of aragonite against the peak height of the 3.27 Å peak. The presence of gypsum and calcite was determined but their amounts were not estimated. The strontium content was obtained by X-ray fluorescence analysis;

* Geological Survey of Israel, Jerusalem
** A commercial sodium hypochlorite solution.

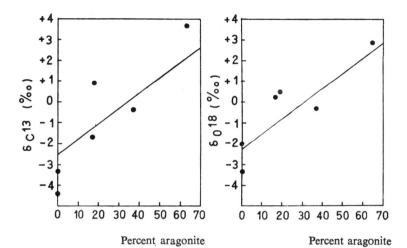

Figure 1
Relationship between aragonite
content and δC^{13} of carbonates
for nearshore sediments of the
Dead Sea.

Figure 2
Relationship between aragonite
content and δO^{18} of carbonates
for nearshore sediments of the
Dead Sea.

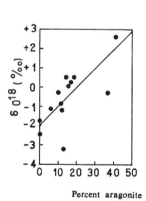

Figure 3
Relationship between aragonite
content and δO^{18} of carbonates
for offshore sediments of the
Dead Sea.

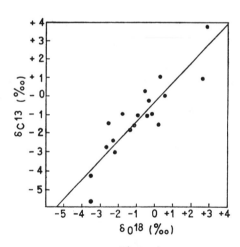

Figure 4
Relationship between δC^{13} and δO^{18}
for carbonate sediments of the
Dead Sea.

281

Carbon and oxygen isotope measurements were made with a gas mass spectrometer of the Nier-McKinney type. The values were converted to the P.D.B. scale using Craig's (1953) correction factors. The reproducibility of the $\delta C^{13}/C^{12}$ and $\delta O^{18}/O^{16}$ values is within 0.1 per mil.

RESULTS

Figures 1 and 2 illustrate the relationship between the aragonite content of nearshore sediments and their $\delta C^{13}/C^{12}$ and $\delta O^{18}/O^{16}$ ratios. These figures show a linear relationship between the aragonite content and isotope composition of the carbonates (aragonite and calcite) present in the sediments. The correlation lines in Figures 1 and 2 have been computed by the method of least squares. Aragonite has been taken as the independent variable and the isotope ratios as the dependent variable. The coefficient of correlation (r) for the relationship between aragonite content and carbon isotope ratio is 0.92 which shows that the correlation is significant, and it follows that 84.6% (r^2) of the variation in the carbon isotope ratios is explained by its dependence on the aragonite content. The coefficient of correlation (r) for the relationship between aragonite content and oxygen isotope ratios is 0.90. This correlation is significant; statistically, 82% (r^2) of the variation in the oxygen isotope ratio is explained by the variation in the aragonite content. An increasing aragonite content of the sediment is accompanied by enrichment of the heavier carbon and oxygen isotopes. The sediment with the highest aragonite content shows the greatest enrichment in the heavier isotopes. Figure 3 shows the relationship between the aragonite content and the $\delta O^{18}/O^{16}$ ratio for offshore sediments. The relationship is approximately linear, with increasing aragonite content being accompanied by enrichment in the heavier oxygen isotope. The coefficient of correlation (r) is 0.78. Figure 4 is a plot of $\delta C^{13}/C^{12}$ against oxygen $\delta O^{18}/O^{16}$ for both nearshore and offshore sediments. This plot, which displays a linear relationship with a correlation coefficient (r) of 0.89, indicates that with increasing enrichment in the heavier carbon isotopes there is a corresponding increase in the heavier oxygen isotopes.

Table I shows the mineralogical and geochemical compositions of pairs of black and white layers (varves) of the sediments studied. This table shows that white layers 1. contain abundant aragonite, 2. are enriched in the heavier isotopes, whereas black layers are enriched in the lighter isotopes, and 3. have a much higher strontium content than the black layers.

DISCUSSION

The ratio $\delta C^{13}/C^{12}$ in limestones is related to the environment of deposition and to post-depositional diagenetic changes. The Recent carbonate sediments of the Dead Sea have not undergone diagenetic changes by post-depositional contacts with fresh water. Therefore the carbon isotope ratios must reflect the environmental conditions under which the carbonate sediments were deposited.

TABLE I
COMPOSITION OF SEDIMENT PAIRS
(WHITE AND BLACK LAYERS)*

		Color of layer	Aragonite (%)	Strontium (p.p. m)	δC^{13}	δO^{18}
Pair 1.		White	63	4,000	+3.61	+2.81
		Black	N.D.**	890	−3.13	−2.05
Pair 2.		White	40	4,000	+0.81	+2.54
		Black	N.D.**	1,000	−1.08	−1.82

* These samples contain gypsum.
** N.D. = none determined. The carbonate mineral in these samples is calcite; other samples of black layers analysed were found to contain both calcite and aragonite (Neev 1964).,

Carbonate precipitated in a non-marine environment tends to be enriched in the lighter carbon isotope (C^{12}) as compared to marine limestone. In non-marine carbonates C^{13} can be concentrated by strong evaporative processes in which the lighter carbon isotope is preferentially removed as part of CO_2. Figure 1 shows that with increasing aragonite content a more or less progressive enrichment in C^{13} takes place in Dead Sea carbonate ($r = 0.92$). In fact, enrichment in C^{13} has taken place to the extent that the carbonate samples have typical "marine" values.

Marine water has a greater O^{18} content than fresh water and, as with carbon isotopes, fresh water carbonates are deficient in the heavier (O^{18}) isotope as compared to marine carbonates. However, enrichment in O^{18} in non-marine carbonates can result from strong evaporation. Figure 2 indicates that for nearshore sediments, as the aragonite content increases, the carbonate shows a tendency to be enriched in the heavier oxygen isotope ($r = 0.90$). This essentially confirms the inference already made from the distribution of the carbon isotopes that the aragonite is formed under conditions of intense evaporation. In Figure 3, aragonite content and oxygen isotope composition have been plotted for sediments from deeper water. This figure indicates that also in offshore sediments aragonite enrichment is for the most part accompanied by concentration of the heavier oxygen isotope ($r = 0.78$). A statistically significant correlation was obtained between the aragonite content and the isotopic composition of the carbonates in the samples, despite the presence of a notable gypsum content, and it is estimated that the correlation would hold even better if only the carbonate isotopes in the samples had been used.

These data and interpretations support the conclusions of Neev (1963, 1964) who suggested that high temperatures were the mechanism that triggers the mass precipitation of aragonite, i.e. the whitening. The conclusions of Neev and those of the present study differ from those of Cloud (1962) and Wells and Illing (1964).

Cloud and Wells and Illing worked in areas which abound with organisms, and their explanation of photosynthetic CO_2 uptake as the triggering device for widespread aragonite precipitation is very convincing. In the Dead Sea there are hardly any organisms capable of photosynthesis. The trigger mechanism which leads to CO_2 expulsion and mass precipitation of aragonite must be inorganic (maximum temperature). The differences between the processes proposed for the Bahamas and Persian Gulf and those for the Dead Sea demonstrate that aragonite precipitation can be triggered off by more than one mechanism, depending on the environmental conditions. They may also explain why the subject of aragonite formation in the Bahamas is so controversial, since both organic and inorganic processes may be responsible for aragonite precipitation.

In another hypersaline lake, Salt Flat of West Texas, U.S.A., layers of isotopically-heavy and isotopically-light carbonate sediments occur in alternating beds similar to those of the Dead Sea (Friedman, 1966). The light layers, as in the Dead Sea, are composed for the most part of calcite which has resulted from the bacterial decomposition of gypsum in which the lighter isotopes are drawn from the carbon source (Feeley and Kulp, 1957, p. 1844). Neev has suggested a similar bacterial origin for the Dead Sea calcite. Lighter calcite, both for the Dead Sea and Salt Flat, Texas, occurrences, possibly reflects bacterial fractionation. At Salt Flat the isotopically heavy carbonate is dolomite and not aragonite as in the Dead Sea.

This study is essentially deductive. The approach has been to relate the isotopic composition to the aragonite content in the studied samples and to explain the observed correlations. To confirm these explanations it is necessary to measure the isotopic composition of Dead Sea waters, of aragonite and calcite actually precipitating, and formed diagenetically, and of detrital limestones which may have served as a source for the Dead Sea carbonates. Such a study would be more ambitious than the scope of the present investigation, but would lead to important information on sedimentary processes in the Dead Sea.

Table I shows that the strontium content increases with an increase in aragonite. This is not unexpected for it has been shown both in laboratory precipitates and in nature that aragonite generally contains more strontium than calcite (Zeller and Wray, 1956; Wray and Daniels, 1957). The ionic radius of strontium is larger than that of calcium and it tends to favor the more open lattice of the aragonite unit cell.

ACKNOWLEDGEMENTS
Thanks are extended to I. R. Kaplan for critical reading of the manuscript. Carbon and oxygen isotope measurements were made by William M. Sackett and R. L. Ames at Pan American Petroleum Corporation's Research Center. David Neev's personal interest and stimulating discussions are gratefully acknowledged. He arranged the field trip to the Dead Sea, including boat transportation, and provided additional samples for study. His enthusiasm made this study possible.

REFERENCES

BLOCH, M.R., LITMAN, H.Z., AND ELAZARI-VOLCANI, B., 1944, Occasional whiteness of the Dead Sea, *Nature*, **154**, 402.

CLOUD, P.E., JR., 1962, Environment of calcium carbonate deposition west of Andros Island, Bahamas, U.S. Geol. Survey Professional Paper, 350, pp. 138.

CRAIG, H., 1953, The geochemistry of the stable carbon isotopes, *Geochim. Cosmochim. Acta*, **3**, 53–92.

FEELEY, H.W., AND KULP, J.L., 1957, The origin of Gulf Coast Salt Dome sulfur deposits, *Bull. Amer. Assoc. Petrol. Geol.*, **41**, 1802–1853.

FRIEDMAN, G.M., 1966, Occurrence and origin of Quaternary dolomite of Salt Flat, West Texas, *J. sedim. Petrol.*, **36**, 263 263–267.

GINSBURG, R.N., 1956, Environmental relatlonships of grain size and constituent particles in some south Florida carbonate sediments, *Amer. Assoc. Petrol. Geol.*, **40**, 2384–4227.

LOWENSTAM, H.A., AND EPSTEIN, S., 1957, On the origin of sedimentary aragonite needles of the Great Bahama Bank, *J. Geol.*, **65**, 364–375.

NEEV, D., 1963, Recent precipitation of calcium salts in the Dead Sea, *Bull. Res. Counc. Israel*, **11G**, 153.

NEEV, D., 1964, The Dead Sea, Recent Sedimentary Processes, Unpublished Ph.D. thesis, Hebrew University of Jerusalem.

PURDY, E.G., 1963, Recent calcium carbonate facies of the Great Bahama Bank, 2. Sedimentary facies, *J. Geol.*, **71**, 472–497.

RASUMNY, JANINE, 1961, (Publ. 1962), The solubility of Neocomian formations from the Sdome region (Israel) in various aqueous solutions: *C. R. Congr. Natl. Soc. Savantes, Sect. Sci.*, **86**, 387–388.

REVELLE, R. AND FAIRBRIDGE, R.W., 1957, Carbonates and carbon dioxide, *Mem. geol. Soc. Ame.*, **67**, 239–295.

SHALEM, N., 1949, Whiting of the waters of the Dead Sea, *Nature*, **164**, 72.

WELLS, A.J., AND ILLING, L.V., 1964, Present-day precipitation of calcium carbonate in the Persian Gulf, *in* L.M.J.U. Van Straaten, Editor, *Deltaic and Shallow Marine Deposits, Proc. 6th International Sedimentological Congress, Amsterdam and Antwerp* 1963, 429–435.

WRAY, J.L., AND DANIELS, F., 1957, Precipitation of calcite and aragonite, *J. Amer. Chem. Soc.*, **79**, 2031–2034.

ZELLER, E.J., AND WRAY, J.L., 1956, Factors influencing precipitation of calcium carbonate, *Amer. Assoc. Petrol. Geol.*, **40**, 140–152.

11

Reprinted from pages 428–429, 432–433, and 439–454 of *Jour. Sed. Petrology* **36:**428–454 (1966), by permission of the publisher

GENESIS OF RECENT LIME MUD IN SOUTHERN BRITISH HONDURAS[1]

R. K. MATTHEWS
Brown University, Providence, Rhode Island

ABSTRACT

In view of the abundance of lime mud in the geologic record, it is paradoxical that studies of Recent carbonate sediments generally have concentrated on the origin of sand-size carbonate particles rather than on the origin of the finer-grained constituents. Only on the Great Bahama Bank has the genesis of lime mud received paramount attention; so much attention, in fact, that the words "aragonite needles" have become virtually synonymous with "lime mud" in the minds of many geologists. Preliminary investigation of Recent lime mud from Southern British Honduras, however, revealed a paucity of aragonite needles. An investigation was therefore undertaken to ascertain the nature and origin of the lime mud in this area.

Mineralogical, chemical, and petrographic point-count data were gathered. The strontium content of the carbonate mud fraction of lagoon samples increases systematically toward carbonate shoals. The mineralogical composition of the carbonate mud fraction of lagoon samples averages 25 percent high-strontium aragonite, 24 percent low-strontium aragonite, 44 percent high-magnesium calcite, and 7 percent low-magnesium calcite. Petrographic data suggest that the high-strontium aragonite is primarily coral debris admixed with lesser amounts of *Halimeda* debris. Similarly, low-strontium aragonite consists primarily of mollusc debris; and high-magnesium calcite, of Foraminifera debris. The data suggest that the Shelf Lagoon mud consists of transported shoal-derived debris and *in situ*-produced mollusc debris and hyaline Formainifera debris.

Physical breakage and abrasion in agitated environments are considered the dominant processes of lime mud production on the carbonate shoals; whereas the major factors in the *in situ* production of lagoonal lime mud appear to be: (1) the inherently fragile nature of the shells of molluscs and tests of hyaline Foraminifera of the lagoon environment, (2) the removal of binding organic matter from mollusc shells, (3) the weakening of larger skeletal particles by the activity of boring micro-organisms, and (4) the mastication, ingestion, and perhaps even simple movement of sediment by the vagrant benthos.

The results of this study indicate that lime mud may originate in a variety of ways. While it may be difficult or impossible in the geologic record to recognize ancient analogues of the various types of lime mud that can be recognized in the Recent, an awareness of the possible multiple origins of lime mud serves to increase our understanding of the genesis of ancient lime mudstones.

INTRODUCTION

Even a cursory examination of most limestone stratigraphic sequences suffices to demonstrate the volumetric importance of lime mud in the stratigraphic record, for such sequences contain mud in varying proportions ranging from the mud matrix surrounding the sand-size constituents of many calcarenites to the lime mud comprising the bulk of calcilutites. Indeed, if a general impression be admitted as evidence, it would seem that the quantity of lime mud in the stratigraphic record is far greater than that of lime sand. It is therefore paradoxical that studies of Recent carbonate sediments generally have concentrated on the origin of sand-size carbonate particles rather than on the origin of the finer-grained constituents. In fact, it is only on the Great Bahama Bank that lime mud genesis has received paramount attention, for here the considerable areal extent of Recent lime mud deposits has suggested to many workers a recent analogue of the many calcilutites that occur in the geologic record. Bahamian lime muds consist largely of aragonite needles, the origin of which

[1] Manuscript received September 8, 1965; revised March 7, 1966.

is still being debated (for example, Lowenstam and Epstein, 1957; Cloud, 1962). But there are other types of Recent lime mud that are just as likely to be represented in the geologic record. For example, on Alacran Reef Hoskins (1963) and Folk and Robles (1964) have reported a lime mud consisting predominantly of coral grit and *Halimeda* fragments, although they do not find lime mud to be an abundant sediment type in their study areas. In Southern British Honduras a lime mud containing both calcite and aragonite but virtually devoid of aragonite needles is an extremely common sediment type. The genesis and geologic significance of this lime mud is the subject of the present report.

GENERAL FEATURES OF THE STUDY AREA

Physiographic Setting

British Honduras is located on the southeast side of the Yucatan Peninsula. The country is bounded to the north by Mexico and to the west and south by Guatemala. The east side of the Yucatan Platform is bounded by a submarine escarpment throughout its entire length from the Yucatan Channel to the Gulf of Honduras. Along the Mexican portion of this coastline, the

286

escarpment lies close to the mainland; whereas in British Honduras a relatively shallow shelf eight to thirty miles wide and one hundred-twenty miles long separates the mainland from the submarine escarpment.

That portion of the shelf which lies north of the city of Belize (fig. 1) is hereafter referred to as Northern British Honduras. The sediments of this extremely shallow area have been investigated by Pusey (1964).

That portion of the British Honduras shelf which lies south of the city of Belize is hereafter referred to as Southern British Honduras (fig. 1), and it is in this area that the present study was undertaken.

Bathymetry and Water Movements in Southern British Honduras

A shallow platform, termed herein the Barrier Platform, exists along the eastern margin of the shelf of Southern British Honduras. Water depths over the platform vary from a few feet to as much as five fathoms. The width of the platform varies from one to seven miles; the platform being broadest just south of Belize (fig. 1) and narrowing abruptly just south of Twin Cays (16°50′ N, 88°05′ W). The platform is bounded on the east by the submarine escarpment mentioned previously. The western edge of the Barrier Platform is characterized by an abrupt descent into ten to twenty-five fathoms of water. This is the eastern margin of the Shelf Lagoon. In west to east profile, the Shelf Lagoon deepens gradually from the mainland to the Barrier Platform, attaining its greatest depth near the base of the Barrier Platform. The maximum depth of the Shelf Lagoon increases systematically from ten fathoms in the north to thirty-five fathoms in the south. North of Twin Cays there are only a few carbonate shoals rising out of the lagoon, while south of Twin Cays numerous shoals appear in the Shelf Lagoon. The shoals illustrated in figure 1 are covered by less than five fathoms of water and frequently have cays developed on them.

Little data are available on the direction of currents on the shelf of Southern British Honduras. Hydrographic Office Charts indicate a general southerly flow of surface waters in the Shelf Lagoon and an easterly flow between Hunting Cays (16°05′ N, 88°15′ W) and Cabo Tres Puntas, Guatemala.

[*Editors' Note:* Material has been omitted at this point.]

FACIES RELATIONSHIPS

The structural control of facies distribution in Southern British Honduras has been discussed by Purdy and Matthews (1964). In broadest terms, the recent marine sediments of Southern British Honduras may be divided into the following three categories: (1) reef complex, (2) Shelf Lagoon lime mud, and (3) near-shore terrigenous clastics.

The reef complex consists almost exclusively of coral reef and reef associated debris of sand-size and larger. Some samples in the lee of island contain as much as fifty percent lime mud, but by and large lime mud is a minor constituent of reef complex sediments. Reef complex sediments occur primarily beneath the shallow waters of the Barrier Platform and of the shoals which rise out of the Shelf Lagoon, although sediments of this facies also occasionally appear in the Shelf Lagoon in areas bordering the Barrier Platform or shoals.

The sediment of the Shelf Lagoon is typically seventy to ninety-five percent mud with *in situ* molluscs and Foraminifera comprising the majority of the coarse fraction. Near the Barrier Platform and in those areas of the lagoon where shoals abound, seventy to ninety percent of the mud fraction is lime mud; the remainder being terrigenous silt and clay.

Westward from the areas of abundant lime mud accumulation the sediment is predominantly terrigenous clastics. Typically there is a gradual east to west transition from lime mud to terrigenous mud, then from terrigenous mud to terrigenous sand as the shore-line of the mainland is approached.

At distances less than one-tenth of a nautical mile from a shoal, lagoon mud often contains coarse debris of shoal origin. Such samples could be described correctly as "coral and algal fragments in a mud matrix." At greater distances from a shoal, sand-size shoal debris is commonly not present in the lagoon mud. Here, mollusc remains and Foraminifera are the obvious "grain types," and such samples would be described accurately as "mollusc fragments and

Foraminifera in a mud matrix." An important distinction remains to be made. however, in the description of these two sediment types. In the former, the "grains" are transported to the site of deposition; in the latter, they are in place. With increasing distance from a shoal, the size of the transported "grains" diminishes until they become indistinguishable from "matrix." It is the origin of this "matrix" that constitutes the focal point of the present investigation.

CONSTITUENT PARTICLE COMPOSITION OF LIME MUD IN SOUTHERN BRITISH HONDURAS

The abundance of lime mud in the lagoon of Southern British Honduras increases systematically toward the Barrier Platform (fig. 2). Thus it is a reasonable working hypothesis that at least some lime mud in the lagoon is derived from the Barrier Platform. The problem becomes one of how to demonstrate whether this is or is not the case. The most direct approach to a solution of this problem is to identify constituent particles in the lagoon mud which can have their origin only on the shoal areas and to simply point-count their abundance in the lagoon mud.

With this in mind, grain-mounts of the silt fraction of lagoon samples and shoal samples were examined in detail at a magnification of 400X and point-counted for constituent particle types. Previous petrographic determinations of constituent particles in recent carbonate sediments (Ginsburg, 1956; Purdy, 1963; Hoskins, 1963; and Pusey, 1964 for example) have dealt primarily with the identification of sand-size particles in thin-section. With decreasing particle size. many of the simple criteria applicable to larger fragments are no longer available to the investigator. For example, shape and shell architecture are common criteria for the identification of sand-sized mollusc fragments; however, silt-size mollusc debris may appear to be single crystals, thus rendering such criteria of limited value. Similarly, sand-sized fragments of coral are easily recognized by the diagnostic extinction of their trabecluae, but this type of extinction can be quite difficult to distinguish from molluscan undulatory extinction in silt-sized particles. As a final example, sand-sized segments of *Halimeda* are quite easily recognized by their "Swiss-cheese" appearance (Purdy, 1963) produced by the abundance of utricles, but silt-sized *Halimeda* particles seldom possess evidence of even a single utricle.

Some silt-sized constituent particles are, however. easily recognized. Tunicate spicules, alcyonarian spicules, and echinoid fragments, for example, are quite readily identified. Moreover, the wall structure of Foraminifera (Wood, 1948) affords distinction among porcelaneous, radial hyaline, and granular hyaline Foraminifera.

To facilitate identification, many of the common organisms of the study area were crushed, and the silt-size debris thus produced was examined with a petrographic microscope. Descriptive categories were erected on the basis of this examination preparatory to making point-counts of the silt fraction of lagoon and shoal samples.

[*Editors' Note:* Material has been omitted at this point.]

Discussion of Petrographic Data

Point-count analysis of the silt fraction of lagoon samples was undertaken as the most direct approach to demonstrating that the lime mud in the Shelf Lagoon originated from the Barrier Platform and other carbonate shoal areas, an hypothesis suggested by the distribution pattern of lime mud in the Shelf Lagoon (fig. 2). Toward this end, point-count analysis has been only partially successful. Table 1 compares the point-count data for the silt fraction of lagoon samples and shoal samples.[2] Visual inspection of these data indicates no profound difference in constituent particle composition. In fact, the similarity between the two sets of data suggests a common origin for the silt fraction in both localities. These data, however, do not prove that the majority of the material has originated on the carbonate shoals, for only Coral Debris, Alcyonarian Spicules, and Tunicate Spicules can be said to originate exclusively on the shoals, and these point-count categories account for only about ten percent of the silt fraction of the lagoon samples. However, a considerable volume of coral debris may be contained in the point-count category Coral or Mollusc Debris, and, similarly, a large volume of *Halimeda* may be contained in the point-count

[2] A table listing the results of point-count, X-ray, and Sr analyses has been deposited as Document No. 8827 with the American Documentation Institute, Auxiliary Publications Project, C/O Library of Congress, Washington 25, D. C. Copies may be secured by citing the Document number and remitting $1.25 for photo copies or $1.25 for 35 mm microfilm.

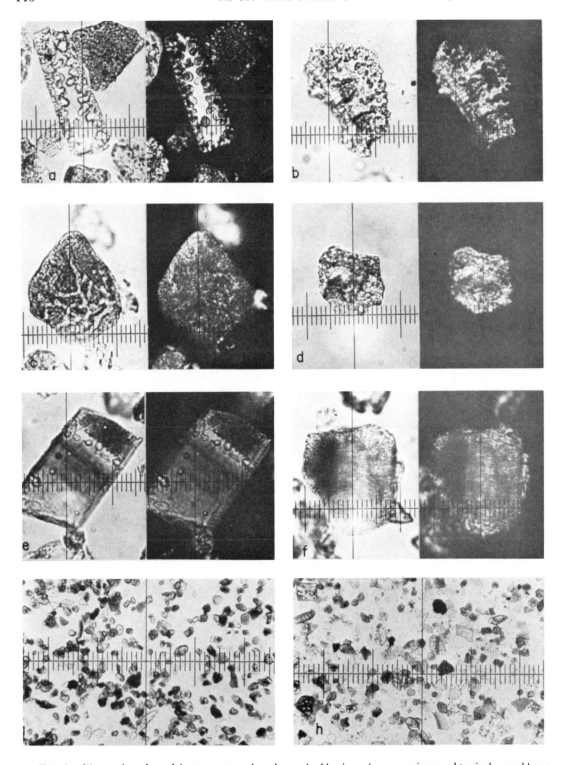

FIG. 5.—Illustration of particle type categories: the work of boring micro-organisms and typical assemblages of particle types.

Continued on following page.

TABLE 1.—*Comparison of point-count data for silt-size carbonate from lagoon samples and carbonate shoal samples from Southern British Honduras*

	Lagoon Samples[a]			Carbonate Shoal Samples[b]		
	Mean	Observed Range	Standard Deviation	Mean	Observed Range	Standard Deviation
Coral Debris	3.6	1–9	2.7	6.9	1–14	3.9
Mollusc Debris	4.4	1–12	2.9	2.0	0–4	1.4
Coral or Mollusc Debris	29.8	17–52	8.7	35.9	15–51	11.7
Cryptocrystalline Particles	25.0	16–42	7.3	24.0	11–43	8.1
Calcified Green Algae[c]	2.5	0–5	1.3	4.6	1–14	3.6
Porcelaneous Foram Debris	9.2	2–16	3.0	8.5	3–22	5.0
Radial Hyaline Foram Debris	8.4	1–18	4.2	3.5	0–9	2.8
Whole Radial Hyaline Forams	3.6	0–10	2.6	1.6	0–9	2.6
Granular Hyaline Foram Debris	2.2	0–6	1.7	0.7	0–3	0.9
Echinoderm Debris	0.7	0–3	0.8	0.5	0–2	0.6
Alcyonarian Spicules	1.3	0–4	1.2	2.1	0–5	1.8
Tunicate Spicules	4.0	0–12	2.8	2.4	0–4	1.3
Other	5.5	2–10	2.3	6.5	2–16	3.6

[a] Twenty-six samples used in calculation.
[b] Thirteen samples used in calculation.
[c] *Halimeda* excluded.

category Cryptocrystalline Particles. Indeed, if it could be demonstrated that these point-count categories contain largely coral and *Halimeda*, it would then appear that the majority of the silt fraction of the lagoon samples was derived from the carbonate shoals, for corals and *Halimeda* live almost exclusively on the shoals.

Clearly other lines of evidence are needed. Specifically, the point-count category Cryptocrystalline Particles must be segregated into *Halimeda* debris, porcelaneous Foraminifera debris, red algae debris, cryptocrystalline mol-lusc debris and cryptocrystalline coral debris; and the point-count category Coral or Mollusc Debris must be segregated as one or the other. Furthermore, the point-count data provide knowledge only of the silt fraction; the finer particle-sizes being too small to investigate optically. Thus, some method must be devised to translate knowledge of the silt fraction into the knowledge of the total mud fraction.

It would appear that a combination of mineralogical data and strontium data may be used to accomplish the desired refinements outlined

Continued from preceding page.

Particles a through f are illustrated at a magnification of 500× in plane polarized light (left) and between crossed nicols (right) as they appear mounted in Canada Balsam. One small division on the scale equals 2.5 microns.

Photomicrographs g and h are magnified 100× in plane polarized light. One small division on the scale equals 13 microns.

a. Mollusc Debris from the silt fraction of sample 295. Note the effects of boring algae or other micro-organisms. Though this prism probably originated through skeletal disintegration of a mollusc shell, further particle size reduction might result from the activity of boring algae or other micro-organisms.

b. Mollusc or Coral Debris from the silt fraction of sample 85. Note that the particle is riddled with microborings. It seems likely that the slightest physical pressure might result in further particle size reduction.

c. Porcelaneous Foraminifera Debris from the silt fraction of sample 85. Note the microborings two to five microns in diameter.

d. Mollusc or Coral Debris from the silt fraction of sample 84. Note one micron diameter microborings.

e. Mollusc Debris from the silt fraction of sample 85. The upper right end of this mollusc prism is so riddled by one-half micron diameter microborings that it appears cryptocrystalline.

f. Transition from Mollusc or Coral Debris on the right to Cryptocrystalline Particle on the left, from the silt fraction of sample 226. It is evident that this mollusc or coral particle is becoming cryptocrystalline. A few possible microborings appear along the left margin of the particle.

g. Assemblage of particle types from the silt fraction of sample 227. Note the generally granular nature of this near-shoal assemblage rich in coral debris and cryptocrystalline particles.

h. Assemblage of particle types from the silt fraction of sample 165. Note the generally platey nature of this lagoon assemblage rich in Foraminifera debris.

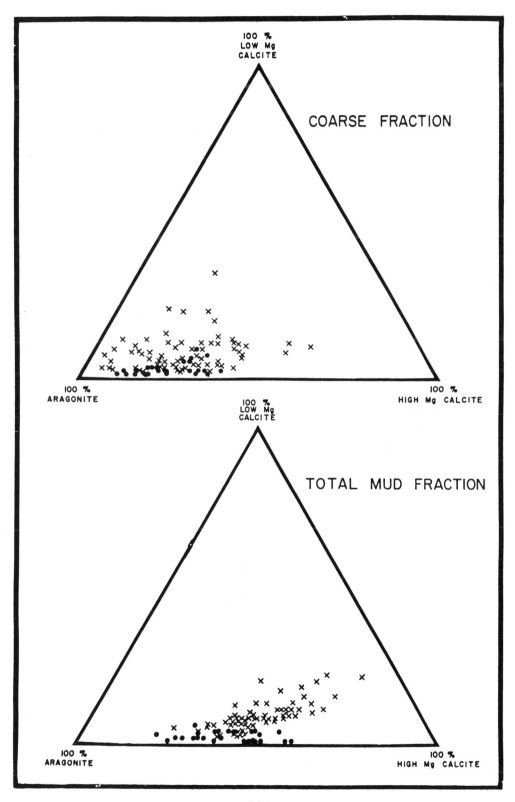

above. If the point-count category Cryptocrystalline Particles can be shown to consist largely
of high-magnesium calcite, *Halimeda*, mollusc,
and coral debris (aragonitic skeletons; Chave,
1962b) can be excluded as possible constituents
of this category. Similarly, if this category can be
shown to consist largely of aragonite, porcelaneous Foraminifera and red algae (high-magnesium
calcite skeletons; Chave, 1962b) can be excluded
as possible constituents of this category.

The data of Odum (1957) suggest that corals
and calcified green algae may be set apart from
all other organisms common in the study area on
the basis of their high strontium content. Thus,
if the point-count category Coral or Mollusc
Debris can be shown to consist largely of aragonite with a high strontium content, the category
may be identified as consisting largely of coral
debris rather than mollusc debris. Similarly, if
the aragonite in the point-count category Cryptocrystalline Particles can be shown to have a
high strontium content, the aragonite may be
identified as that of calcified green algae or corals
rather than molluscs.

Finally, if sufficient mineralogical and chemical similarity can be demonstrated between the
silt fraction and the total mud fraction, then
constituent particle composition of the silt fraction may be extrapolated to infer constituent
particle composition of the total mud fraction.

Toward the purposes outlined above, mineralogical composition and strontium content were
determined for the coarse fraction, silt fraction
and total mud fraction of lagoon and shoal
samples.

Mineralogical Data

The carbonate mineralogy of 106 samples from
Southern British Honduras is summarized in
figures 6 and 7 and table 2 and 3.[2] The data indicate an enrichment of high-magnesium calcite in
the mud fraction.

Figure 6 compares the mineralogy of the
coarse fraction and the total mud fraction. The
total mud fraction of the shoal samples appears

TABLE 2.—*Comparison of carbonate mineralogy
of size fractions of selected samples from
Southern British Honduras*

Sample		Aragonite	High Mg Calcite	Low Mg Calcite
120				
	>62 μ	72	26	2
	20–62 μ	69	29	2
	< 5 μ	48	46	6
135				
	>62 μ	49	41	10
	20–62 μ	63	35	2
	< 5 μ	34	57	9
152				
	>62 μ	31	58	11
	20–62 μ	54	40	6
	< 5 μ	27	60	13
292				
	>62 μ	75	22	3
	20–62 μ	66	31	3
	< 5 μ	39	49	12

to be enriched with high-magnesium calcite at
the expense of aragonite; moreover, in the lagoon
samples this enrichment is even more pronounced. These data suggest that either shoal-
derived high-magnesium calcite is being selectively transported to the lagoon or *in situ* production of high-magnesium calcite is occurring in
the lagoon.

Figure 7 compares histograms of the mineralogical composition of the coarse fraction, silt
fraction, and total mud fraction of shoal samples. Table 2 compares the mineralogical composition of the coarse fraction, silt fraction, and
clay fraction of four lagoon samples located close
to carbonate shoals. The data in both illustrations suggest that the enrichment of high-magnesium calcite in the mud fraction is primarily at
the expense of aragonite.

If it is considered that the production of lime
mud results primarily from particle size reduction of sand sized materials, then it follows that
the relative enrichment of high-magnesium calcite in the mud fraction results from differential

← ⋘

FIG. 6.—Carbonate mineralogy of the coarse fraction and total mud fraction of 106 samples from Southern
British Honduras. Shoal samples are represented by solid dots and lagoon samples by crosses. Note the general
shift toward higher concentrations of high-magnesium calcite in the total mud fraction as compared to the coarse
fraction.

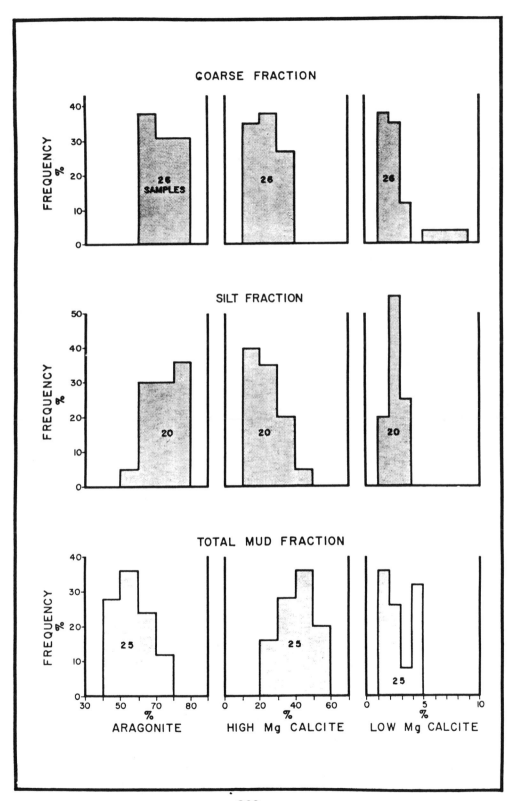

TABLE 3.—*Comparison of carbonate mineralogy of the mud fraction of lagoon samples[a] and carbonate shoal samples from Southern British Honduras*

	Lagoon Samples[b]			Carbonate Shoal Samples[c]		
	Mean	Observed Range	Standard Deviation	Mean	Observed Range	Standard Deviation
High-Strontium Aragonite	31.7[d] (16.8[e])	9–69 (0–61)	11.7 (13.4)	46.5[d] (35.3[e])	15–69 (0–61)	12.3 (15.5)
Low-Strontium Aragonite	17.1[d] (32.0[e])	0–32 (9–47)	7.1 (8.5)	9.9[d] (21.0[e])	0–28 (0–42)	8.8 (11.6)
High-Magnesium Calcite	44.0	25–56	6.3	41.2	21–59	10.2
Low-Magnesium Calcite	7.3	3–13	2.4	2.5	1–6	1.4

[a] Includes only those samples that are less than 4.0 miles from a shoal and contain greater than 40% carbonate in the mud fraction.
[b] Sixty samples used in calculations.
[c] Twenty-two samples used in calculations.
[d] Based on $Sr = 2.0°$ $_{oo}$ for low Sr aragonite.
[e] Based on $Sr = 3.0°$ $_{oo}$ for low Sr aragonite.

rates of particle size reduction. If it is considered that the production of lime mud results primarily from physical chemical precipitation of mud sized calcite carbonate, then one must hypothesize the physical chemical precipitation of high-magnesium calcite. Although Purdy (1964) has reported the replacement of skeletal high-magnesium calcite by cryptocrystalline high-magnesium calcite, the precipitation of mud-sized high-magnesium calcite is not well documented and is not considered as a serious possibility in this study. Thus, while enthusiastic supporters of the Bahama Bank model may wish to argue that some of the aragonite in the mud fraction in Southern British Honduras consists of aragonite needles (undoubtedly *some* aragonite needles *are* present), it is not solely the origin of the aragonite which should concern us here. The unusual abundance of high-magnesium calcite in the mud fraction equally commands our attention. This contrasts sharply with the situation on the Great Bahama Bank (Lowenstam and Epstein, 1957; Cloud, 1962; and Friedman, 1964) where physical chemical precipitation and/or skeletal disintegration has resulted in a predominantly aragonitic lime mud.

Samples of the lagoonal lime mud of Southern British Honduras contain 30 to 70 percent high-magnesium calcite, 20 to 60 percent aragonite, and 5 to 25 percent low-magnesium calcite. From these data some inferences can be drawn with regard to the constituent-particle composition of this mud. The high-magnesium calcite debris is most likely derived from the high-magnesium calcite skeletons of organisms that inhabit the shelf of Southern British Honduras. Such organisms include porcelaneous Foraminifera (table 4 and Blackman and Todd, 1959), some radial hyaline Foraminifera (Blackman and Todd, 1959), red algae (table 4 and Johnson, 1961), alcyonarians (Chave, 1954 and 1962b), and echinoderms (table 4 and Chave, 1954 and 1962b). Thus debris from some combination of these organisms accounts for 30 to 70 percent of the mud-size carbonate in the lagoon samples from Southern British Honduras.

Similarly the aragonite content of the lime mud of Southern British Honduras is most likely

←◀◀◀◀

FIG. 7.—Histograms of the mineralogical composition of size fractions of carbonate shoal samples. Note that the relative abundances of aragonite and high-magnesium calcite are approximately the same in the coarse fraction and silt fraction; whereas in the total mud fraction, aragonite is diminished and high-magnesium calcite is more abundant.

contributed chiefly by the skeletons of corals, calcified green algae, most molluscs (see table 4), and bryozoans (Chave, 1962b). Thus debris from some combination of corals, calcified green algae, molluscs, and bryozoans accounts for 20 to 60 percent of the mud-size carbonate in the lagoon samples from Southern British Honduras.

Following the same line of reasoning, 5 to 25 percent of the mud-size carbonate in the lagoon samples is contributed by skeletons of low-magnesium calcite from radial hyaline Foraminifera (Blackman and Todd, 1959), granular hyaline Foraminifera (Blackman and Todd, 1959) and molluscs (Chave, 1962b).

To be able to make a good estimate of the constituent particle composition of the lime mud in Southern British Honduras, there remains only the problem of distinguishing molluscan

TABLE 4.—*Mineralogy and strontium content of the calcareous skeletons of some marine organisms common in Southern British Honduras*

Organism	Remarks	Sr (°/oo)	Carbonate Mineralogy	Mole % $MgCO_3$ in Calcite
ALGAE				
Amphiroa rigida	Fresh, Sta. 71	1.6	calcite	22
Lithothamnium sp.	Fresh, Sta. 226	1.7	calcite	21
Halimeda opuntia	Fresh, locality unknown	8.5	aragonite	
Halimeda sand	Sta. 340	7.9	aragonite	
Penicillus dumetosus	Locality unknown	7.9	aragonite	
Rhipocephalus phoenix	Tarpum Cay	8.0	aragonite	
FORAMINIFERA				
Archais compressus	Sta. 149	3.2	calcite	15
assemblage[a]	Sta. 338	2.9	calcite	15
assemblage[a]	Sta. 306	2.2	calcite	14, 4[b]
assemblage[a]	Sta. 301	(0.7)[c]	calcite	4
CORALS				
Acropora cervicornis	Fresh	7.3	aragonite	
A. cervicornis	Worn fragment, Sta. 193	6.3	aragonite	
A. cervicornis	Fragment in mud, Sta. 129	6.8	aragonite	
A. cervicornis	Worn fragment, Sta. 292	7.0	aragonite	
A. cervicornis	Slightly abraded, from beach at Laughing Bird Cay	7.8	aragonite	
A. cervicornis	Worn fragment, Sta. 209	7.2	aragonite	
A. palmata	Fresh	7.5	aragonite	
A. palmata	Worn fragment, from beach at Laughing Bird Cay	7.5	aragonite	
A. palmata	24 feet subsea in Colson Cay bore hole	7.8	aragonite	
Montastrea cavernosa	Fresh	7.3	aragonite	
MOLLUSCS				
Pelecypods				
Diplodonta candeana	Worn shells, Sta. 132	1.8	aragonite	
Nuculana sp.	Fresh, Sta. 165	2.0	aragonite	
Tellina sp.	Fresh, Sta. 292	1.5	aragonite	
Gastropods				
Cerithium litteratum	Fresh, Sta. 151	1.8	aragonite	
Conus pygmaeus	Fresh, Sta. 141	1.3	aragonite	
Strombus gigas	Fresh	1.2	aragonite	
Pteropod assemblage	Sta. 171A	(1.8)	aragonite	
ECHINODERMS				
Mellita sexiesperforata	Fresh, Sta. 141	1.4	calcite	16
Comatulid		1.5	calcite	15
TUNICATES				
Didemnum spicules		6.5	aragonite	

[a] Water was added to dry sediment sample and those Foraminifera tests which floated were collected for analysis.
[b] Two calcites present.
[c] Value questionable, very small sample.

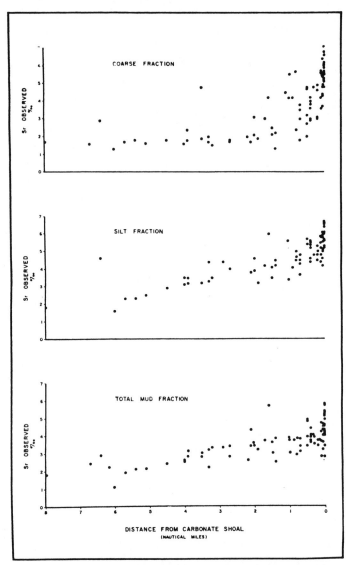

FIG. 8.—Strontium content of the carbonate fraction of Southern British Honduras samples plotted against distance from nearest carbonate shoal. Note that the strontium content of the coarse fraction falls off rapidly to a base level concentration of less than 2.0 parts per thousand; whereas the silt fraction and total mud fraction show a regular decrease in strontium with increasing distance from carbonate shoal.

aragonite from that of corals and *Halimeda*. To accomplish this distinction, strontium determinations were made.

Strontium Data

Analyses of the skeletons of the common marine organisms of Southern British Honduras (see table 4) suggest that only corals, calcified green algae, and tunicate spicules have strontium contents greater than 3.2 parts per thousand. The strontium content of corals and calcified green algae ranges from 6.3 to 8.5 parts per thousand. Since the abundance of tunicate spicules is volumetrically small, a mud fraction containing greater than 3.5 parts per thousand strontium may be considered to contain some debris of corals and/or calcified green algae while a mud fraction containing 8.0 parts per thousand strontium may be considered to consist almost entirely of coral and/or calcified green algae debris. It is evident from figure 8 that the strontium values for the silt fraction and total mud fraction of lagoon samples are sufficiently high to suggest the presence of considerable quantities of coral and/or calcified green algae debris. The strontium values also decrease systematically with increasing distance from carbonate shoals. Inasmuch as living corals and calcified green algae are characteristically absent in the lagoon, this systematic decrease suggests the transportation of high-strontium aragonite debris from carbonate shoals into the adjacent lagoon. (There are a few instances where corals and calcified algae were found living in relatively deep water (up to 25 fathoms) on small bathymetric highs in the extreme southern portion of the study area but such occurrences are rare. Extensive sampling, coring, and dredging has revealed the lagoon floor to be typically a soupy mud with no macroscopic flora and no corals.) Thus, examination of the strontium data, viewed in the light of known distribution of living corals and calcified green algae, leads to the conclusion that fragments from corals and/or calcified green algae are major constituents of the lagoon mud and further that these fragments are derived from carbonate shoals, the Barrier Platform being the largest example of such a shoal.

If coral and calcified green algae debris finds its way in significant quantities from the carbonate shoals to the Shelf Lagoon, then other shoal-derived mud-sized debris must also be present in the lagoon sediment. There remains only to combine point-count, mineralogical, and strontium data into a single best-estimate of constituent particle composition of the lime mud fraction.

Lime Mud Data Synthesis

In the preceding discussion it was concluded that most of the aragonite in the lagoon mud occurs in the form of fragments of corals, calcified green algae, molluscs, and bryozoans. Combination of mineralogical data and strontium data permits distinction of a high-strontium aragonite, which consists primarily of fragments of corals and calcified green algae, and low-strontium aragonite, which consists primarily of fragments of molluscs and bryozoans. Inasmuch as corals and calcified green algae contain about 7.4 parts per thousand strontium (table 4) and the skeletons of all other common organisms in the study area contain less than 3.2 parts per thousand strontium, the abundance of coral and calcified green algae debris in a sample may be estimated as follows:

Let

A = observed strontium content of the sample, in parts per thousand.

b = average strontium content of all constituents other than corals and calcified green algae, in parts per thousand.

7.4 = average strontium content of corals and calcified green algae, in parts per thousand.

and

X = that fraction of the sample which is composed of high-strontium aragonite and therefore approximates the abundance of corals and calcified green algae in the sample.

Then:

$$A = 7.4X + b(1 - X)$$

$$X = \frac{A - b}{7.4 - b}$$

The fraction of the sample which consists of low-strontium aragonite may be estimated by subtracting the percentage of high-strontium aragonite from the total aragonite content of the sample as determined by X-ray diffraction.

The only serious source of error in this calculation lies in the choice of a proper value for b. Examination of table 4 suggests that a value $b = 2.0$ would approximate the strontium content of coralline algae, mollusc, and echinoderm debris; whereas a value $b = 3.0$ would be required to approximate the strontium content of the Foraminifera which occur on the carbonate shoals. Consequently, two calculations were made for each sample, one using $b = 2.0$ and the other using $b = 3.0$. Thus, the fraction of the sample which is high-strontium aragonite should lie somewhere between these two calculated estimates. Table 3 compares the chemical-mineralogical composition of the total mud fraction of carbonate shoal samples that are less than four miles from a shoal and contain greater than forty percent carbonate in the mud fraction. These data indicate that the mud fraction of shoal samples contain more high-strontium aragonite than the mud fraction of lagoon samples, and, conversely, the mud fraction of lagoon samples contains more low-strontium aragonite than the mud fraction of the shoal samples. These data indicate either selective transport of low-strontium aragonite from the shoals into the lagoon, or in situ production of mud-sized low-strontium aragonite in the Shelf Lagoon.

There remains only to combine the chemical-mineralogical data with the petrographic data to arrive at a final best-estimate of the constituent particle composition of the carbonate mud fraction. The point-count data are regarded, for reasons presented previously, as the least accurate data gathered in this study. Therefore, the chemical-mineralogical data are relied upon as the major categories among which the point-count data may be distributed. The point-count data were combined with the chemical-mineralogical data in the following manner:

1. The chemical-mineralogical data for the silt fraction provide four headings (high-strontium aragonite, low-strontium aragonite, high-magnesium calcite, and low-magnesium calcite) among which the point-count categories are classified as appropriate.

2. Granular Hyaline Foraminifera Debris, Radial Hyaline Foraminifera Debris, and Whole Radial Hyaline Foraminifera are first assigned to the low-magnesium calcite group. If these three point-count categories total more than the percentage of low-magnesium calcite indicated for the sample, and this is the usual case, the excess percentage of Foraminifera is assumed to represent radial hyaline Foraminifera, and this percentage is then transferred to the high-magnesium calcite group. This procedure is justified on the basis that radial hyaline foraminifera may form tests of either low-magnesium calcite or high-magnesium calcite (Blackman and Todd, 1959).

3. Porcelaneous Foraminifera Debris, Echinoderm Debris, Alcyonarian Spicules, and any excess of Radial Hyaline Foraminifera resulting from the calculation described above are totaled, and this figure is compared with the percent of high-magnesium calcite indicated for the sample. This total is never sufficient to satisfy the observed percentage of high-magnesium calcite. Consequently this deficit is made up by assigning an appropriate percentage of the point-count category Cryptocrystalline Particles to the high-magnesium calcite group. This procedure is justified by the fact that this point-count category undoubtedly consists at least in part of irregular fragments of porcelaneous Foraminifera and/or red algae.

4. Mollusc Debris and whatever percentage is needed from the category Coral or Mollusc Debris are used to satisfy the sample data for the abundance of low-strontium aragonite.

5. Tunicate Spicules, Calcified Green Algae Debris (exclusive of *Halimeda*), the remainder of the category Coral or Mollusc Debris (this assumes that only coral fragments remain after step 4) and the remainder of the category Cryptocrystalline Particles (assumes only *Halimeda* debris remains) constitute the high-strontium aragonite group.

6. Finally, the importance of each of the constituents within a chemical-mineralogical group is adjusted to fit the chemical-mineralogical data of the total mud fraction as follows:

$$\frac{s}{S} = \frac{t}{T}$$

Where

S = any chemical-mineralogical group for the silt fraction, expressed in percent of the total carbonate in the silt fraction.

s = any constituent in the chemical-mineralogical group S, expressed in percent of the total constituents in the silt sample.

T = the same chemical-mineralogical group for the total mud fraction, expressed in percent of the total carbonate in the total mud fraction.

and

t = the same constituent in the chemical-mineralogical group T, expressed in percent of the total constituents in the total fraction.

The resulting estimates of constituent particle composition of the total mud fraction are summarized in table 5. It should be noted that two major uncertainties arise from the estimation procedure out-lined above. First, the identity of the "cryptocrystalline particles" which are assigned as high-magnesium calcite is not certain. Although these particles are undoubtedly high-magnesium calcite, because only the point-count category Cryptocrystalline Particles could possibly accommodate the excess high-magnesium calcite so consistently apparent when the point-count data are compared with the mineralogical data, it is not certain whether these particles are porcelaneous Foraminifera debris or red algae debris. In view of the relative hardness of coralline algae as compared to Foraminifera (personal observation) and in view of the fact that the magnesium calcite peak typical of coralline algae ($d_{(104)} = 2.9850$, or less) was not observed in any samples of the mud fraction, it is likely that this material represents primarily porcelaneous Foraminifera debris or "Melobesia" debris (Pusey, 1964). (The writer was unable to isolate sufficient "Melobesia" to accurately determine the magnesium content of its calcite in Southern British Honduras.)

The second major uncertainty exists with respect to the relative abundance of coral debris as compared to *Halimeda* debris. In step 4 above, low-strontium aragonite is satisfied from the categories Mollusc Debris and Coral or Mollusc Debris. It is probable, however, that low-strontium aragonite molluscs give rise to Cryptocrystalline Particles through alteration. If this is the case, then the outlined computation procedure has resulted in a division of high-strontium aragonite that tends to over-estimate the abundance of *Halimeda* debris. Thus, the figures presented for the abundance of *Halimeda* debris,

TABLE 5.—*Comparison of estimated abundance of constituent particles in the total mud fraction of lagoon samples and carbonate shoal samples from Southern British Honduras*

	Lagoon Samples[a]			Carbonate Shoal Samples[b]		
	Mean	Observed Range	Standard Deviation	Mean	Observed Range	Standard Deviation
Mollusc Debris	23.6	5–35	8.6	9.4	0–18	5.5
Coral Debris	14.0	0–55	12.5	33.1	17–53	12.2
Halimeda Debris	7.9	0–22	5.1	11.3	2–16	3.6
Other Codiaceae Debris	1.4	0–4	1.0	2.7	0–8	2.3
Granular Hyaline Foram Debris	2.1	0–6	1.8	0.2	0–1	0.4
Radial Hyaline Foram Debris[c]	15.3	1–30	7.4	7.4	0–17	5.2
Porcelaneous Foram Debris	12.0	2–28	5.0	14.0	7–27	6.3
Porcelaneous Foram Debris or Red Algae Debris	13.6	0–30	9.1	13.6	0–23	7.4
Echinoderm Debris	1.0	0–3	1.0	1.2	0–3	1.2
Alcyonarian Spicules	1.6	0–5	1.7	2.8	0–10	3.1
Tunicate Spicules	2.3	0–8	1.7	2.0	0–3	1.0

[a] Twenty-five samples used in calculation.
[b] Nine samples used in calculation.
[c] Includes whole radial hyaline forams. In both lagoon and shoal samples, debris exceeds whole forams by about 3:1.

are, if anything, too high; and the figures presented for coral debris, if anything, are too low.

GENESIS OF LIME MUD IN SOUTHERN BRITISH HONDURAS

It is evident from the foregoing that the debris of organisms common on the carbonate shoals bulks large in the lagoon mud. Some groups of these organisms live only on the carbonate shoals; therefore, their debris must originate in the shoal environment and be transported into the lagoon. Other groups of organisms are represented both on the shoals and in the lagoon and the debris from these organisms may thus originate either on the shoals or in the lagoon. Most of the skeleton-bearing organisms inhabiting the shoals are taxonomically distinct from their relatives living in the lagoon; however, this distinction can seldom be made in the petrographic identification of silt-size carbonate debris. Thus, in the present study, corals and *Halimeda* are the only major sediment contributors that may be considered as exclusive inhabitors of shoal areas. This being the case, it is necessary to compare the relative abundance of coral debris and *Halimeda* debris in the mud fraction of shoal samples with the relative abundance of these same constituents in the mud fraction of lagoon samples to gain some idea of the contribution of shoal-derived fines to the lagoon mud. Examination of the mean values reported in table 5 suggests that about half of the lagoon mud is of shoal derivation (Coral Debris plus Halimeda Debris = 44 percent for shoal samples and only 22 percent for lagoon samples). It should be noted, however, that the calculations for lagoon samples take

into account only those samples which contain greater than forty percent carbonate in the mud fraction. The data indicate that the carbonate mud fraction of lagoon samples nearer to the mainland contains virtually no lime mud of shoal origin. To make an accurate evaluation of the relative importance of shoal-derived lime mud versus *in situ* lagoonal lime mud requires involved calculations concerning the relative abundance of these components in each sample, the aerial distribution of mud characterised by each sample, and the vertical thickness of mud represented by each sample. The resulting numbers are, at best, rough figures and do not contribute greatly to recognition of the principles of lime mud genesis which Southern British Honduras has to offer us. Suffice it to say that a large volume of the lime mud in the Shelf Lagoon had its origin on the carbonate shoals and that a large volume of lime mud appears to have formed by *in situ* processes in the Shelf Lagoon itself.

Thus, Southern British Honduras offers us two models for the genesis for lime mud: the production of mud sized debris on carbonate shoals, and the *in situ* production of lime mud in the quiet, relatively deep waters of a protected continental shelf. The fact that both models are in operation in Southern British Honduras today may be viewed as coincidental.

Lime Mud Production on the Carbonate Shoals

Processes of particle-size reduction in modern carbonate environments have been discussed by several workers. In general, four processes have been recognized and these may be listed as follows: (1) Skeletal disintegration by removal of binding organic matter (Lowenstam, 1955; Lowenstam and Epstein, 1957; and Purdy, 1963). (2) Skeletal disintegration resulting from the activity of boring organisms (Ginsburg, 1957; Cloud, 1959; Neumann, 1963; Goreau and Hartman, 1963; and Pusey, 1964). (3) Particle-size reduction effected through mastication and ingestion of invertebrate skeletons by organisms (Cloud, 1959 and Folk and Robles, 1964). (4) Particle-size reduction effected through physical breakage and abrasion of skeletal material in agitated environments (Cloud, 1959; Chave, 1962a; Folk and Robles, 1964). The following discussion attempts to evaluate the importance of each of these processes on the carbonate shoals in Southern British Honduras. It will be noted that this is primarily a discussion of processes of particle-size reduction rather than of the more limited topic of lime mud production. Given the composition of the lime mud of Southern British Honduras, there is no *a priori* reason to suspect that particle-size reduction processes which produce carbonate sand are any different from particle-size reduction processes which produce lime mud. Any process of particle-size reduction which is suggested to produce sand-size carbonate material may also be expected to produce a certain amount of lime mud.

Lowenstam (1955) has demonstrated that many calcified green algae disintegrate into aragonite needles upon removal (oxidation) of the organic matter which binds the needles together in the living specimen. Purdy (1963) has noted a similar disintegration of molluscs and certain coelenterates into prismatic or granular particles upon removal of organic matter by hydrogen peroxide. In contrast to the Great Bahama Bank where aragonite needles are the predominant mud-size carbonate particle type, aragonite needles are not an abundant constituent of the mud in Southern British Honduras. Granular particles and to a lesser extent prismatic particles, however, are common in the silt fraction of samples from Southern British Honduras. While the granular particles may be of diverse origin, it would appear likely that the prismatic particles result largely from the disintegration of mollusc shells containing large amounts of binding organic matter as suggested by the experiments of Purdy (1963). Removal of binding organic matter may account for the origin of prismatic particles and perhaps also of some granular particles.

The boring activities of certain algae, sponges, and molluscs have been noted by several authors (Ginsburg, 1957; Cloud, 1959; Neuman, 1963; Goreau and Hartman, 1963; and Pusey, 1964). Cloud (1959) and Neumann (1963) note particularly the weakening effect these organisms can have on large masses of reef or limestone. Cloud (1959) concludes: "Such activities themselves provide little new sediment, but they commonly render the rock so weak and punky that it crushes or breaks with the lightest tap of a hammer. In this state it is, of course, easily broken by hydraulic wedging or by the impact of loose fragments of rock or broken coral masses that may be rolled or bounced across it at time of storm." Thus, Cloud points out that boring organisms serve primarily to facilitate mechanical breakup. Ginsburg (1957, p. 84) mentions this "weakening of the reef by organic erosion" but also notes the work of boring organisms on smaller skeletons such as molluscs. Pusey (1964, p. 72) notes further that sand-sized mollusc fragments may be reduced to a delicate mesh by boring organisms. Thus, boring organisms may be expected to weaken skeletons and thereby facilitate particle-size reduction over an enormous range of original material sizes. At one end of this size spectrum, whole blocks of reef may be so weakened as to finally break into large boulders; while at the other end of the spectrum, sand-sized mollusc fragments may be so weakened as to give rise to mud-size particles. This full range of boring activity is undoubtedly present on the carbonate shoals of Southern British Honduras, but the relative importance of this activity to the production of lime mud is difficult to evaluate.

Cloud (1959) attaches major significance to the destructive role of certain organisms in the production of reef-associated sediments. Fish in particular are credited with producing vast quantities of sediment as a result of their ingesting living coral and coralline algae. To illustrate his point, Cloud (1959, p. 399) attempts to calculate the sediment contribution ascribable to fish and concludes they reduce to sediment 1,100 to 1,600 metric tons of living coral and coralline algae per square mile per year. Cloud further calculates that this represents an annual rate of sedimentary accretion of 0.2 to 0.3 millimeters per year from the activity of fish.

Cloud's calculations imply that a maximum of 2 meters of fish-produced sediment would accumulate during Holocene time. In Southern British Honduras, however, 20 to 30 meters of carbonate sand have accumulated on the Barrier Platform alone; to say nothing of the shoal-derived mud which has accumulated in the Shelf Lagoon. Thus either fish in Southern British Honduras are ten to twenty times more active than provided by Cloud's calculations, or Cloud's conclusion as to the relative importance of fish as sediment producers is not applicable to

Southern British Honduras. As no reason can be offered to explain such a large contrast between fish activity in Saipan and fish activity in Southern British Honduras, it is concluded that fish are relatively minor agents in effecting particle-size reduction in Southern British Honduras.

Physical breakage and abrasion of skeletal material in agitated environments is also discussed by Cloud (1959). He considers that the greatest production of sediment by inorganic factors alone occurs along the reef front and that this sediment is deposited as fore-reef talus. Cloud (1959, p. 400) concludes that "sediments of direct organic origin and those produced by the browsing activities of reef fish are . . . most conspicuous in lagoons and shoals" and thus suggests that agitation of sediment on shoals is relatively unimportant in effecting particle-size reduction.

In contrast, the experiments of Chave (1962a and 1964) suggest that many carbonate skeletons may undergo particle-size reduction in agitated environments. Folk and Robles (1964) reach a similar conclusion from study of a natural environment.

Although the operation of processes of particle-size reduction in agitated environments appears qualitatively well documented, quantitative data on the relative importance of purely physical processes in the production of lime mud is virtually non-existent. In this study it is concluded by default that such processes are indeed quantitatively important; that is, the mechanism is obviously in operation, and other mechanisms which have been suggested seem inadequate to explain the large volume of sediment observed.

In Situ Lime Mud Production in the Shelf Lagoon

Table 5 compares the estimated abundance of constituent particles in the total mud fraction of lagoon samples and carbonate shoal samples. Note that lagoon samples contain considerably more mollusc debris and radial hyaline Foraminifera debris than do shoal samples. These data suggest either selective transport of mollusc and radial hyaline Foraminifera debris from the shoals to the lagoon or in situ production of mollusc and radial hyaline Foraminifera debris in the lagoon itself. If one chooses to defend the idea of selective transport, one must explain why both mollusc debris and radial hyaline Foraminifera debris react to the transporting mechanism in the same fashion. Mollusc debris is predominantly granular in shape, quite similar to coral debris; whereas radial hyaline Foraminifera debris is platey, much like porcelaneous Forami-

nifera debris. Thus it seems unlikely that mollusc debris and radial hyaline Foraminifera debris have anything in common that causes them to be selectively transported to the lagoon from the carbonate shoals. Moreover, molluscs and radial hyaline Foraminifera are abundant inhabitants of the Shelf Lagoon. If processes can be suggested by which the skeletons of these organisms may become comminuted, then clearly in situ lagoonal production of this excess mollusc and Foraminifera debris is the most tenable hypothesis.

The four factors considered most important in the in situ production of Shelf Lagoon lime mud are as follows: (1) the inherently fragile nature of the shells of molluscs and tests of Foraminifera that inhabit this environment; (2) the removal of the binding organic matter from mollusc shells (Purdy, 1963); (3) the weakening of larger skeletal particles by the activity of boring microorganisms; and (4) the mastication, ingestion, and perhaps even simple movement of sediment by the vagrant benthos.

Small, thin-shelled molluscs and hyaline Foraminifera are the predominant carbonate-secreting organisms of the Shelf Lagoon. Mollusc shells are commonly fragmented, and fragments of radial hyaline Foraminifera are twice as abundant as whole radial hyaline Foraminifera in the silt fraction of lagoon samples (table 1). Processes of wave and current agitation appear inadequate in this environment to produce the observed breakage of skeletal remains. Biological processes must account for the vast majority of the particle-size reduction which occurs in the Shelf Lagoon.

CONCLUSION

If this investigation has served no other purpose, it has demonstrated that lime mud may have multiple origins. Even if one argues that recrystallization may render it difficult or impossible to distinguish in the geologic record among the various types of lime mud that can be recognized in the Recent, an awareness of the possible multiple origins of lime mud may aid in interpreting the genesis of ancient lime mudstones.

The example of lime mud production on the Barrier Platform and deposition in the Shelf Lagoon calls attention to the possible importance of transportation in the development of facies in carbonate sediments. Indeed, much of the discussion of the reef-complex shoals as lime mud producers can be extended to any carbonate shoal. Carbonate environments which are subjected to wave and/or current agitation should be expected to produce lime mud, and in most cases it should be expected that this mud will be transported away from its agitated site of origin and deposited in quiet water elsewhere.

In contrast, the example of *in situ* lime mud production in the Shelf Lagoon indicates that lime mud can form *in situ* in quiet water without the presence of calcified green algae or abnormal salinity conditions which must be postulated if one envisions an "aragonite needle" type of lime mud.

It would appear that Recent environments offer the geologist at least four possible modes of origin of lime mud in ancient lime sediments: (1) Production of aragonite needles by physical precipitation from waters of abnormally high salinity and carbonate saturation as suggested by Cloud (1962), Wells and Illing (1964) and Friedman and Neev (1966); (2) Production of aragonite needles by *post mortem* disintegration of calcified green algae (Lowenstam and Epstein, 1957); (3) Production of mud-sized skeletal debris by predominantly physical processes of particle-size reduction in agitated environments (Chave, 1962a and 1964; Folk and Robles, 1964; and this study); and (4) Production of mud-sized skeletal debris other than aragonite needles by predominantly biological processes of particle-size reduction in quiet-water environments (Purdy, 1963; and this study).

Not only do the available data suggest that lime mud may form in a spectrum of environments by a variety of mechanisms, but they also suggest that fine-grain carbonate sediments may vary greatly with respect to particle-size distribution, particle shape, mineralogy and minor element chemistry. The possible importance of these original sediment parameters in neomorphism (Folk, 1965) in fine-grain carbonates has

scarcely been mentioned. Truly, much study is needed if ancient lime muds are to be understood in terms of their presumed counterparts in the Recent.

ACKNOWLEDGEMENTS

This paper is a shortened version of a .Ph.D. thesis submitted to the Department of Geology of Rice University. The study was supervised by Professor Edward G. Purdy as part of an overall British Honduras research project financially supported by the Humble Oil & Refining Company, the Jersey Production Research Company, and Imperial Oil Limited. Dr. J. J. W. Rogers kindly served as thesis committee chairman in the absence of Dr. Purdy.

The friendly cooperation of persons in the British Honduras Government is gratefully acknowledged. Mr. W. Ford Young of Belize, British Honduras, was an invaluable help in expediting field preparations.

The International Department of Phillips Petroleum Company kindly contributed a blueline copy of a photomosaic of the Port of Honduras. Without this coverage, our map of this area would have been highly inaccurate.

A debt of gratitude is owed to Mr. John Mc Crevey and Dr. Walter C. Pusey, III, for their assistance with portions of the laboratory work, and to Dr. James Lee Carter for operating the Atomic Absorption Spectrophotometer. The drafting is the work of Harrison E. Jones of Houston, Texas.

Finally, a debt of gratitude is owed to Dr. Gerald M. Friedman and Dr. Robert N. Ginsburg for their valuable assistance in editing the original manuscript of this paper.

REFERENCES CITED

AMERICAN SOCIETY FOR TESTING MATERIAL, 1960, X-ray powder data file: Am. Soc. Testing Material, Philadelphia, Pennsylvania.

BLACKMON, P. D. AND TODD, RUTH, 1959, Mineralogy of some Foraminifera as related to their classification and ecology: Jour. Paleontology, v. 33, p. 1–15.

CHAVE, K. E., 1954, Aspects of biogeochemistry of magnesium: Jour. Geology, v. 62, p. 266–283 and 587–599.

———— 1962a, Carbonate skeletons to limestones; problems: New York Acad. Sci. Trans., v. 23, p. 14–24.

———— 1962b, Factors influencing the mineralogy of carbonate sediments: Limnology and Oceanography, v. 7, p. 218–223.

———— 1964, Skeletal durability and preservation, *in* Imbrie, John and Newell, Norman, D., eds., Approaches to paleoecology. John Wiley & Sons, Inc., New York, 432 p.

CLOUD, P. E., JR., 1959, Geology of Saipan, Mariana Islands, part 4, Submarine topography and shoal water ecology: U. S. Geol. Survey Prof. Paper 280-K, p. 361–445.

———— 1962, Environment of calcium carbonate deposition west of Andros Island, Bahamas: U. S. Geol. Survey Prof. Paper 350, 138 p.

FOLK, R. L., 1965, Some aspects of recrystallization in ancient limestones, *in* Pray, L. C., and Murray, R. C., eds., Dolomitization and limestone diagenesis: Soc. Econ. Paleontologists and Mineralogists Spec. Paper 13, 180 p.

FOLK, R. L., AND ROBLES, ROGELIO, 1964, Carbonate sands of Isla Perez, Alacran Reef Complex, Yucatan: Jour. Geology, v. 72, p. 255–292.

FRIEDMAN, G. M., 1964, Early diagenesis and lithification in carbonate sediments: Jour. Sedimentary Petrology, v. 34, p. 777–813.

FRIEDMAN, G. M., AND NEEV, DAVID, 1966, On the origin of aragonite in the Dead Sea: Israel Journal Earth
    Sciences, in press.
GINSBURG, R. N., 1956, Environmental relationships of grain size and constituent particles in some south Flor-
    ida carbonate sediments: Am. Assoc. Petroleum Geologists Bull., v. 40, p. 2384–2427.
———— 1957, Early diagenesis and lithification of shallow water carbonate sediments in south Florida, in
    LeBlanc, R. J., and Breeding, J. G., eds., Regional aspects of carbonate deposition—A symposium: Soc.
    Econ. Paleontologists and Mineralogists. Spec. Pub. No. 5, 178 p.
GOLDSMITH, J. R., GRAF, D. L., AND JOENSUU, O. I., 1955, The occurrence of magnesium calcites in nature:
    Geochim. et Cosmochim. Acta, v. 7, p. 212–230.
GOREAU, T., AND HARTMAN, W. D., 1963, Boring sponges as controlling factors in the formation and maintenance
    of coral reefs in R. F. Sognnaes, ed., Mechanisms of hard tissue destruction. Am. Assoc. Advancement Sci.,
    Washington, D. C., 764 p.
HATCH, F. M., RASTALL, R. H., AND BLACK, MAURICE, 1938, The petrology of the sedimentary rocks. 3rd. ed.,
    Allen and Unwin, London, 383 p.
HOSKINS, C. M., 1963, Recent carbonate sedimentation on Alacran Reef, Yucatan, Mexico: Natl. Acad. Sci.,
    Natl. Research Council, Pub. No. 1089, 160 p.
HURLBUT, C. S., JR., 1955, Dana's manual of mineralogy. John Wiley and Sons, Inc., New York, 530 p.
JOHNSON, J. H., 1961, Limestone-building algae and algal limestones: Johnson Publishing Co., Boulder, Colo-
    rado, 297 p.
LOWENSTAM, H. A., 1954, Factors affecting the aragonite:calcite ratios in carbonate secreting marine organisms:
    Jour. Geol. v. 62, p. 284–322.
———— 1955, Aragonite needles secreted by algae and some sedimentary implications: Jour. Sedimentary Petrol-
    ogy, v. 25, p. 270–272.
LOWENSTAM, H. A., AND EPSTEIN, S., 1957, On the origin of sedimentary aragonite needles of the Great Bahama
    Bank: Jour. Geology, v. 65, p. 364–375.
MATTHEWS, R. K., 1965, Genesis of Recent lime mud in Southern British Honduras: Ph.D. thesis, University
    Microfilm No. 65-10345, Rice Univ., Houston, Texas, 148 p.
NEUMANN, A. C., 1963, Processes of recent carbonate sedimentation in Harrington Sound, Bermuda: Ph.D.
    thesis, Lehigh University, Bethlehem, Pennsylvania, 157 p.
ODUM, H. T., 1957, Biogeochemical deposition of strontium: Public Inst. Marine Sci., v. 4. no. 2, p. 38–114.
PURDY, E. G., 1963, Recent calcium carbonate facies of the Great Bahama Bank, 1. Petrography and reaction
    groups, 2. Sedimentary facies: Jour. Geology, v. 71, p. 334–355, 472–497.
PURDY, E. G., AND MATTHEWS, R. K., 1964, Structural control of recent calcium carbonate deposition in British
    Honduras (abs.): Geol. Soc. America, Spec. Pub. 82, p. 128.
PUSEY, W. C., 1964, Recent calcium carbonate sedimentation in Northern British Honduras: Ph.D. thesis, Rice
    Univ., Houston, Texas, 247 p.
TAFT, W. H., AND HARBAUGH, J. W., 1964, Modern carbonate sediments of southern Florida, Bahamas, and
    Espiritu Santo Island, Baja California; a comparison of their mineralogy and chemistry: Stanford Univ.
    Pub., Geol. Sci., v. 8, no. 2, 133 p.
WELLS, A. J., AND ILLING, L. V., 1964, Present day precipitation of calcium carbonate in the Persian Gulf, in
    Van Straaten, L. M. J. V., ed., Deltaic and shallow marine deposits. Elsevier Publishing Co., Amsterdam
    464 p.
WOOD, A., 1948, The structure of the wall of the test in the Foraminifera; Its value in classification: Geol. Soc.
    London Quart. Jour., v. 104, p. 229–255.

12

Reprinted from *Jour. Sed. Petrology* **37**:633-648 (1967), by permission of the publisher

THE PRODUCTION OF LIME MUD BY ALGAE IN SOUTH FLORIDA[1,5]

K. W. STOCKMAN,[2] R. N. GINSBURG,[3] AND E. A. SHINN[4]

ABSTRACT

We have made estimates of the annual production of fine aragonite mud ($<15\mu$) by algae with fragile skeletons from sea floor observations in two areas of modern lime mud accumulation, Florida Bay and the nearshore part of the Florida Reef Tract. Comparing these rates of production with rates of mud accumulation, we calculate that one genus, *Penicillus*, is a major contributor of fine aragonite mud.

Penicillus sp. is one of several lightly calcified green and red algae that disintegrate post-mortem and produce fine crystals ($<15\mu$) of aragonite and calcite. The annual contribution of fine aragonite mud from two species of this alga was estimated from a combination of the following observations: (1) a year's surveillance of permanent bottom stations representative of conditions within the major areas of mud accumulation, (2) counts of plant abundance made by divers, and (3) determination of the weight of aragonite per plant for representative specimens.

Production of fine aragonite mud ($<15\mu$) by *Penicillus* sp. was compared with average rates of sediment accumulation for the equivalent size grades. The rates of sediment accumulation were determined indirectly from C^{14} age determinations of subsurface samples. Considering *only production within the areas of maximum mud accumulation*, we calculated that since the areas were flooded by rising sea level 4000 to 10,000 years ago, the present rate of production by *Penicillus* sp. could account for all the fine aragonite mud in the inner Florida Reef Tract and one-third of the same material in northeastern Florida Bay.

The production of aragonite mud by *Penicillus* sp. provides a base line for evaluating the production of other similar algae. In the reef tract and the southeastern and western margins of Florida Bay, two other related green algae together are more abundant than *Penicillus* sp., in northeastern Florida Bay, one alga is more abundant than *Penicillus*.

Another source of fine lime mud is the biological and mechanical breakdown of resistant skeletons, mollusks, algae, corals, etc. This breakdown gives recognizable skeletal silt ($15-62\mu$), but appreciable finer particles ($<15\mu$) must also be produced.

Both organic sources of fine lime mud, fragile algae and skeletal breakdown, are present in abundance in the large wave- and current-swept areas seaward of the mud accumulations. The mud which they generate cannot accumulate in the agitated environment in which it is produced. The mud from one of these "source areas," the one west and southwest of Florida Bay, is very probably a major contributor to the large mud banks in western Florida Bay. The other agitated source area, the outer reef tract, may contribute to the mud accumulation in the nearby inner reef tract, but because the outer reef tract is so close to deeper water, an unknown but significant amount of mud is probably carried seaward to the Straits of Florida and beyond.

The production of lime mud by all these skeletal sources is believed to be more than adequate to explain the amount of lime mud sediment in south Florida. The similarity in the conditions of accumulation of these Recent lime muds and many ancient lime muds (textures, structures, and faunal variety) suggests that plant and animal skeletons, particularly the fragile ones, have been major sources of fine lime sediment in the past.

INTRODUCTION

The purpose of this report is to present direct evidence of the large amount of fine lime mud, $<15\mu$ (1/64 mm), produced by the post-mortem disintegration of the fragile skeletons of algae. This algal mud and the mud produced by the biological and mechanical trituration of more resistant skeletons are believed to be the source of most, if not all, the recent lime mud in south Florida sediments.

The origin of Recent and ancient lime muds is a familiar geologic problem. How much of a given lime mud is derived from plant and animal skeletons, and how much is derived from nonskeletal precipitation, either physicochemically or biochemically?

Lime mud is usually used for the size fraction, $<62\mu$ (1/16 mm). In Recent lime muds and in many ancient ones, the particles between 15 and 62μ (1/64-1/16 mm) are recognizable grains of skeletons of various plants and animals, excluding fecal pellets that are aggregates of finer particles. It is the finer fraction, $<15\mu$, whose source cannot be identified with the microscope. When this fine lime mud is only a small percent of the total sediment and the much larger re-

[1] Contribution No. 475, Shell Development Company.

[2] Present address, 9450 Caribbean Blvd., Kendall, Florida.

[3] Present address, The Dept. of Geology, The Johns Hopkins University, Baltimore, Maryland.

[4] Present address, Shell Development Co., Houston, Texas.

[5] Manuscript received November 22, 1966; revised February 17, 1967.

FIG. 1. Electron photomicrograph of the fine fraction of a lime mud sample taken a few miles west of Andros Island, Bahamas.

mainder is mostly skeletal debris, the mud is usually considered to be the product of biological and mechanical erosion of the skeletal debris. However, when fine lime mud predominates and little or no skeletal debris is present, as is frequent in ancient carbonates, there are no guidelines for interpreting the origin of the mud.

To develop some basis for interpreting ancient fine lime mud, geologists have for some time studied Recent lime sediments. Until recently these studies have been largely indirect; that is, they have been concerned with the mechanisms of nonskeletal precipitation that could produce fine lime mud. In the most thoroughly studied area of Recent lime muds, the area west of Andros Island on the Great Bahama Bank, several workers studied the chemistry of the water and concluded that physicochemical precipitation is responsible for the fine mud (Gee *et al.*, 1932; Smith, 1940; Cloud and Barnes, 1948). The most complete treatment of this interpretation is the paper by Cloud (1962), who concluded that approximately three-fourths of the fine aragonite mud is a nonskeletal precipitate, largely physicochemical. Other workers (Drew, 1914; Lipman, 1924; Bavendamm, 1932, Oppenheimer, 1961 and Greenfield, 1963), studied the role of bacteria and suggested that their metabolism and decay can produce precipitation of fine aragonite mud.

More recently, Lowenstam has attempted a more direct approach to the origin of the Bahamian muds. In 1955 he first reported that the fragile skeletons of several of the common green algae disintegrate spontaneously post-mortem to prismatic crystals of aragonite 2–10μ long. These algal crystals (fig. 2), resemble the ara-

gonite crystals in the muds west of Andros (fig. 1). Later Lowenstam and Epstein (1957) showed that an algal source was consistent with the oxygen isotope ratios of the sediment. Some workers (Cloud, 1962, pp. 97, 98; Purdy, 1963) have questioned the assumed temperatures and salinities on which Lowenstam's and Epstein's (1957) conclusion is based; Cloud (1962, pp. 96–98) doubts that the abundance of algae is sufficient to explain the quantity of fine aragonite mud in the sediments.

In this report we shall describe the results of another direct approach to the origin of fine lime mud, a quantitative estimate of the production by algae with fragile skeletons. For this study we chose the most common fragile green alga in south Florida, *Penicillus* sp., and made direct sea floor observations of its annual production.

The time required for a *Penicillus* plant to grow and die, its life span, was determined from periodic observations for a year of fixed submarine stations established in representative areas of the two major environments, the Florida Reef Tract and Florida Bay. The average plant abundance over a large area was estimated from numerous bottom observations. The weight of aragonite per plant was determined by acid dissolution of several representative collections. From these three estimates—life span, average abundance, and weight of aragonite per plant—the annual rates of mud production by this plant can be calculated. These rates of mud produc-

FIG. 2a.—Electron photomicrograph of the aragonite crystals from the head of *Penicillus* sp.

tion by *Penicillus* are then compared with the rate of mud accumulation based on radiocarbon age determinations of subsurface sediments.

PENICILLUS: MORPHOLOGY AND CALCIFICATION

Penicillus capitatus and *Penicillus pyriformis* are only two of several green algae whose fragile skeletons are composed of aragonite. However, because these are the only species that occur in both Florida Bay and the Florida Reef Tract, *Penicillus* was chosen for this study.

Penicillus is an erect alga composed of continuous filaments that are organized into a holdfast, stem, and head (fig. 3). The holdfast consists of a ball of filaments intertwined about themselves and the sediment to which they are firmly attached. The filaments in the holdfast are not calcified, but when they pass up into and form the stem, they are bundled by a sheath of aragonite. When these same filaments separate to become the head, each of them is encased in a delicate porous sheath of aragonite. The sheaths of both the stem and of the head filaments are so fragile that when a plant is treated with Clorox, the pressure of a dissecting needle breaks them easily into crystals mostly less than 20μ. Most of the crystals in the stem are prisms (fig. 2b): in the head, prisms and irregular crystals are present (fig. 2a).

Penicillus sp. reproduces vegetatively and sexually. Vegetative reproduction is by rhizomes that extend laterally through the sediment (fig. 4); therefore, *Penicillus* commonly occurs in clumps on the sea floor. The sexual phase of reproduction has not been described.

STUDY AREA AND ABUNDANCE OF *PENICILLUS*

The south Florida region is divided into two major environments, restricted Florida Bay and the open reef tract (fig. 5). The hydrography, sediments, and faunas of these two environments are discussed by Ginsburg (1956) and are summarized by Ginsburg (1964).

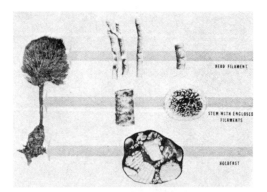

FIG. 3.—Morphology of *Penicillus* sp.

Reef Tract

The reef tract is divided into two subenvironments (fig. 5); (1) an outer part characterized by twice-daily tidal flushing, coral-algal reefs, and lime sands and (2) an inner part that is less frequently flushed by tidal exchange and commonly has a relatively dense cover of turtle grass and sediment that contains appreciable quantities of lime mud.

The distribution of *Penicillus* in the outer reef tract is irregular; plants are absent on coral-algal reefs and are generally rare on the lime sands that are moved frequently by waves and currents. Off the reefs and away from areas of moving sand, which together constitute an estimated 50 percent of the area, *Penicillus* plants are large and locally abundant.

Penicillus is present throughout the inner part of the reef tract. Locally, plants are very abundant. Most of a narrow zone approximately 100 ft wide along the windward margin of Rodriguez and Tavernier banks (Ginsburg, 1964) has up to 60 plants/m². Bottom observations indicate that some of the deeper areas have hundreds of

FIG. 2b.—Electron photomicrograph of the aragonite crystals from the stem of *Penicillus* sp.

FIG. 4.—Underwater view of *Penicillus* sp. showing vegetative reproduction. The rhizomes that connect plants were exposed by waving the sediment away with a swim fin.

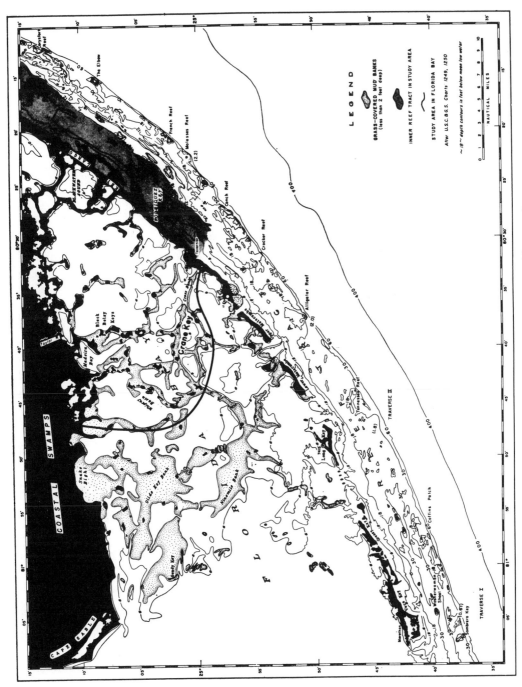

FIG. 5.—Study area in northeastern Florida Bay and inner reef tract.

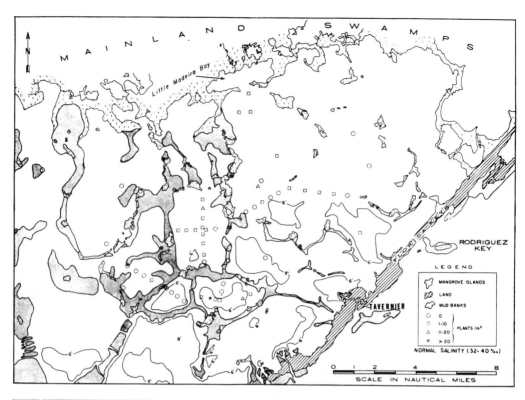

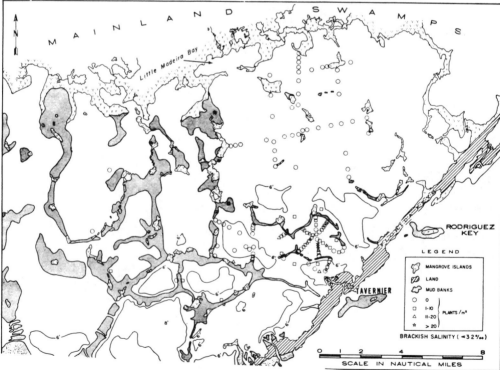

FIG. 6.—Study area in northeastern Florida Bay showing observations of plant abundances, during periods of normal salinity (top) and brackish salinity (bottom).

plants/m². However, these areas of maximum abundance are of limited lateral extent, and the more usual abundance is from 5 to 15 plants/m². The average abundance for the inner reef tract is estimated at 8 plants/m².

Florida Bay

Florida Bay (fig. 5) is a large triangular-shaped area in which linear near-surface mud banks surround irregular-shaped areas of deeper water (4–8 ft deep) known locally as lakes. Hydrographically, the bay can be divided into three zones—a marginal zone in which flushing keeps salinity near that of normal sea water for this region, a runoff zone most frequently affected by runoff of fresh water from the mainland, and an interior zone in which salinity fluctuates from brackish to supersaline, depending on the amount of runoff (McCallum and Stockman, 1966).

The present study was limited initially to the northeastern part of the bay (fig. 5). *Penicillus* is present throughout the study area during periods of normal or super-saline salinity, but it disappears from the interior and runoff zones when the salinity is brackish. More recently,

field observations of plant abundance in the western and southwestern parts of the bay indicate much greater plant abundances than in the study area. In the lakes that make up approximately 90 percent of the study area, the average abundance ranges from 0 to 30 plants/m². Locally, in a narrow zone along the protected margin of some of the mud banks, the plant abundances are high (76–108 plants/m²). The average abundance for the lakes that form 90 percent of the Florida Bay study area was estimated to be 2 plants/m². This estimate was based on numerous observations of abundance during periods of both brackish and normal salinity (fig. 6) weighted according to area.

It is very difficult to make a reliable estimate of the average abundance on the banks, 10 percent of the study area. Some banks, especially those in the southeast part of Florida Bay, had very abundant plants (several tens of plants/m²) in a narrow zone on the protected margin and sparse plants (several plants/m²) on the exposed margin. Other narrow banks had from a few to several plants/m² all across them. Still other banks, particularly those in the western and northwestern parts of the study area, had no

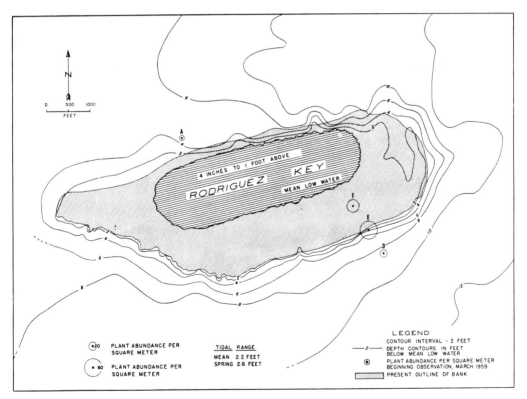

FIG. 7.—Location of permanent stations on Rodriguez Bank, inner reef tract, diameter of circles proportional to plant abundance.

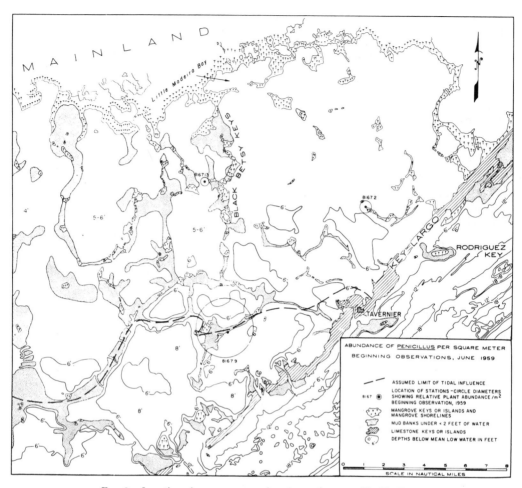

FIG. 8.—Location of permanent stations in northeastern Florida Bay.

plants. Because the observations are not numerous enough to evaluate the variations, the average abundance is assumed to be the same as that of the lakes (2 plants/m²). This figure is probably well below the real average.

DETERMINATION OF LIFE SPAN

To determine the life span of *Penicillus*, it was necessary to establish permanent stations on the bottom and observe the growth and death of individual plants for an extended period. To span the range of different environmental conditions under which *Penicillus* lives, twenty-three stations were selected after the initial reconnaissance. In the reef tract sixteen stations were established at four locations around Rodriguez Bank (fig. 7), and in Florida Bay seven stations were established at three different localities (fig. 8). At each station one or more square plots 0.5 m on a side (area, 0.25 m²) were established by driving stakes at the corners of the plots.

These stakes were used as fixed reference points for a frame with an underwater camera attached (fig. 9) or a marked counting frame (fig. 10). Approximately every 20 days for a year, each of these stations was visited, the frames were oriented from the stakes, and *Penicillus* within the plots was observed. For the clear-water stations in the reef tract it was possible to take underwater photographs of the bottom and compare the number and position of plants on successive visits. In the murky water of Florida Bay the number and positions of plants were determined visually with the counting frame.

At the end of the year's observations the number of plants that grew, died, and contributed their skeletons to the sediment was known for each of the plots. The average annual life spans for individual stations varied from 30 to 63 days. The average life span for each of the major environments was determined graphically by plotting average plant abundance vs total an-

FIG. 9.—Diver with frame and attached camera for photographing station.

Counting grid used in Florida Bay − 1/4 m.²

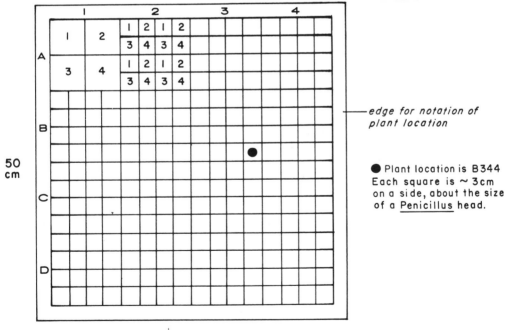

FIG. 10.—Counting grid used in Florida Bay.

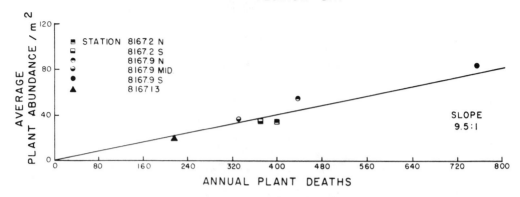

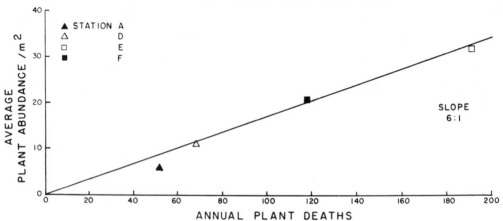

FIG. 11.—Plot of average plant abundance vs annual plant deaths.

nual deaths for each station (fig. 11). This graphic estimate of the average life span was used because the estimate of life spans for each station is largely a function of the frequency of observation and because the graphic method shows that plant abundance does not influence life span, that is, the points can be connected by a straight line.

The slopes of the lines in figure 11 are the average annual number of crops of *Penicillus* in the major enivronments. Thus, for Florida Bay (slope, 9.5:1) the average life span in round numbers is about 40 days, (360/9.5) and for the inner reef tract (slope, 6:1) the average life span in round numbers is 60 days (360/6).

WEIGHT OF ARAGONITE PER PLANT

Several representative plants were collected from each of the stations established for the determination of life spans. The plants were care-

fully cleaned to remove sediment and attached organisms and were dried, and the amount of aragonite was determined by the loss of weight after dissolution in dilute acid. The results are shown in table 1.

The lowest figure for Florida Bay, 0.17 g, is taken as the average for plants in the lakes because of the observed similarity in size. The average weight for plants in the inner reef tract was taken as 0.52 g, the average weight of carbonate in plants collected near stations A and D.

RATES OF MUD PRODUCTION AND ACCUMULATION FROM *PENICILLUS*

The preceding sections have shown how the estimate of life spans, average abundances, and average weights of aragonite/plant were determined. With these three figures the annual production of mud by *Penicillus* can be calculated.

	Crops/year	Weight/Plant	Average Abundance	Annual Production
		(g)	(plants/m²)	(g/m²)
Florida Bay Study Area	9.5	0.17	2	3.23
Inner Reef Tract	6	0.52	8	25

These weights of solid aragonite are converted to sediment volume and rare of accumulation with 2.9 as the specific gravity of aragonite and 60 percent as the average porosity of Recent lime muds (see fig. 12).

marine sedimentation at a given locality by determining the depth below present sea level of the lithified Pleistocene floor.

If for example, the Pleistocene rock floor is 10

	Weight	Volume	Vol as Porous Sediment	Rate of Accumulation	
	(g/m²)	(cm³/m²)	(cm³/m²)	(cm/yr)	(cm/1000 yr)
Florida Bay	3.23	1.1	2.8	0.0003	0.3
Inner Reef Tract	25	8.6	23	0.0023	2.3

COMPARISON OF RATES OF MUD PRODUCTION BY PENICILLUS WITH RATES OF MUD SEDIMENTATION

To evaluate the figures for rates of production of fine aragonite mud by *Penicillus*, it is necessary to compare then with rates of accumulation on the sea floor. Estimating rates of accumulation in this region is difficult, because significant variations exist from place to place. For example, the rates of accumulation on banks in both Florida Bay and the inner reef tract are much higher than in adjacent areas. For this reason, and because lime mud is shifted in suspension from one place to another, the average rates of accumulation were estimated for the two main areas, northeastern Florida Bay and the inner reef tract. Using such large areas removes the large local variations in rates. Comparison of rates of accumulation with the rates of fine aragonite mud production will indicate how much of the sea floor mud could be produced by *Penicillus*.

Rate of Sediment Accumulation

The rates of sediment accumulation for Recent sediments are usually calculated from radiocarbon dates of sediment cores. Because this method requires more dates than are available, another method was used for the present study.

The basis of the method is an estimate of the rate of relative rise of sea level in this region during the past 10,000 years. Radiocarbon dating of subsurface samples of Recent sediments that accumulated at or very near sea level gives a curve of rate of relative rise of sea level (fig. 13). Because the Recent unconsolidated sediments lie on lithified Pleistocene limestones, it is possible to use figure 13 to date the start of

ft below present sea level, figure 13 shows that marine sedimentation could not have begun before 4300 years ago. Knowing when marine sedimentation began and the present thickness of unconsolidated sediment, one can calculate an average rate of accumulation.

Florida Bay.—From numerous probings

TABLE 1.—*Penicillus deaths and aragonite production*

Environment	Station	Aragonite Weight of Representative Plant (grams)	Deaths per m²/yr
Florida Bay	8167.2N	0.23	400
	8167.2S	0.23	372
	8167.9N	0.23	436
	8167.9M	0.24	756
	8167.9S	0.17[1]	332
	8167.13	0.20	216
Inner Reef Tract	A1	0.507	84
	A2	0.507	8
	A3	0.507	44
	A4	0.507	72
	D1	0.525	48
	D2	0.525	60
	D3	0.525	92
	D4	0.525	76
	A–D[2]	0.52	58
	E1	0.662	228
	E2	0.662	296
	E3	0.662	256
	F1	0.337	100
	F2	0.337	112
	F3	0.337	164
	F4	0.337	100

[1] This weight was used as the average for plants in interior Florida Bay.

[2] The average of A1–A4 and D1–D4 is taken as the average weight for plants in the inner reef tract.

313

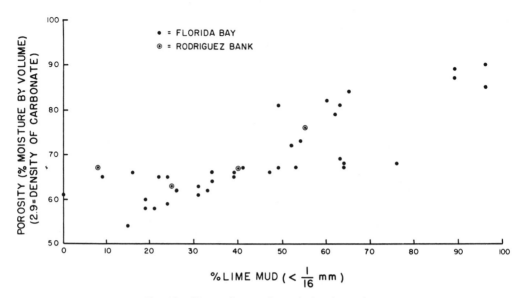

FIG. 12.—Percent lime mud vs calculated porosity.

throughout the study area the average depth of the Pleistocene rock floor below present sea level was found to be 188 cm. Figure 13 shows that marine sedimentation began at this level 3000 years ago. The average thickness of Recent unconsolidated sediment in the lakes is 16 cm,

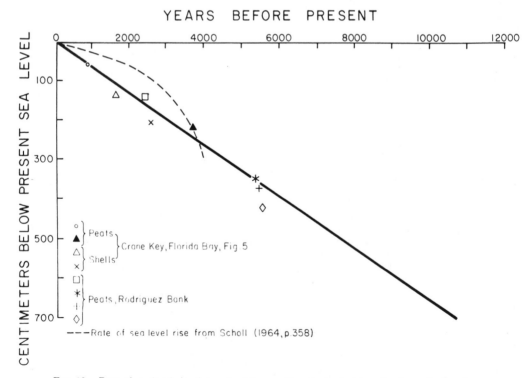

FIG. 13.—Rate of sea level rise determined from radiocarbon age determinations of subsurface sediments that accumulated near sea level.

and that on the banks is 90 cm. Therefore, the average rates of accumulation are 5.3 cm/1000 years for the lakes and 33 cm/1000 years for the banks. Combining these two figures according to the relative proportions of banks (10 percent) and lakes (90 percent) gives an over-all average rate of accumulation of 8 cm/1000 years. This rate is for total sediment, which includes particles produced by organisms other than *Penicillus*. An appreciable amount is sand and coarse silt size recognizable debris of mollusks, Foraminifera, and other organisms. Of the total whose origin cannot be determined with the microscope, 13 percent (based on X-ray analyses of numerous samples from Florida Bay by J. F. Burst and on pipette analyses of five surface samples made during this study) is fine aragonite mud equivalent in size to the disintegration products of *Penicillus*. Therefore, the average rate of accumulation of fine aragonite mud in the Florida Bay study area is 1.0 cm/1000 years.

Inner reef tract.—On the basis of several sparker profiles by M. M. Ball, in the inner reef tract the average depth below sea level of the Pleistocene rock floor is 667 cm. From figure 13, extrapolated, marine sedimentation at this level began 10,200 years ago. The average thickness of Recent unconsolidated sediment is 228 cm; thus, the rate of total sediment accumulation is 22 cm/1000 years. It is estimated from size analysis by Ginsburg (1956) that the percentage of fine aragonite mud in the surface sediment is 10 percent. Assuming that the subsurface sediments are similar to those now on the surface the calculated rate of accumulation of fine aragonite mud is 2.2 cm/1000 years.

These two rates of accumulation of fine aragonite mud, 1.0 cm/1000 years for northeastern Florida Bay and 2.2 cm/1000 years for the inner reef tract, can now be compared with the rates of mud accumulation produced by *Penicillus* in the two areas. For Florida Bay the rate for *Penicillus* mud is 0.3 cm/1000 years compared with 1.0 cm/1000 years. If the present rate of production is assumed to apply for the past 3000 years, *Penicillus* could have produced one-third of the fine aragonite mud in Florida Bay. For the inner reef tract the rate of *Penicillus* mud production is 2.3 cm/1000 years compared with a rate of accumulation of 2.2 cm/1000 years. If the present rate of accumulation is assumed to apply to the past 10,000 years in the inner reef tract, *Penicillus* could have produced all of the fine aragonite mud in this area.

EVALUATION OF RESULTS

The estimates of rates of production by *Penicillus* are believed to be conservative for several reasons: (1) The figures for average abundance are probably minimum ones. (2) The estimates do not include mass mortality of plants from unfavorable conditions, temperature, salinity, storms, or water level. (3) The estimates do not include plants that grew and died in less than 20 days, the period between observations, nor do they include parts of plants removed by browsing animals and replaced by later growth.

The assumption that the average abundance of *Penicillus* during the past several thousand years has been the same as it is today is open to question. Fresh-water runoff, which inhibits *Penicillus*, was probably more extensive in Florida Bay a few thousand years ago than it is today. On the other hand, the abundance of *Penicillus* may well have been higher during the time when Florida Bay and the inner reef tract were shallower than they are today.

In the estimation of sea floor rates of accumulation of fine aragonite mud, the most critical assumptions are (1) the rate of relative rise of sea level (fig. 13), (2) the coincidence of the start of marine sedimentation with the flooding of the Pleistocene rock floor by rising sea level, and (3) that the amount of fine aragonite mud in the subsurface sediments is the same as that in the surface sediments.

The curve showing relative rise of sea level (fig. 13) is generally similar to that given by Shepard (1961) based on samples from widely separated localities. More recently, Scholl (1964, p. 358) has given the results of a detailed study of the rate of sea level rise for southwest Florida. Scholl's curve of sea level rise (1964 fig. 8) shown by the dashed line on figure 13 is considerably different from ours. The use of Scholl's curve to estimate rate of sediment accumulation in the range from present sea level to 3 meters below would give slower rates than those estimated from our own straight line in figure 13. If the rates of sediment accumulation are in fact slower the calculated percentage of fine aragonite mud produced by *Penicillus* would be larger. Because Scholl had no samples from depths greater than 2 meters below present sea level the extrapolation of his curve beyond 3 meters below sea level is not justified.

The assumption that the beginning of recent marine sediment accumulation was coincident with the initial flooding of the rock floor is open to question. If the time lag between flooding and the start of sediment accumulation is significant, the calculated rates of accumulation are low and the estimated contribution from *Penicillus* is too high.

The proportion of fine aragonite mud in the subsurface sediments is assumed to be the same as that of the surface sediments. For Florida Bay, examination of numerous cores from the

"Lakes" show that this assumption is valid. However, the assumption may not be valid for the inner reef tract, where the only long cores available, those from Rodriguez Bank, show a marked increase in lime mud with depth.

The comparison of rates of production by *Penicillus* with rates of accumulation assumes that both Florida Bay and the inner reef tract are sediment traps that retain lime mud produced within them. In addition, as will be suggested below, we believe that these traps collect lime mud from adjacent current-swept areas. However, aerial and surface observations show that during the ebb or seaward-moving tide the water is most invariably more turbid than on the flood (Ball, Stockman, and Shinn, 1966). Some of this turbid water eventually reaches the Straits of Florida and is carried away from the shelf into deep water. The amounts of these losses are unknown, but unless they are large compared with the volumes in the sediment traps, they will not seriously affect the comparison of production and accumulation.

It is impossible to evaluate quantitatively all the various sources of errors in the estimates of rates of production and accumulation. We believe that the most reliable figures are those for the life span, the average plant abundance and the weight of aragonite/plant. The least reliable figures are probably the rates of accumulation of fine aragonite mud. Considering all the possible errors we believe that our estimates of the proportion of fine aragonite mud produced by *Penicillus* have an accuracy of $\pm 50\%$—Interior Florida Bay, $33\% \pm 16$, Inner Reef Tract, 100% ± 50.

OTHER SOURCES OF SKELETAL MUD

The preceding treatment was limited to the production of fine aragonite mud by *Penicillus* living within the areas of maximum mud accumulation, Florida Bay and the inner reef tract. Three other sources of lime mud derived from organic skeletons contribute to these accumulations: (1) Three other green algae with fragile skeletons similar to *Penicillus* disintegrate post-mortem to produce fine aragonite mud. At least one widespread and abundant red alga disintegrates to calcite mud. (2) Biological and mechanical erosion of skeletons of algae, mollusks, corals, and Foraminifera produces lime mud. (3) All the algae with fragile skeletons—*Penicillus*, the three other similar green algae, and the red alga—occur in the current-swept sandy bottom area adjacent to Florida Bay and the inner reef tract.

The three green algae that have fragile skeletons of aragonite similar to *Penicillus* are *Rhipocephalus* sp., *Udotea* sp., and *Acetabularia* sp.

Bottom observations indicate that *Rhipocephalus* sp. and *Udotea* sp. together are more abundant in the reef tract and around the outer margin of Florida Bay than *Penicillus* sp. *Acetabularia* is more abundant than *Penicillus* in the northeast interior of Florida Bay. *Rhipocephalus* sp. and *Udotea* sp. are in the same family as *Penicillus*, and it is reasonable to assume that their life spans are similar to that of *Penicillus*. *Acetabularia* sp. is a dasycladacean alga whose life span is not known but from the work of Rupp (1966) on a related genus, *Chalmassia* sp., it is probably much longer than *Penicillus* sp. Taken together these three lightly calcified green algae certainly produce as much lime mud as does *Penicillus* and probably much more.

The abundances and life spans of red algae with fragile skeletons of calcite have not been studied. However, one of these, *Melobesia* sp., is present throughout the reef tract and over a large part of the Florida Bay study area. *Melobesia* sp. almost completely covers the blades of turtle grass, *Thallassia testudinum*, with very fragile crusts. This heavy incrustation disintegrates when the grass blades fall. Because the fast-growing turtle grass is everywhere abundant in the inner reef tract and Florida Bay, this one alga must be a major contributor to the fine calcite mud fraction of the sediment. Many other red algae with relatively fragile skeletons are present in the reef tract, and their post-mortem disintegration and decay also add calcite mud to the sediments.

Almost all of the sediment particles larger than 125 microns (1/8 mm) are fragments of plant and animal skeletons whose source organism can be identified (Ginsburg, 1956). In the range 15–125 microns (1/64–1/8 mm) the majority of the particles can be recognized as skeletal debris, although it is difficult to determine the source organism. Matthews (1966) studied the origin of the particles from 20μ to 62μ in the lagoonal sediments of Southern British Honduras and was able to identify the skeletal source of 3/4 of the particles in this size fraction. The abundance of this skeletal silt suggests that even smaller particles (<15 microns) are produced during the size reduction of skeletons. It is impossible to estimate the proportion of fine lime mud formed in this way, but on the basis of the abundance of skeletal silt, it is probably appreciable.

The production of lime muds from organic skeletons is not limited to the areas of mud accumulation. *Penicillus*, the three related green algae, and fragile red algae are locally abundant in the outer reef tract, where the sediments are predominantly sand size. Intense biological

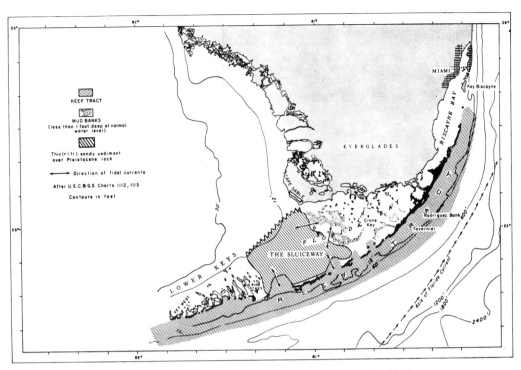

FIG. 14.—Chart showing the area of mud accumulation in Florida Bay
and the adjacent current-swept Sluiceway.

erosion of skeletons occurs in these sandy sediments and the coral-algal reefs which they surround. The total production of skeletal lime mud may exceed the amount produced in the inner reef tract.

In the outer reef tract, lime mud can be put in suspension by the wave action produced by winds of 20 mph or more and moved by the tidal currents. When the tide is falling and the currents flow seaward, this mud moves out into the Straits of Florida and eventually accumulates in deeper water (Ball, Shinn, and Stockman, 1966). When the tide is rising and the currents flow landward, the lime mud is carried shoreward and can accumulate in the inner reef tract or the margins of Florida Bay.

E. M. Brownson (personal communication) found that *Penicillus* is common throughout the current-swept area southwest of Florida Bay he termed The Sluiceway (fig. 14). The Recent sediments in The Sluiceway are only inches thick and are predominantly sand size. The skeletal debris in the sediments is strongly corroded and blackened. The water is commonly turbid with suspended mud in The Sluiceway. Frequent "whitings" occur in The Sluiceway and in Florida Bay. These streaks of very turbid water, white from the air, are produced by bot-

tom-feeding fish, mullet. Fishermen locate the schools of mullet by the whitings which they produce, and can even determine whether the schools are feeding by the character of the whiting. When this muddy water is carried onto the southwestern Florida Bay mud banks (fig. 14) by the flood tide, its suspended load can settle out on and between these grass-covered banks. The presence of trace quantities of clay mineral from the eastern Gulf of Mexico suite and quartz silt from southwest Florida in the sediments of western Florida Bay offer support for this inward transportation of lime mud produced in The Sluiceway and the adjacent open Gulf shelf. It is impossible to estimate the amount of mud that is trapped in Florida Bay from these adjacent areas of production. However, because the size of the mud banks increases enormously from the interior toward these potential source areas (fig. 5), it seems likely that shoreward movement of mud made on the current-swept sea floors is a major contributor to the margin of Florida Bay.

The preceding discussion of skeletal sources of lime mud other than *Penicillus* is largely qualitative. However, it is evident that the lime mud contributed from "other sources" far exceeds the contribution by *Penicillus* living in

317

the areas of mud accumulation. If the figures for production by *Penicillus* presented above are accurate within a factor of 3, the total production of skeletal mud is more than enough to account for the amount of lime mud sediment. It is therefore concluded that skeletal breakdown is the major source of lime mud in the south Florida area. The spontaneous disintegration of fragile algal skeletons within and outside the areas of accumulation is believed to be the predominant source for particles $<15 \mu$.

COMPARISON WITH OTHER DATA

We have been unable to find published data on the life span of *Penicillus* or related genera from other areas. Cloud (1962, p. 95) does give some information on the abundance of *Penicillus* and a related genus *Rhipocephalus* from an area of 25 m² near the middle of the northwestern Great Bahama Bank. He found a plant abundance of 1.5 plants/m² with an average of about 0.3 gms of carbonate/plant. These figures are similar to those from Florida Bay, 2 plants/m² and 0.17 gms/plant.

Matthews (1966) studied the origin of the lime mud in Southern British Honduras. He did not make estimates of rates of production but instead identified the source skeletons for the silt fraction, 20–62 μ and deduced the skeletal sources for the finer particles using mineralogy and Sr content. Matthews concluded (1966, p. 450) that all the mud is produced by the breakdown of skeletons and their debris.

IMPLICATIONS FOR THE ORIGIN OF ANCIENT LIME MUDS

To what extent can the breakdown of calcareous skeletons explain ancient lime muds? At present no direct method exists for determining whether or not ancient lime muds are derived from organic skeletons. We can only test this possibility by examining how well it fits the geologic facts of ancient lime muds.

One test is to compare the rates of production with average rates of accumulation in the geologic record. The observed rates of production by *Penicillus* are given on page 645. If we double these figures as a conservative estimate of the total production by fragile green algae, ignoring contribution from outside the areas of accumulation and breakdown of more resistant skeletons, we have, in round numbers, a range of 5–45 meters/million years. For comparison, Kay (1955, p. 678) gives a range of 1–6 meters/million years for the rate of deposition of post-Proterozoic sediments in North America. Kay (ibid., table 1) gives rates of deposition for Paleozoic miogeosynclines of 75–120 meters/million years. Admittedly, this comparison is subject to much

question; its only value is to show that the observed production of aragonite mud today could in time produce appreciable sediment thicknesses, thicknesses comparable to those we know from the geologic past.

Another test is to compare the conditions of accumulation of Recent lime muds with those of the ancient muds. The general similarity of occurrence indicated by similar textures and structures offers support for the argument that ancient muds, like the Recent muds of south Florida, were derived at least in large part from the breakdown of skeletal organisms. Many ancient lime muds, like Recent muds, as evidenced by mud cracks, accumulated in areas that were only intermittently covered by water. Other ancient lime muds have the same scarcity of skeletal debris and the churned appearance found in some Florida Bay sediments. Still others have the same mixtures of mud and skeletal sand that occur in the inner reef tract. Thus, it appears that the Florida region includes a representative range of settings similar to those of the past. If Recent lime muds accumulating in this range of conditions are of skeletal origin, it is reasonable to infer that ancient muds may have had a similar origin.

The lack of fossils or fossil fragments in ancient lime muds is frequently cited as evidence for inorganic precipitation. However, *Penicillus* and the other fragile green algae disintegrate completely when they die; thus, even in Recent sediments, they do not occur as recognizable skeletons.

The presence of interbedded lime mud and evaporites is also cited as evidence for inorganic precipitation of the lime mud. If the lime mud has been produced close by, this might be true. However, mud areas are sites of accumulation for transported material as well as in situ material. Therefore, lime mud associated with evaporites could have originated elsewhere as a product of skeletal breakdown.

There is every reason to believe that plants and animals with skeletons as fragile as *Penicillus* have existed since Cambrian times. Codiacean algae and red algae are known as far back as the Ordovician, and it seems likely that some of them had fragile skeletons. Animals with fragile skeletons were undoubtedly in ancient seas, that is, juveniles of most groups, Foraminifera, bryozoans, calcareous sponges, etc. Mechanical breakdown of skeletons is also evident in many ancient limestones. Only heavier skeletons are unbroken, and fragments of skeletons are the rule. All these observations support the likelihood that disintegration and breakdown of fragile and semi-resistant skeletons were major sources for many ancient lime muds.

ACKNOWLEDGEMENTS

We are indebted to R. Michael Lloyd for the original idea of this study and for his continued interest in the work and his advice on the report. Our colleague, Mahlon M. Ball, provided data on the reef tract and made numerous helpful suggestions during the preparation of the manuscript. Donald V. Higgs was a most thorough critic of the manuscript and his incisive queries helped to clarify many points. We are grateful to Shell Development Company for permission to publish this paper and for sponsoring its publication. Gerald M. Friedman and Willis W. Tyrrell, Jr., reviewed the manuscript and made some helpful suggestions for improving it.

REFERENCES

BALL, M. M., SHINN, E. A., AND STOCKMAN, K. W., 1966, The Geologic Effects of Hurricane Donna in South Florida: (submitted to Journal of Geology).

BAVENDAMM, W., 1932, Die mikrobiologische Kalkfallung in der Tropischen See: Arkiv für Mikrobiologie, Zeitschrift für die Erforschung der pflanzlichen Mikroorganismen, v. 3, p. 205–276.

CLOUD, P. E., JR., AND BARNES, V. E., 1948, Paleoecology of the Early Ordovician Sea in Central Texas: Natl. Res. Coun., Div. Geol. Georg., Report of The Committee on a Treatise on Marine Ecol. and Paleoecol., No. 8, p. 29–83.

———, 1962, Environment of Calcium Carbonate Deposition West of Andros Island, Bahamas: U. S. Geol. Surv. Prof. Paper 350, 138 p.

DREW, G. H., 1914, On the Precipitation of Calcium Carbonate in the Sea by Marine Bacteria and on the Action of Denitrifying Bacteria in Tropical and Temperate Seas: Carnegie Inst. Washington Pub. 182, Papers from the Tortugas Lab., v. 5, p. 7–45.

GEE, HALDANE, et al., 1932, Calcium Equilibrium in Sea Water: Scripps Inst. Oceanog. Bull., Tech Series, v. 3, p. 145–190.

GINSBURG, R. N., 1956, Environmental Relationships of Grain Size and Constituent Particles in Some South Florida Carbonate Sediments: Am. Assoc. Petr. Geol. Bull., v. 40, p. 2384–2427.

———, et al., 1964, South Florida Carbonate Sediments: Guidebook Field Trip No. 1, Geol. Soc. America, Ann. Meeting 1964, Inst. Mar. Sci. Univ. of Miami, Florida

GREENFIELD, L. J., 1963, Metabolism and Concentration of Calcium and Magnesium and Precipitation of Calcium Carbonates by a Marine Bacterium: N. Y. Acad. of Sci. Annal., v. 109, p. 23–45.

KAY, MARSHALL, 1955, Sediments and Subsidence Through Time: Geol. Soc. America, Sp. Paper 62, p. 665–684.

LIPMAN, C. B., 1924, A Critical and Experimental Study of Drew's Bacterial Hypothesis of CaCO₃ Precipitation in the Sea: Carnegie Inst. Washington Pub. 340, Papers from the Dept. Marine Biol., v. 19, p. 181–191.

LOWENSTAM, H. A., 1955, Aragonite Needles Secreted by Algae and Some Sedimentary Implications: Jour. Sedimentary Petrology p. 270–272.

———, AND EPSTEIN, SAMUEL, 1957, On the Origin of Sedimentary Aragonite Needles of the Great Bahama Bank, Jour. Geology, v. 65, p. 364–375.

MATTHEWS, R. K., 1966, Genesis of Recent Lime Mud in Southern British Honduras: Jour. Sedimentary Petrology, v. 36, p. 428–454.

McCALLUM, J. S. AND STOCKMAN, K. W., 1964, Water Circulation, Florida Bay, in Ginsburg, R. N., South Florida Carbonate Sediments: Guidebook Field Trip #1, Geol. Soc. Amer. Ann. Meeting, 1964, Inst. of Mar. Sci., Univ. of Miami, Florida, p. 11–13.

NEWELL, N. D., et al., 1959, Organism Communities and Bottom Facies, Great Bahama Bank: Am. Mus. Nat. History Bull., v. 117, p. 193.

OPPENHEIMER, C. H., 1961, Note on the Formation of Spherical Aragonite Bodies in the Presence of Bacteria From the Bahama Bank: Geochim et Cosmochim. Acta. v, 23, p. 295–296.

PURDY, E. G., 1963, Recent Calcium Carbonate Facies of the Great Bahama Bank. 2. Sedimentary Facies: Jour. Geology, v. 71, p. 472–497.

RUPP, A. W., 1966, Origin, Structure and Environmental Significance of Recent Fossil Calcispheres: Program for Geol. Soc. America Annual Meeting Abstract, p. 186.

SCHOLL, D. W., 1964, Recent Sedimentary Record in Mangrove Swamp and Rise in Sea Level Over the South Western Coast of Florida: Marine Geology, v. 1, p. 344–366.

SHEPARD, F. P., 1961, Sea Level Rise during the Past 20,000 Years, in Pacific Island Terraces: Eustatic, Annals of Geomorphology N. S. Supp. v. 3, p. 30–35.

SMITH, C. L., 1940, The Great Bahama Bank, I. General Hydrographic and Chemical Factors; II. Calcium Carbonate Precipitation: Jour. Marine Res., v. 3, p. 147–189.

13

Reprinted from *Marine Geology* **12**:123-140 (1972)

CARBONATE PRODUCTION BY CORAL REEFS[1]

KEITH E. CHAVE, STEPHEN V. SMITH AND KENNETH J. ROY[2]

Department of Oceanography and Hawaii Institute of Geophysics, University of Hawaii, Honolulu, Hawaii (U.S.A.)

(Received August 18, 1971)

ABSTRACT

Coral reef $CaCO_3$ production may be conveniently divided into three categories: potential production, gross production, and net production.

The first, potential production, is the amount of $CaCO_3$ produced per unit area of reef surface covered by an organism or a colony of organisms. Most calcareous phyla have potential productions near 10^4 (g $CaCO_3/m^2$)/year. Some values may go as high as 10^5 (g/m²)/year.

The second, gross production, is the amount of $CaCO_3$ produced by the reef community per unit area of sea floor. Hypothetical reef associations have gross productions ranging from $4 \cdot 10^2$ to $6 \cdot 10^4$ (g $CaCO_3/m^2$)/year. Environments consisting of combinations of these associations may produce between $3 \cdot 10^3$ and $6 \cdot 10^4$ (g $CaCO_3/m^2$)/year. The gross production of entire reef systems is apparently near 10^4 (g $CaCO_3/m^2$)/year.

The third, net production, the amount of $CaCO_3$ retained by the reef system, is largely a function of changing sea level. Presently net production is about 10^3 (g $CaCO_3/m^2$)/year. More rapid rates of sea level rise several thousand years ago probably were accompanied by greater net (and gross) production—up to more than $2 \cdot 10^4$ (g $CaCO_3/m^2$)/year—or by reef drowning.

The above rate estimates are largely hypothetical. However, they appear in reasonable agreement with the data which are available for real reef systems.

INTRODUCTION

No two reefs are identical; they differ in fauna and flora, in morphology, and in oceanographic climate. Yet, all reefs are similar, being magnificent complex communities of organisms able "to erect their own hard substrate . . . (and) create their own shore habitat which they are capable not only of maintaining but also expanding in the wake of destructive physical forces", (LOWENSTAM, 1950, p.433).

---

[1] Supported by ONR contract NR 083-603 and NSF (Sea Grant) grants GH-26 and GH-93. Hawaii Institute of Geophysics Contribution No. 432.

Invited paper, Symposium on "Sedimentation of Marine Organisms", Commission for Marine Geology and International Association of Biological Oceanography, E. Seibold and J. Hedgpeth, Conveners. "The Ocean World". Joint Oceanographic Assembly. Presented September 15th, 1970, in Tokyo.

[2] Present address: Institute of Sedimentary and Petroleum Geology, 3303 33rd St., N.W., Calgary 44, Alta., Canada.

The term reef, as used in this paper, includes all of the macro- and micro-environments of the system related to the reef community—lagoon, reef flat, algal ridge, outer slope, and so forth.

The similarities which do exist among reefs should make them susceptible to modeling. In this paper, we will first create hypothetical models of reefs, based upon carbonate production estimates for individual organisms; then we will compare the models with what is known about real reef communities and their geologic histories. Because comprehensive review articles about coral reefs are available (e.g., STODDART, 1969) we will make no attempt to present a survey of the coral reef literature.

The discussion will involve three terms. These are defined as:

(1) Potential Production (P). The amount of $CaCO_3$ produced by a single organism or colony of organisms per unit area of reef surface covered by that organism or colony. (g $CaCO_3/m^2$)/year.

(2) Gross Production (G). The amount of $CaCO_3$ produced per unit area of reef. G is obtained by summing the product of P of each organism in a given reef habitat, times the proportion of the reef area covered by the organism.

$$G = 0.01 \times \sum_{i=1}^{n} (P_i \times \% \text{ cover}) \text{ (g } CaCO_3/m^2)/\text{year}$$

(3) Net Production (N). The amount of $CaCO_3$ permanently retained by the reef system. $N = G - (\text{mechanical} + \text{chemical} + \text{biological loss})$. Gains through chemical diagenesis (e.g., LAND and GOREAU, 1970) will not be considered, because the rates of these gains are not known well enough. If a reef is to exist for a long period of time during rising sea level, N must be positive. (g $CaCO_3/m^2$)/year.

POTENTIAL PRODUCTION

Potential production is the amount of $CaCO_3$ produced by a single organism, or colony of organisms, per unit area of surface covered by that organism. This definition provides a unit of measurement by which the carbonate producing abilities of organisms with vastly different sizes and life histories can be compared. For the purpose of building models of reefs in this paper, P is estimated to the nearest order of magnitude. More exact estimation of P is limited by the availability of data and by the likely ability of organisms to produce differently in different environments. Detailed calculations might be undertaken for particular reefs or even portions of reefs for which adequate data were available.

It is assumed in the calculations that all calcareous organisms have a density of 1 g/cm³ of organism. This value is equivalent to about 37% $CaCO_3$ by volume. Probably few calcareous organisms differ from this proportion by as much as a factor of two, so this value is adequate for our order of magnitude calculations.

Potential production, P, will be calculated by one of two methods. In the first method P is calculated as the product of the organism turnover rate (year^{-1}), and the mass of the organism per unit area it covers (g/m^2). This calculation is most useful for organisms of definite size and definable longevity. For organisms of indefinite extent, P is most readily expressed as the CaCO$_3$ mass per volume of organism (1.0 g/cm^3 or 10^6 g/m^3) times the upward growth rate (cm/year or m/year). Each procedure yields equivalent units—(g CaCO$_3$/m^2)/year.

Phyla of calcareous organisms frequently encountered in reef environments include corals, red Algae, green Algae, mollusks, echinoderms, and Foraminifera. Our estimates of P, by phylum, are summarized in Table I.

Corals, the most prominent calcareous organisms on reefs, have been extensively studied, so their potential production is better known than is that of

TABLE I

ESTIMATES OF POTENTIAL PRODUCTION P BY PHYLUM

Organism	A. Size (m^2)	B. Mass (g)	C. Upward growth (m/year)	D. Turnover (1/year)	P. P (g CaCO$_3$/m^2)/year
Corals					
Acropora	—	—	10^{-1}	—	10^5
Other	—	—	10^{-2}	—	10^4
Green Algae					
Halimeda	10^{-3}	10^1	—	10^0	10^4
Penicillus	10^{-3}	10^0	—	10^1	10^4
Red Algae					
Lithothamnioid	—	—	10^{-2}	—	10^4
Articulated	10^{-3}	10^0	—	10^1	10^4
Mollusks					
Micro	10^{-6}	10^{-3}	—	10^1	10^4
Macro	10^{-3}	10^2	—	10^{-1}	10^4
Echinoderms	10^{-3}	10^2	—	10^{-1}	10^4
Foraminifera	10^{-6}	10^{-3}	—	10^1	10^4

$P = B/A \times D = $ g/m^2 $\times$ 1/year $= $ (g/m^2)/year

or

$P = C \times$ density $= $ m/year $\times$ g/m^3 $= $ (g/m^2)/year

[1] The order of magnitude of potential production (P) can be calculated in either of the two manners shown in this table. The density used here is assumed to be 10^6 g m^3.

the other organisms. Fig. 1 shows typical upward growth rates for a variety of common coral genera and the references from which these rates were derived. The growth rates range from more than 100 mm/year for *Acropora* and *Montipora* to less than 10 mm/year for some slower growing genera. This range of upward growth rates yields P values of 10^4 to 10^5 (g $CaCO_3/m^2$)/year (Fig. 1 and Table I). The lower rate is apparently very common, with the rapid rate being restricted to a few genera.

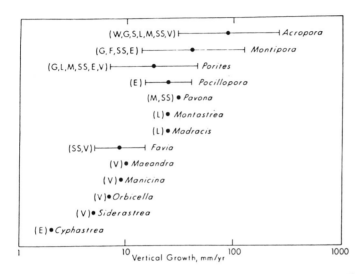

Fig.1. Vertical growth rates of typical reef-building corals. The letters in parentheses refer to the following publications: W = WOOD-JONES (1910); G = GUPPY (1889); S = SHINN (1966); L = LEWIS et al. (1968); M = MAYOR (1924b); SS = STEPHENSON and STEPHENSON (1933); V = VAUGHAN (1915); F = FINCKH (1904); E = EDMONDSON (1929).

In contrast to the available data on corals, relatively little is known about the growth of calcareous Algae, which often are a conspicuous portion of the reef community. Therefore the approximations to algal potential production are necessarily cruder than the coral potential production estimates.

Two genera of calcareous green Algae are locally abundant on coral reefs. *Penicillus*, common in the Atlantic but absent in the Pacific, has turnover rates of 6–9 year^{-1} in Florida Bay (STOCKMAN et al., 1967). Data in COLVINAUX et al., (1965) suggest that *Halimeda*, which is found on coral reefs throughout the world, has a turnover of about 2 year^{-1}. Plants of both genera commonly have cross-sectional areas of 10^{-3} m, corresponding to a square 2 to 5 cm on a side. *Penicillus* is assumed to weigh 10^0 g/plant, while the somewhat heavier *Halimeda* plants are assumed to weigh 10^1 g. These values yield similar P values of 10^4 (g $CaCO_3/m^2$)/ year (Table I).

The red Algae are another important phylum of calcareous Algae. It is convenient to divide these Algae into the encrusting, or lithothamnioid, coralline

Algae and the branching, or articulated, coralline Algae. JOHNSON (1961) cites upward growth rates of 1 cm/year, or 10^4 (g $CaCO_3/m^2$)/year for the litho-thamnioid Algae. Little is known about tropical articulated Algae, but the data for temperate water species (SMITH, 1970 and in press), as well as analogy with the calcareous green Algae, suggest that they also have P values over 10^4 (g $CaCO_3/m^2$)/year.

Mollusks are conveniently divided into two categories for calculation of P—micromollusks and macromollusks. E. Alison Kay (personal communication, 1970) suggests that the turnover rate of micromollusks is about 10 year^{-1}. If these mollusks have cross-sectional areas of about 10^{-6} m^2 (1 mm diameter), and a mass of about 10^{-3} g, then their P value will be on the order of 10^4 (g $CaCO_3$ g/m^2)/year (Table I). Various estimates suggest that macromollusks live from a few to more than 10 years (turnover about 10^{-1} year^{-1}) (FRANK, 1968; SMITH, 1970 and in press). If these mollusks have an area of about 10^{-3} m^2 and a mass of about 10^2 g, then they will also produce 10^4 (g $CaCO_3/m^2$)/year (Table I).

Echinoderms commonly live several years in temperate waters (SMITH, 1970 and in press), and are here assumed to have similar longevities in the tropics (turnover 10^{-1} year^{-1}). If their cross-sectional areas average 10^{-3} m^2, and they have a mass of about 10^2 g, then they also have P values of 10^4 (g $CaCO_3/m^2$)/year (Table I).

Benthic Foraminifera have been reported to have a turnover rate of up to 10 year^{-1} (MYERS and COLE, 1957). The size of these foraminifers is approximately 10^{-6} m^2; their mass is about 10^{-3} g, so they too yield P values of 10^4 (g $CaCO_3/m^2$)/year (see Table IV).

These order of magnitude calculations for production rates of common calcareous reef organisms, perhaps surprisingly, point out one thing: common taxa of tropical marine organisms appear to have similar potential carbonate production abilities, to within an order of magnitude. Only a few genera of corals are obvious exceptions to this suggested generality. If this generality proves to be supported as more reliable data become available, then most phyla of calcareous organisms have the ability to produce $CaCO_3$ sediment as rapidly as the modern coral reef community. This possibility is of considerable importance to interpreting ancient carbonate deposits.

GROSS PRODUCTION

The gross production of $CaCO_3$ in the different reef environments could range from that of a sandy reef flat without calcareous Algae, and with a few living micromollusks and Foraminifera, producing less than 10^2 (g $CaCO_3/m^2$)/year, to an *Acropora* thicket producing more than 10^5 (g $CaCO_3/m^2$)/year.

Simple models of coral reefs can be constructed by inserting potential production figures into typical reef habitats, and then combining these habitats

Fig.2. Hypothetical sand flat with calcareous green Algae and mollusks as the major CaCO₃ producers. As a scale, the tuft-like *Penicillus* plants are typically about 10 cm tall. Drawing by Helen Tong.

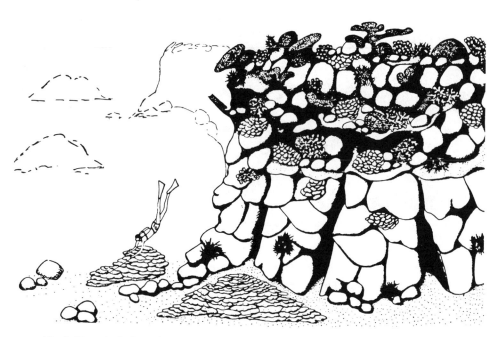

Fig.3. Hypothetical coral mound, with corals as the major CaCO₃ producers. The diver serves as a scale. Drawing by Helen Tong.

Fig.4. Hypothetical view of the algal ridge with encrusting coralline Algae as the major CaCO₃ producers. Typical vertical relief shown in this diagram is less than 1 m. Drawing by Helen Tong.

Fig.5. Hypothetical area of extensive coral coverage, with corals as the major CaCO₃ producers. Diver for scale. Drawing by Helen Tong.

into reef environments, and finally constructing total reef systems. We have chosen four typical habitats from which to construct our model reefs. These are:

(*1*) Shallow sand flat with calcareous Algae (Fig.2), a common sheltered back reef or lagoon habitat.

(*2*) Coral mounds on sand (Fig.3), a common semi-sheltered habitat.

(*3*) Algal ridge (Fig.4), the windward breaker zone habitat.

(4) Complete coral coverage (Fig.5), common on the outer slopes of reefs below strong wave action.

The gross production of these four habitats, calculated from potential production figures (Table I), is given in Table II–V. Sand flats are likely to produce about $4 \cdot 10^2$ (g $CaCO_3/m^2$)/year. Both coral mounds on sand and the algal ridge habitat apparently can produce 10^4 (g $CaCO_3/m^2$)/year. Complete coral cover may produce $6 \cdot 10^4$ (g $CaCO_3/m^2$)/year.

TABLE II

GROSS PRODUCTION OF A SAND FLAT WITH CALCAREOUS ALGAE

Organism	% Coverage	P	G
		(g $CaCO_3/m^2$)/year	
Penicillus	1	10^4	10^2
Halimeda	1	10^4	10^2
Macromollusks	1	10^4	10^2
Micromollusks	1	10^4	10^2
Foraminifera	0.1	10^4	10^1
		Total	410
		or $4 \cdot 10^2$	

TABLE III

GROSS PRODUCTION OF CORAL MOUNDS ON SAND

Organism	% Coverage	P	G
		(g $CaCO_3/m^2$)/year	
Acropora	10	10^5	10^4
Other coral	20	10^4	$2 \cdot 10^3$
Halimeda	1	10^4	10^2
Macromollusks	1	10^4	10^2
Micromollusks	1	10^4	10^2
Echinoderms	1	10^4	10^2
Foraminifera	1	10^4	10^2
		Total	13,000
		or $1 \cdot 10^4$	

TABLE IV

GROSS PRODUCTION OF ALGAL RIDGE

Organism	% Coverage	P	G
		(g CaCO$_3$/m^2)/year	
Coralline Algae	80	10^4	8 · 10^3
Coral other than Acropora	10	10^4	10^3
Barren	10	0	0
		Total or 9 · 10^3	9,000

TABLE V

GROSS PRODUCTION OF AREA WITH COMPLETE CORAL COVERAGE

Organism	% Coverage	P	G
		(g CaCO$_3$/m^2)/year	
Acropora	50	10^5	5 · 10^4
Other coral	50	10^4	5 · 10^3
		Total or 6 · 10^4	55,000

TABLE VI

GROSS PRODUCTION ESTIMATES FOR FIVE REEF ENVIRONMENTS

Environment	% Habitat (Table no.)	G (g CaCO$_3$/m^2)/year
Lagoon	50% Table II	
	50% Table III	5 · 10^3
Reef flat	50% Table II	
	25% Table III	
	25% Barren	3 · 10^3
Algal ridge	100% Table IV	9 · 10^3
Upper slope	100% Table V	6 · 10^4
Lower slope	10% Table V	
	20% Table III	
	20% Table II	
	50% Barren	8 · 10^3

From combinations of these four habitats, model reef environments can be constructed, and estimates of G for each can be made. Table VI summarizes such estimates for five reef environments. The production of these environments appears likely to range from $3 \cdot 10^3$ (g $CaCO_3/m^2$)/year on the reef flat to $6 \cdot 10^4$ (g $CaCO_3/$ m^2)/year on the upper slope seaward of the algal ridge. The lagoon, the algal ridge, and the lower portion of the outside reef slope can all produce between $5 \cdot 10^3$ and $9 \cdot 10^3$ (g $CaCO_3/m^2$)/year.

From these estimates of G for typical reef environments, we can calculate G for model reef systems. Table VII shows the calculation for a typical atoll 15 km in diameter. Similar calculations for barrier and fringing reefs are given in Table VIII and IX respectively. All three reef types can produce approximately 10^4 (g $CaCO_3/m^2$)/year.

TABLE VII

GROSS PRODUCTION ESTIMATE OF A CIRCULAR ATOLL

Environment	Width (km)	% Area	G (g $CaCO_3/m^2$)/year
Lagoon	15	77	$4 \cdot 10^3$
Reef flat	0.5	11	$3 \cdot 10^2$
Algal ridge	0.1	3	$3 \cdot 10^2$
Upper slope	0.1	3	$2 \cdot 10^3$
Lower slope	0.3	7	$6 \cdot 10^2$
		Total	$7 \cdot 10^3$

TABLE VIII

GROSS PRODUCTION ESTIMATE OF A BARRIER REEF COMPLEX

Environment	Width (km)	% Area	G (g $CaCO_3/m^2$)/year
Lagoon	1	50	$3 \cdot 10^3$
Reef flat	0.5	25	$8 \cdot 10^2$
Algal ridge	0.1	5	$5 \cdot 10^2$
Upper slope	0.1	5	$3 \cdot 10^3$
Lower slope	0.3	15	$1 \cdot 10^3$
		Total	$8 \cdot 10^3$

TABLE IX

GROSS PRODUCTION ESTIMATE OF A FRINGING REEF COMPLEX

Environment	Width (km)	% Area		G (g CaCO$_3$/m^2)/year
Reef flat	0.5	50		$2 \cdot 10^3$
Algal ridge	0.1	10		$9 \cdot 10^2$
Upper slope	0.1	10		$6 \cdot 10^3$
Lower slope	0.3	30		$2 \cdot 10^3$
			Total	$1 \cdot 10^4$

The models presented are highly oversimplified. Data in the literature, used in calculating P, are very sparse and in many cases are open to question. Potential production estimates lie in the vicinity of 10^4–10^5 (g CaCO$_3$/m^2)/year. Gross production estimates for model reef systems lie between $7 \cdot 10^3$ and $10 \cdot 10^3$ (g CaCO$_3$/m^2)/year. The test of our models lies in the real world: in estimates of real, gross, and net production, and in estimates of the stresses placed on the reef system by eustatic changes in sea level and tectonism.

NET PRODUCTION

Potential and gross production of carbonate by reef organisms and systems are principally biological considerations of how an organism, or a community of organisms, grows and calcifies. Net production of carbonate, on the other hand, is largely of geological interest, dealing with how much material the reef retains and what kind and size of structure is built.

Although for brief periods a reef, or part of a reef, may lag behind, or grow faster than rising sea level, over long periods a reef must stay abreast of sea level, or die. Subsidence combined with sea level rise can add to the community production requirements. During stillstands or times of falling sea level, a community able to keep its structure repaired against wave damage can probably remain a reef.

According to most authors, sea level has risen very slowly over the past several thousand years. A mean worldwide rise of about 1 mm/year since about 5,000 B.P. can be derived from the sea level curve of MILLIMAN and EMERY (1968). This rate is similar to the sea level rise over the last 100 years (FAIRBRIDGE, 1966). Considering initially only reefs in tectonically stable areas, we can calculate the demand which this rate of sea level rise imposes on the net production of a reef. If the accumulating material has a porosity of 50%, a net production of $1 \cdot 10^3$ (g CaCO$_3$/m^2)/year is necessary to keep up with sea level rise. Clearly, our model reefs can produce this much, with an order of magnitude to spare. The spare gross production can produce lateral accretion and lagoon filling, can be lost to the surrounding sea as fine detritus in suspension, or can escape as bottom sediment moving downslope. Loss of material from the system may be more important than lateral accretion, as evidenced by a plume of reef-derived carbonates observed in the water column 120 miles west of Jamaica (K. E. Chave and S. V. Smith, unpublished data), and reef sands on deep slopes south of Bermuda (CHAVE et al., 1962).

From about 15,000 to 7,000 B.P., sea level rose much more rapidly than the present rate. MILLIMAN and EMERY's (1968) sea level curve suggests a rate of up to 17 mm/year. This rate demands an N value of about $23 \cdot 10^3$ (g CaCO$_3$/m^2)/year, considerably higher than the G values for our hypothetical reefs.

Several occurrences during rapid sea level rise are possible. First, of course, our models may be too conservative; further refinement of values of P will resolve this. Second, many reefs may have drowned. Drowned reefs are well known (DILL, 1969, and others). Third, the morphology of the reefs may have changed; slower growing areas may have deepened, while the faster growing areas kept up, or nearly kept up with the rising sea. Enough of the reef community may have survived deepening to allow renewed growth of the reef as a whole as the sea level rise slowed. Fourth, the community structure may have altered to accommodate a higher G.

In tectonically active areas, where subsidence over short periods can add significantly to the deepening due to sea level rise, many reefs must drown.

CARBONATE PRODUCTION OF REAL WORLD REEFS

So far we have ignored information on the carbonate production of real world reefs. Since much of this information is as speculative as are our models, we hesitate to call the production itself real world—only the reefs.

Fig.6 and Table X summarize the available estimates of carbonate production by real coral reefs. Since most of the estimates are presented in terms of upward growth, the production has been calculated by assuming the reef materials have a porosity of 50%. Three useful guideposts have also been put on Fig.6. The fastest known rate of coral growth, 250 mm/year (LEWIS et al., 1968), can be considered to be the upper limit of possible CaCO$_3$ production. The approximate present rate of sea level rise (0.6–1.0 mm/year) and the maximum Holocene rate (17 mm/year) are also shown.

Most of the estimates fall into one of three clusters: about $25 \cdot 10^3$ (g CaCO$_3$/m^2)/year, about $7 \cdot 10^3$ (g CaCO$_3$/m^2)/year, and about $2 \cdot 10^3$ (g CaCO$_3$/m^2)/year. It is tempting to force these clusters to be identified as potential, gross, and net production. Although this interpretation is not entirely true, it is close enough to the truth to warrant consideration.

Points *1–5*; ranging from 22 to $35 \cdot 10^3$ (g CaCO$_3$/m^2)/year, might be

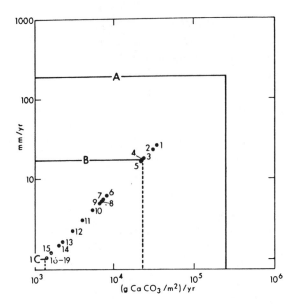

Fig.6. CaCO₃ production rates estimated for "real-world" coral reefs. The conversion between (g CaCO₃/m²)/year and mm/year is made on the assumption of 50% porosity for reef sediments. The numbers refer to the reef areas listed in Table X. Line *A* is our estimate of maximum potential production, based on the maximum known rate of coral growth. Line B is the approximate maximum rate of Holocene sea level rise, and line C is the approximate present rate of sea level rise.

expected to be estimates of *P*—essentially growth rates of corals. Indeed, points *1*, *4*, and *5* are estimates of *P* (GARDINER, 1903, Fiji; MAYOR, 1924, Samoa; VAUGHAN, 1915, Florida–Bahamas; respectively). Each of these authors interpreted coral growth to equal reef growth. Points 2 and 3 are intended as estimates of *G*. Point *2* (ODUM and ODUM, 1955) for the upward growth of an inter-island reef at Eniwetok Atoll, is the product of the coral standing crop times the coral growth rate reported by MAYOR (1924) from Samoa. ODUM and ODUM converted Mayor's growth rate estimate into per cent area increase, giving a new growth rate of 80 mm/year rather than 24 mm/year rate as given by Mayor. This method implies that the area of coral cover on the reef is increasing, rather than static. Although the Odum and Odum model might be realistic, we prefer the more conservative hypothesis that coral cover will remain at 20% of the reef areas, and that mean growth will be 5 mm/year (20% of 24 mm), or 6,800 (g CaCO₃/m²)/year, which is Point 9. Recent data by SMITH (in preparation) suggest that this latter value is closer to the correct gross production and perhaps is still somewhat too high. We suspect the same argument will hold for Point *3*, SARGENT and AUSTIN's (1949) production estimate from Rongelap Atoll. However, their standing crop data are presented in such a manner that the conversion cannot be made. Thus, with these exceptions, the first cluster may actually represent true values of *P*.

TABLE X

ESTIMATES OF CORAL REEF CaCO$_3$ PRODUCTION

Reef no.	Reef location	Reef environment	Production[*1] $(g/m^2)/year$	$mm/year$	Method	Apparent production type	Reference
1	Fiji	general	(35,000)	26	coral growth	P	Gardiner (1903)
2	Eniwetok	inter-island reef flat	31,000	16	coral growth × standing crop	G	Odum and Odum (1955)
3	Rongelap	reef flat	24,000	14	coral growth × standing crop	G	Sargent and Austin (1949)
4	Samoa	general	(23,000)	9–24	coral growth	P	Mayor (1924)
5	Florida–Bahamas	general	(22,000)	6–25	coral growth	P	Vaughan (1915)
6	Florida	Pleistocene reef	(8,100)	4–8	coral growth and in-filling	N	Hoffmeister and Multer (1964)
7	Hawaii	reef flat	(7,400)	5.5	C^{14} dating	N	Easton (1969)
8	Fiji	channel across reef	(7,000)	5.2	coral growth × standing crop	G	Gardiner (1903)
9	Eniwetok	inter-island reef flat	6,800	5	coral growth × standing crop	G	This paper; revision of no. 5 with more conservative estimate of growth than used by Odum and Odum (1955)
10	Bikini	lagoon	(5,400)	4	C^{14} dating	N	Emery et al. (1954)
11	Cocos–Keeling	lagoon	(4,100)	3	coral growth × standing crop	G	Guppy (1889)
12	Hawaii	barrier reef system	3,000	(2)	growth × standing crop	G	Smith et al. (1970)
13	Fiji	channel across reef	2,200	(1.6)	fill in dredged channel	N	Gardiner (1903); alternate of no. 8
14	Pacific	general	(2,000)	1.5	standing crop × growth	G	Dana (1872)
15	Jamaica	slope	(1,600)	1.2	C^{14} dating	N	Goreau and Land (in press)

TABLE X (continued)

Reef no.	Reef location	Reef environment	Production[*1] (g/m²)/year	Production[*1] mm/year	Method	Apparent production type	Reference
16	Hawaii	reef community on lava flow	1,400	(1)	accumulation on dated lava flow	N	OOSTDAM (1963)
17	Bikini	general	(1,400)	0.9–1.3	rate of sea level change	N	EMERY et al. (1954)
18	Bermuda	boiler reefs	(1,400)	1	C14	N	SCHROEDER and GINSBURG (1971)
19	Samoa	reef flat	1,400	0.8[*2]	coral growth × standing crop	G	MAYOR (1924)

[*1] A porosity of 50% has been used as a conversion factor between mm/year and (g/m²)/year for those values given in only one of these two units. We calculated the value in parenthesis from the companion value in all instances.

[*2] Mayor's paper gives this figure to be 8 mm/year rather than 0.8 mm/year.

The next cluster of points (6 through 10) lies between 5 and $8 \cdot 10^3$ (g CaCO$_3$/ m^2)/year. Point *8* is an estimate of *G* (GARDINER, 1903, Fiji). Point *9* has been previously discussed, and is *G*. Points *6*, *7*, and *10* are estimates of *N* (HOFF-MEISTER and MULTER, 1964, Florida; EASTON, 1969, Hawaii; EMERY et al., 1954, Bikini Atoll; respectively). These are determined for fossil reefs. The high values of *N* can be attributed to reef growth during periods of more rapid sea level rise, where *N* can approach *G*. The second cluster may all truly represent *G*.

Points *11* and *12* do not fall into any of the clusters. Point *11* is an estimate of *G* by GUPPY (1889) from Cocos–Keeling lagoon. He reported that the lagoon is shoaling locally. Point *12* is from SMITH et al. (1970) for a reef in a polluted bay on Oahu, Hawaii. While this reef has a *G* value of approximately $3 \cdot 10^3$ (g CaCO$_3$/ m^2)/year, it is apparently losing more material than it produces; thus its *N* value is negative. This reef will disappear if this condition persists.

The lowest group of points (*13* through *19*) lies between 1 and $2 \cdot 10^3$ (g CaCO$_3$/m^2)/year. Except for points *14* and *19* these figures truly represent values of *N*. Points *13* and *15* through *18* are taken from GARDINER, 1903, Fiji; GOREAU and LAND, in press, Jamaica; OOSTDAM, 1963, Hawaii; EMERY et al., 1954, Bikini; SCHROEDER and GINSBURG, 1971, Bermuda; respectively. Point *14* comes from the general, and poorly supported, statement by DANA (1872) that Pacific reefs can grow about 1.5 mm/year, based on CaCO$_3$ production by corals and other calcareous organisms. This should be *G*, but appears to be conservative. Point 19 is another estimate of *G* from Mayor (1924a and b) for Samoa. He cal-culated production from coral growth and standing crops. Unfortunately, he incorrectly converted 1,400 (g CaCO$_3$/m^2)/year to 8, rather than 0.8 mm/year upward growth and went to considerable effort to explain how this large amount of material was lost from the reef flat. Perhaps no material is being lost from the reef, or possibly that reef is actually eroding. Therefore, the last cluster of points may actually indeed be values of *N*.

With a certain amount of qualification, there appears to be a reasonable similarity between the hypothetical model reefs and the real world reefs. Most of the variability in production estimates of these real world reefs can be explained in terms of our potential, gross, and net production models.

SUMMARY

In summary, it can be seen that the range of coral reef growth rate estimates reflects, in part, the diversity of production types. If reef growth is taken to equal individual component growth (potential production), then reefs can grow at rates well in excess of 10^4 (g CaCO$_3$/m^2)/year (> 7 mm/year). Such unrealistic estimates can be revised downward by considering the area covered by each of the com-ponents as well as their growth rate (gross production). With the exception of dense, rapidly growing coral thickets, gross production is rarely likely to exceed

10^4 (g $CaCO_3/m^2$)/year (7 mm/year). Present rates of sea level rise are less than 1 mm/year, so net production (the amount of material retained by the reef system) will generally be limited to near 10^3 (g $CaCO_3/m^2$)/year.

In this study we have not pursued the various implications of this clarified coral reef $CaCO_3$ production terminology. However, it seems obvious to us that our clarification will prove useful in furthering geological, geochemical, and biological studies of coral reef processes.

ACKNOWLEDGMENT

This paper is dedicated to Dr. Thomas F. Goreau, a friend and a world-renowned reef expert, who died recently and suddenly, at much too young an age. Tom taught us much that we know about reefs. We wish that Tom were here to review this paper. He would be the first to say, "You are wrong, because..." and we sincerely wish that he were here to say this.

REFERENCES

CHAVE, K. E., SANDERS, H. L., HESSLER, R. R. and NEUMANN, A. C., 1962. Animal–sediment interrelationships of the Bermuda slope and adjacent deep sea. *Rep. ONR Contract Nonr-1135(02)*, *Bermuda Biol., Stn.*, 17 pp., unpublished.

COLINVAUX, L. H., WILBUR, K. M. and WATABE, N., 1965. Tropical marine algae: growth in laboratory culture. *J. Phycol.*, 1: 69–78.

DANA, J. D., 1872. *Corals and Coral Islands*. Dodd and Mead, New York, N.Y., 398 pp.

DILL, R. F., 1969. Submerged barrier reefs on the continental slope north of Darwin, Australia. *Geol. Soc. Am., Abstr.*, 7: 264–266.

EASTON, W. H., 1969. Radiocarbon profile of Hanauma Reef, Oahu (abstract). *Geol. Soc. Am. Spec. Pap.*, 121: 86.

EDMONDSON, C. H., 1929. Growth of Hawaiian corals. *B. P. Bishop Mus., Bull.*, 58: 1–138.

EMERY, K. O., TRACEY JR., J. I., and LADD, H. S., 1954. Geology of Bikini and nearby atolls: part 1, geology. *Geol. Surv. Prof. Pap.*, 260-A: 1–265.

FAIRBRIDGE, R. W., 1966. Mean sea level changes, long-term-eustatic and other. In: R. W. FAIRBRIDGE (Editor), *Encyclopedia of Oceanography*. Reinhold, New York, N.Y., pp. 479–485.

FINCKH, A. E., 1904. The Atoll of Funafuti: biology of the reef-forming organisms at Funafuti Atoll. *R. Soc. Lond., Rep. Coral Reef Comm.*, 6: 125–150.

FRANK, P. W., 1968. Growth rates and longevity of some gastropod molluscs on the coral reef at Heron Island. *Oecologia*, 2: 232–250.

GARDINER, J. S., 1903. *The Fauna and Geography of the Maldive and Laccadive Archipelagoes*. Cambridge Univ. Press, Cambridge, 423 pp.

GOREAU, T. F. and LAND, L. S., in press. Fore-reef morphology and depositional processes.

GUPPY, H. B., 1889. The Cocos–Keeling Islands. Part III. *Scott. Geogr. Mag.*, 5: 569–588.

HOFFMEISTER, J. E. and MULTER, H. G., 1964. Growth rate estimates of a Pleistocene coral reef of Florida. *Geol. Soc. Am. Bull.*, 75: 353–358.

JOHNSON, J. H., 1961. *Limestone-building Algae and Algal Limestones*. Johnson, Boulder, Colo., 297 pp.

LAND, L. S. and GOREAU, T. F., 1970. Submarine lithification of Jamaican reefs. *J. Sediment. Petrol.*, 40: 457–462.

LEWIS, J. B., AXELSEN, F., GOODBODY, I., PAGE, C. and CHISLETT, G., 1968. Comparative growth rates of some reef corals in the Caribbean. *Mar. Sci. Manuscr. Rep., McGill Univ.*, 10: 1–26.

LOWENSTAM, H. A., 1950. Niagran reefs of the Great Lakes area. *J. Geol.*, 58: 430–487.

MAYOR, A. G., 1924a. Causes which produce stable conditions in the depth of the floors of Pacific fringing reef flats. *Dep. Mar. Biol. Carnegie Inst. Wash., Papers*, 19: 27–36.

MAYOR, A. G., 1924b. Growth rate of Samoan corals. *Dep. Mar. Biol. Carnegie Inst. Wash., Papers*, 19: 51–72.

MILLIMAN, J. D. and EMERY, K. O., 1968. Sea levels during the past 35,000 years. *Science*, 162: 1121–1123.

MOBERLY, R., 1963. Coastal geology of Hawaii. *Hawaii Inst. Geophys. Rep.*, 41: 1–216.

MYERS, E. H. and COLE, W. S., 1957. Foraminifera. In: J. W. HEDGEPETH (Editor), *Treatise on Marine Ecology and Paleoecology. Geol. Soc. Am. Mem.*, 67(1): 1075–1082.

ODUM, H. T. and ODUM, E. P., 1955. Trophic structure and productivity of a windward coral reef community on Eniwetok Atoll. *Ecology* 25: 291–320.

OOSTDAM, B. L., 1963. The thickness and rates of growth of corals on top of the historic lava flow near La Perouse Bay, Maui, report of field work. *Unpubl. Rep.*, 68 pp. (Quoted in MOBERLY, 1963).

SARGENT, M. C. and AUSTIN, T. S., 1949. Organic productivity of an atoll. *Am. Geophys. Union Trans.*, 30: 245–249.

SCHROEDER, J. H. and GINSBURG, R. N., 1971. Calcified algal-filaments in reefs: criterion of early diagenesis. *Abstr., Pap. presented to Ann. Meet. Am. Assoc. Pet. Geologists, Houston, Texas, 1971*: 364.

SHINN, E. A., 1966. Coral growth-rate, an environmental indicator. *J. Paleontol.*, 40: 233–240.

SMITH, S. V., 1970. Calcium carbonate budget of the Southern California Continental Borderland. *Hawaii Inst. Geophys. Rep.*, 70–11: 1–174.

SMITH, S. V., in press. Southern California CaCO₃ production. *Limnol. Oceanogr.*

SMITH, S. V., ROY, K. J., CHAVE, K. E., MARAGOS, J. E., SOEGIARTO, A., KEY, G., GORDON, M. J., and KAM, D., 1970. Calcium carbonate production and deposition in a modern barrier reef complex. *Abstr., Pap. presented to Ann. Meet. Geol. Soc. Am., Milwaukee, Wisc., 1970*: 688–689.

STEPHENSON, T. A. and STEPHENSON, A., 1933. Growth and asexual reproduction in corals. *Great Barrier Reef Exped. 1928–1929, Sci. Rep., Br. Mus. (N.H.)*, 3: 167–217.

STOCKMAN, K. W., GINSBURG, R. N. and SHINN, E. A., 1967. The production of lime mud by algae in south Florida. *J. Sediment. Petrol.*, 37: 633–648.

STODDART, D. R., 1969. Ecology and morphology of Recent coral reefs. *Biol. Rev.*, 44: 433–498.

VAUGHAN, T. W., 1915. The geologic significance of the growth-rate of the Floridian and Bahaman shoal-water corals. *Wash. Acad. Sci. J.*, 5: 591–600.

WOOD JONES, F., 1910. *Coral and Atolls*. Lovell, Reeve and Co., London 392 pp.

Part III

COLD-WATER CARBONATES

Editors' Comments
on Papers 14 and 15

14 ASKELSSON
Excerpt from *Contributions on the Geology of the Northwestern Peninsula of Iceland*

15 LEONARD et al.
Excerpts from *Evaluation of Cold-Water Carbonates as a Possible Paleoclimatic Indicator*

A prevailing misconception among geologists is that all carbonate rocks are of shallow-water, low-latitudinal origin and, as such, have been widely used for paleoenvironmental interpretation. While this assumption is to a considerable extent true, exceptions exist. It was Daly as early as 1934 and Askelsson in 1936 (Paper 14) who drew attention to the occurrences of carbonate skeletal sand in shallow marine seas from the west coast of France and south coast of the West Fjord Peninsula of Iceland. Teichert (1958) and Chave (1967) pointed out many places located in higher latitudes where shallow-water sediments contain more than 50 percent calcium carbonate. Since then, Friedman and Sanders (1970), Lees (1975), Nelson (1978), Prasad Rao (1981), Sanders and Friedman (1969), and Paper 15 among others, drew attention to cold- and temperate-water carbonates, and brought to focus that possibly not all ancient carbonates are of tropical, shallow-water origin. Paper 15 is included to reemphasize the importance of higher latitude cold-water carbonates and encourage carbonate sedimentologists to have a fresh look into the problem. The authors of this paper have given an overview of the characteristics of the known occurrences of modern cold-water carbonates and also have given comparative characteristics of cold- and warm-water carbonates. Presently they suggest a pluralistic approach to discriminate cold-water carbonates from their warm-water counterparts.

REFERENCES

Chave, K. E., 1967, Recent Carbonate Sediments—an Unconventional View, *Jour. Geol. Education* **14**:200-204.

Daly, R. A., 1934, *The Changing World of the Ice Age,* Yale University Press, New Haven, 271p.

Friedman, G. M., and J. E. Sanders, 1970, Coincidence of High Sea Level with Cold Climate, and Low Sea Level with Warm Climate: Evidence from Carbonate, Rocks, *Geol. Soc. America Bull.* **81:**2457–2458.

Lees, A., 1975, Possible Influence of Salinity and Temperature on Modern Shelf Carbonate Sedimentation, *Marine Geology* **19:**159–198.

Nelson, C. S., 1978, Temperate Shelf Carbonate Sediments in the Cenozoic of New Zealand, *Sedimentology* **25:**737–771.

Prasad Rao, C., 1981, Criteria for Recognition of Cold-Water Carbonate Sedimentation: Berriedale Limestone (Lower Permian), Tasmania, Australia, *Jour. Sed. Petrology* **51:**491–506.

Sanders, J. E., and G. M. Friedman, 1969, Position of Regional Carbonate/Non-Carbonate Boundary in Nearshore Sediments along a Coast: Possible Climatic Indicator, *Geol. Soc. America Bull.* **80:**1789–1796.

Teichert, C., 1958, Cold- and Deep-Water Coral Banks, *Am. Assoc. Petroleum Geologists Bull.* **42:**2718–2744.

14

CONTRIBUTIONS ON THE GEOLOGY OF THE NORTHWESTERN PENINSULA OF ICELAND

J. Askelsson

This article was translated expressly for this Benchmark volume by G. M. Friedman, Rensselaer Polytechnic Institute, New York, from "Beitrage Zur Geologie der nordwestlichen Halbinsel von Island" in Geol. Fören. Stockholm Förh. **58:***111-112 (1936) by permission of the publisher, Geologiska Foereningen.*

Vast stretches of sand extend along the coast of Iceland, especially on the south coast of the west Fjord Peninsula. Unsuspecting observers may consider the sands to be composed of quartz, but as Keilhack already noted the particles of the sands are composed for the most part of skeletal fragments of pelecypods, especially *Cyprina, Saxicava, Pecten, Mya,* and *Anomia*.

A chemical analysis shows the following:

SiO_2	4.5%
Al_2O_3	1.0%
Fe_2O_3	
CaO	50.5%
MgO	0.5%
NaCl	1.5%
Na_2O	1.6%
CO_2	39.6%
TOTAL	99.2%

The sands are the result of excretion by fish of fragments of molluscan shells as found, for example, in the intestines of the fish *Anarrhiches Lupus L.*, which occurs in abundance near the coast. R. A. Daly has described a similar example from the west coast of France in his book, *The Changing World of the Ice Age* (1934). The calcium carbonate sand stretches far out into the ocean and it is certain that sea level was lower than at present when the sand first formed. At that time, as today, winds picked up the skeletal particles and transported them landward.

REFERENCE

Daly, R. A., 1934, *The Changing World of the Ice Age*, Yale University Press, New Haven, 271p.

15

EVALUATION OF COLD-WATER CARBONATES AS A POSSIBLE PALEOCLIMATIC INDICATOR

JAY E. LEONARD [1], BARRY CAMERON [2], ORRIN H. PILKEY [3] and GERALD M. FRIEDMAN [1]

[1] *Department of Geology, Rensselaer Polytechnic Institute, Troy, N.Y. 12181 (U.S.A.)*
[2] *Department of Geology, Boston University, Boston, Mass. 02215 (U.S.A.)*
[3] *Department of Geology and Marine Laboratory, Duke University, Durham, N.C. 27708 (U.S.A.)*

(Received April 19, 1979; revised and accepted April 15, 1980)

ABSTRACT

The common belief currently shared by many geoscientists concerning the climatic interpretation of limestones is that a warm-water environment is essential. This concept is not necessarily true because the rate and extent of terrigenous sediment dilution, rather than water temperature, is the primary factor determining whether or not a limestone forms at nearshore or continental shelf depths. Because carbonate productivity is lowest in cold climates, however, $CaCO_3$ abundance and the thickness of carbonate accumulations tend to be least at high latitudes. In this regard present-day continental shelves and beaches offer a poor model for comparing cold-water and warm-water carbonates because of the generally emergent continental tectonic framework, recent eustatic sea-level changes, and the presence of ice caps at the modern poles.

Typically, the influence of climate on non-reef continental shelf and beach environments cannot be clearly distinguished by the presence or absence of major taxonomic groups. Faunal diversity and equitability are more sensitive in this regard. The absence of shelf-depth inorganic carbonate precipitates, micrite envelopes, and peloids may also point to the cold-water origin of a rock. Skeletal mineralogy and oxygen isotopes of certain unrecrystallized carbonates may be good paleoclimatic indicators; however, trace elements and physical-textural attributes of the carbonate fraction are probably temperature insensitive.

Previous studies of high-latitude continental shelves have concentrated merely on the abundance of calcareous material and there is seemingly a disproportionate amount of information with respect to low-latitude carbonate studies. Further research on cold-water carbonates may open up new avenues for alternative paleoenvironmental and paleoclimatic interpretations.

INTRODUCTION

Until recently it has been a common assumption of many paleoecologists and sedimentologists that carbonate rocks indicate climatically tropical to

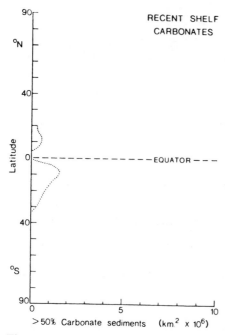

Fig. 1. Latitudinal distribution of modern marine shelf sediments greater than 50% carbonate, i.e., those that are potentially limestones. The bimodal frequency pattern is due to dilution by equatorial terrigenous sedimentation. (Modified from Fairbridge, 1964, p. 473, fig. 12.)

subtropical conditions, where warm oceanic surface waters appear to be saturated or even supersaturated with respect to $CaCO_3$ (Rodgers, 1957; Cayeux, 1970). Moreover, because a significant proportion of modern calcareous sediments appears to be presently forming in areas of warm water (Fig. 1), and because the distribution of marine organisms are strongly temperature-controlled (Fairbridge, 1967; Lees, 1975), many environmental stratigraphers have felt quite comfortable in utilizing a "warm climate" model for the interpretation of carbonate rocks. This classic concept was challenged by Chave (1967) who concluded that (1) carbonate sediments are, in fact, forming in many areas of higher latitude and colder water, (2) high-latitude surface waters may be supersaturated with respect to $CaCO_3$, and (3) non-tropical, colder-water calcareous organisms can calcify rapidly. These three factors indicate that there is a potential for the formation of limestone in colder climates.

Estimates of the percentage of limestone in the stratigraphic record range from 18 to 29%, being second in abundance to shales only (Sanders and Friedman, 1967; Garrells and Mackenzie, 1971; Pettijohn, 1975). They have received disproportionate attention relative to shales, however, because of the (1) great abundance of marine fossils in limestones, (2) large number of petroleum reservoirs in limestones, and (3) more precise paleoenvironmental interpretations possible (Sanders and Friedman, 1967).

Shallow-water and reef carbonates represent about 5% of the area of modern deposition of calcareous sediment, but they comprise about 28% of the total volume of modern marine carbonate sediments (Milliman, 1974). Modern shallow-shelf carbonates are extremely restricted in areal distribution. These carbonates, characteristic of ancient carbonate seas, are limited at present to a few areas such as Campeche and eastern Yucatan, eastern Nicaraguan shelf, western Florida, the Bahamas, the Persian Gulf, the south coast of India, several parts of Indonesia, and several coastal regions of Australia. Interestingly, the areal extent of these modern marine calcareous sediments is on the order of one-tenth or less that of ancient distributions. Ordovician shallow marine carbonates of North America, for example, are nearly continent-wide, whereas those of the Holocene rarely extend for hundreds of miles in any direction (Chave, 1960; Friedman and Sanders, 1978).

The origin of skeletal (especially calcareous) shallow-water benthic marine invertebrates, about 600 million years ago, is one of the great enigmas of earth history and made possible tremendous accumulations of marine carbonates in shallow epicontinental seas. This is the primary reason why limestones are so abundant in post-Precambrian rocks. A second revolution occurred in the mid-Mesozoic with the appearance of tremendous numbers of oceanic calcareous skeletal plankton, such as the planktic forams and coccoliths, which moved the primary locus of $CaCO_3$ deposition to the deep sea.

These two revolutions marked important events in the deposition of clastic limestones; however, the fact remains that there has been a substantially higher ratio of ancient Paleozoic and Mesozoic carbonate rocks compared to modern calcareous sediments. Were past eras generally warmer or perhaps were some of these ancient limestones of 'cold water' origin?

The answer to this question lies in establishing a set of criteria for reinvestigating older limestones. Unfortunately, the Holocene is by no means geologically 'normal,' or typical of the geologic past, because we are now essentially in an interglacial stage of the Quaternary Period, with both climate and sedimentation mechanisms (i.e., glaciers) significantly different from those for most of the distant past (Fairbridge, 1967). The normal mid-latitude temperatures are about 10°C lower today than in the past when there were no major continental ice-caps, such as during the Mesozoic. Oceanic bottom temperatures were likewise warmer in the past by as much as 8—10°C (Fairbridge, 1967). The low-latitude carbonate belt of the circum-Pacific region, for example, has an average present position of 40°N to 30°S, but its width changed from about 90° in the Triassic to about 65° in the Cretaceous (Khudoley, 1974).

With this gradual cooling of the earth since the Triassic, the oceans may have become less saturated or less extensively supersaturated in $CaCO_3$

(Fairbridge, 1967). Tropical shallow-water seawater is saturated by several hundred percent with $CaCO_3$, while cold temperate and polar waters are probably near saturation or even undersaturated at times (Alexandersson, 1972). Also, since the advent in the mid-Mesozoic of calcareous plankton, which extract $CaCO_3$ from seawater for their skeletons, there may also be less $CaCO_3$ dissolved as bicarbonate in the oceans.

Briden and Irving (1964) used paleomagnetic data to locate the paleo-latitude of ancient carbonates, dolomites, evaporites, and reefs. The paleo-latitudes of most Paleozoic limestones of North America fall within 30° of their appropriate paleoequators, but a small proportion lies between 40° and 80°N of their paleolatitudes. For Europe some lie between 40° and 50°N and S paleolatitude, but most occur within 30° of the paleoequator. Those of the Mesozoic and Tertiary are slightly more northerly than those of the Paleozoic. Those for the U.S.S.R. have a maximum at 25°N paleolatitude, but there is a significant proportion between 40° and 70° paleolatitude (Fig. 2a). Most of the lower Paleozoic to lower Carboniferous limestones of Australia occur within 30° of the paleoequator. However, most later lime-stones were deposited at higher mid-southern paleolatitudes, and they are small, local deposits without reefs (Fig. 2b). The question that arises is: how

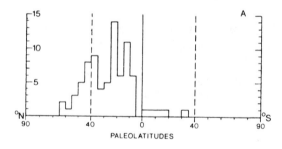

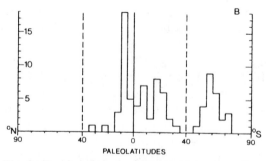

Fig. 2. Equal-angle paleolatitude histograms for carbonates of (a) the U.S.S.R. and (b) Australia. For Australia (b), the carbonates between 40°N and 80°N are post-lower Carboniferous in age. (Modified from Briden and Irving, 1964, pp. 210–211, figs. 15B and 16B.)

many of these higher paleolatitude limestones are of cold-water origin? Few studies in this regard are available, but Nelson (1978), for example, attributes Oligocene shelf limestones of New Zealand to generally cool to warm temperate (60° S to 35° S) climatic conditions.

It is the purpose of this paper to review most of the known shallow-water occurrences of recently-forming 'cold water' calcareous sediments in the western hemisphere and to compare or contrast these sediments with their warmer water counterparts. Little research has been carried out on the problem of distinguishing cold- and warm-water, non-reef carbonates of continental shelves (Lees, 1975; Nelson, 1978). Because most of the pertinent literature is scattered, non-systematic and often buried in studies concerned with other problems, we make no pretense that this report is thoroughly comprehensive. However, we hope that the carbonate sedimentologic characteristics for distinguishing ancient climates, although at present mostly inconclusive, will open up more research in this interesting and rewarding area.

OCCURRENCES OF NON-TROPICAL CARBONATES

Modern beach and nearshore carbonates

Two major types of U.S. Atlantic beaches can be recognized in terms of processes of carbonate sedimentation: barrier beaches and non-barrier beaches.

Beaches south of Long Island, New York, typically contain a mixed molluskan fauna derived partly from onshore transportation of their skeletons from the shelf (Pilkey and Field, 1972), and also by the seaward exposure of lagoonally derived sediments by barrier island migration. The resulting carbonate fraction consists of a mixture of lagoonal, beach, nearshore and open shelf forms. This faunal diversity is particularly apparent in some of the carbonate-rich beaches south of Cape Hatteras (F. Muehlburger and B.W. Blackwelder, personal communication). Comparison of the abundance of shelf and adjacent beach calcareous materials of the entire U.S. Atlantic barrier island coast (Milliman et al., 1972) reveals a similar relationship. South of Cape Hatteras, beach carbonate abundance ranges from 5 to 55% (Giles and Pilkey, 1965). North of the Cape, beach sands rarely contain more than 2—3% calcium carbonate. Cape Hatteras also marks an important abundance boundary of continental shelf carbonates; the average $CaCO_3$ content is higher offshore. In spite of the information available from malacologists on beach faunas, there have been no systematic quantitative studies of species distribution of calcareous skeletal material on U.S. Atlantic barrier beaches.

The carbonate fraction of rocky U.S. Atlantic pocket beaches usually consists of indigenous calcareous forms. Apparently, adjacent relict shelf sediments contribute relatively minor amounts of carbonates, if any. Thus,

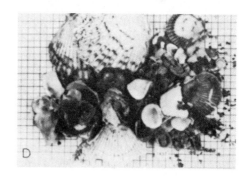

PLATE I

a. Sample No. 6-71-22, Sand Beach, Newport Cove, Acadia National Park, Mt. Desert Island, Maine. Sample taken on beach face at station 2+00 (See Fig. 3). Organic constituents are described in text (fine grid spacing is 1 mm).
b. W.H.O.I. sample No. 1123, collected on Georges Bank (40°59.4'N—67°00.8'W) at a depth of 74 ft (fine grid spacing is 1 mm.)
c. W.H.O.I. sample No. 1237 collected south of Cape Sable, Nova Scotia (43°10.2'N—65°44.3'W), at a depth of 64 ft (fine grid spacing is 1 mm).
d. W.H.O.I. sample No. 1437 collected off Cape Hatteras, N.C. (34°59.2'N—75°31.0'W), at a depth of 44 ft (fine grid spacing is 1 mm.)

the beach fauna along non-barrier coasts usually represents only in situ forms or those living in immediately adjacent environments. Rocky shoreline pocket beaches, such as those described by Raymond and Stetson (1932) and Leonard (1972, 1973), are restricted to cold temperate waters along the U.S. Atlantic coast. Thus, an apparent difference exists between U.S. Atlantic cold- and warm-water beach carbonate fractions; but it is not due to temperature as much as to an accident of geologic framework, i.e., tectonics, whereby rocky beaches are restricted to more northerly latitudes.

On the east coast of North America, Raymond and Stetson (1932) described a highly calcareous (27—67%) beach in cold waters on Little Cranberry Island on the mid-coast of Maine, U.S.A. The major contributors to the calcareous sediment are the blue mussel, *Mytilus*, the barnacle, *Balanus*, and a variety of snails. Other bivalves and echinoderms are of minor importance. The calcareous material also extends into nearby Little Cranberry Harbor.

Leonard (1972, 1973) and Leonard and Cameron (1979) described a highly calcareous pocket beach in rocky Newport Cove on Mt. Desert Island, Maine. The sediment on this beach, which is close to Raymond and Stetson's location has an average carbonate value of 69%. This calcareous percentage, which varies both temporally and spatially, is caused, in part, by seasonal variations of wave climate and energy. The dominant carbonate contributors are barnacles (*Balanus*), with mollusks second (mostly *Mytilus* and *Littorina*), and echinoderms third (sea urchins and sand dollars). Trace amounts of ectoprocts and red crustose coralline algae (*Lithothamnium*) are also present. Mineralogically, the sediment is mainly low-Mg calcite; high-Mg calcite is also present in the form of echinoderm spines and tests (Plate Ia).

On the north coast of Scotland, Raymond and Hutchins (1932) reported coarse beach sands composed of 97% carbonate material. This area, similar in coastal climate to Maine, has calcareous material which is dominated by mollusks.

Keary (1965, 1967), Lees et al. (1968), Lees and Buller (1971, 1972) and Gunatilaka (1977) studied the calcareous beach sediments in the coastal region of Connemara, western Ireland. They found that the sand fraction is dominated by molluskan debris, whereas the sand and gravel fraction is dominated by fragments of *Lithothamnium*. Foraminiferal sand also contains the skeletal debris of bivalves, gastropods, echinoderms and barnacles. Ectoprocts and sponge spicules are present, but they are unimportant in the total mass of the sediment. Although the carbonate content of individual beaches in the Connemara region may exceed 95% (Keary, 1967), the overall beach area has an average carbonate content of about 54%, with a standard deviation of approximately 27.5%. This indicates that, by strictest definition (Pettijohn, 1975; Friedman and Sanders, 1978), the area, if lithified, would contain a lot of limestone and, due to the relatively high standard deviation, various physical and biological agents are differentially concentrating or diluting the calcareous debris.

On the beaches and in the nearshore zones of the Alexander Archipelago, southeastern Alaska (56—58°N), Hoskin and Nelson (1969, 1971) found high percentages, ranging from 37 to 94%, of calcareous sediments of recently forming skeletal origin. The carbonate fraction of these sediments is mostly low-Mg calcite with lesser amounts of aragonite and high-Mg calcite. Texturally, the sediment is mainly sand and gravel; however, trace amounts of carbonate mud were reported. The main constituents of the calcareous sediment are dominated by barnacles, mollusks and echinoderms. Ectoprocts

(Bryozoa) are also important and minor amounts of corals, crustose coralline algae, articulate brachiopods and forams are present. Hoskin and Nelson (1971) suggested that waves erode shells from the rocks and transport this material through boulder bed "grinding mills" which produce shell fragmentation.

Continental shelf carbonates

Because at the present time the sediment cover of most continental shelf areas of the world is not in equilibrium with their terrigenous sediment sources (Emery, 1968), essentially the only sediment forming is the carbonate fraction. Thus, Pilkey (1968) and Milliman et al. (1972) suggested that present-day shelves will be represented by a carbonate-rich layer if preserved in the geologic column. This is probably true for both warm- and cold-water areas. Likewise, during Pleistocene interglacial stages, carbonate sediments accumulated in great thicknesses on the shelves of the continents (Sanders and Friedman, 1967).

Very few studies have been made of the carbonate fractions of continental shelf sediments from cold waters. Most studies involve either beach or near-shore sediments (e.g., Hoskin and Nelson, 1969; Leonard, 1972; Leonard and Cameron, 1979), or are primarily concerned with total abundance (e.g., Hulsemann, 1967). Some data for skeletal components of shelf sediments, primarily from tropical and subtropical environments, are available (e.g., McMaster et al., 1971; Milliman, 1971, 1974).

Generally, the abundance of shelf carbonate fractions plotted by Fairbridge (1964) (Fig. 1) indicates a range of significant accumulation from 30°S to 20°N, with a low at the equator where the high rainfall of the equatorial belt leads to a high terrigenous sedimentation rate. Contrary to Fairbridge, Hayes (1967) believes that 'shell' distribution is not diagnostic of latitude or climate.

Chave (1967) made the first major review of the distribution of modern high-latitude skeletal carbonates. He mentioned eight shelf occurrences from five areas north of 40°N and one area south of 40°S. Also, on a global scale Emery (1968) has suggested that carbonate sedimentation is more important on the western sides of ocean basins relative to the eastern sides. This is due to the normal open-ocean surface current patterns which 'warm up' the western ocean basin margins. Hence, in theory at least, broader expanses of the shelves on either side of the equator along western margins should be covered by $CaCO_3$-rich sediments relative to eastern margins during non-glacial times.

Within this general framework there are local variations in carbonate abundance that are common in colder shelf waters. For example, Georges Bank and the Scotia shelf (Plate Ib and Ic) have a percentage of skeletal $CaCO_3$ which may exceed 50% (Wigley, 1961; Hulsemann, 1967). The continental shelf of Onslow Bay, North Carolina, lying between Cape Lookout

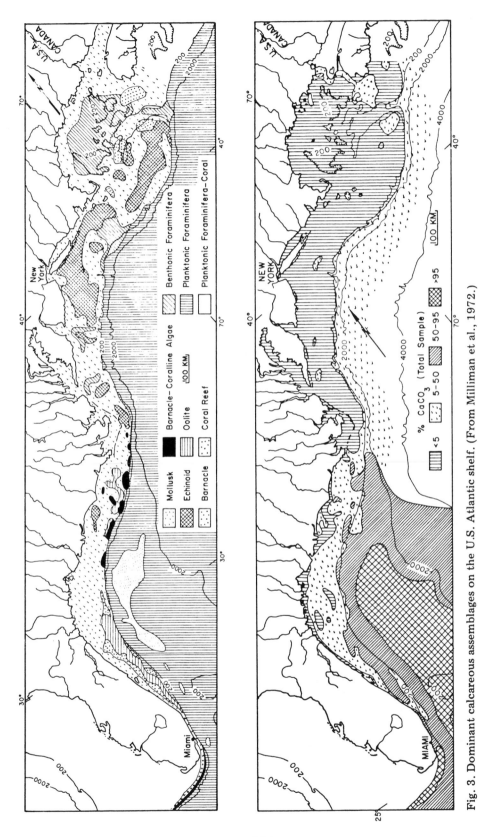

Fig. 3. Dominant calcareous assemblages on the U.S. Atlantic shelf. (From Milliman et al., 1972.)

Fig. 4. Abundance of CaCO₃ in U.S. Atlantic shelf sediments. (From Milliman et al., 1972.)

351

and Cape Fear, has a sediment cover richer in $CaCO_3$ than the adjacent shelf areas to the north or south (Milliman et al., 1968). Similarly, the beaches adjacent to Onslow Bay are relatively carbonate-rich. The reason for this local anomaly appears to be the lack of a terrigenous dilution factor during ice-age transgression and regression of sea level. Both north and south of Onslow Bay relatively large piedmont rivers empty onto the shelf, whereas Onslow Bay has low runoff, with local rivers deriving sediment only from the adjacent coastal plain.

An extensive area of biogenic carbonate sediments also characterizes the Atlantic coastal region of the British Isles and France. These carbonate sediments extend from northeast Scotland to the Isles of Skye in western Scotland and then along the coast of Ireland down to the coast of Bretagne in France. The total area across which carbonate sediments have spread is at least 5000 km^2. Although the thickness of these sediments is not known for certain, it is inferred that they may extend to a depth of 15 m (Gunatilaka, 1977). If sea level more or less reached its present position 3000—4000 years ago, then the rate of generation and accumulation of carbonate sediments was rapid.

In spite of these 'non-tropical' occurrences where the lack of terrigenous influx seems to be the main controlling factor for carbonate accumulation or concentration, temperature also plays an important role on modern continental shelves. An excellent example of temperature control of carbonate abundance is the U.S. Atlantic shelf (Fig. 3) where the Cape Hatteras—North Carolina region marks an important boundary with regard to shelf water temperatures (Emery and Uchupi, 1972). South of Cape Hatteras the environment is subtropical and to the north temperate. Similarly, Cape Hatteras marks a boundary with regard to carbonate abundance (Sanders and Friedman, 1969; Milliman et al., 1972). To the north carbonate material is almost always less than 5% of the shelf sediment, whereas off the southern United States values from 10 to 30% carbonate are more typical (Fig. 4) (Plate Id). The situation is somewhat complicated by the fact that most calcareous shelf material north and south of Cape Hatteras is relict and was deposited under shallower water conditions. For example, heavily bored, broken and polished barnacle valves dominate some carbonate-rich sediments of the southwest Grand Banks, which can reach over 70% $CaCO_3$ at water depths less than 70 m (Muller and Milliman, 1973). Milliman and Emery (1968) and Merrill et al. (1964) found late Pleistocene, shallow-water oysters on the Atlantic continental shelf above 40° latitude. Pilkey et al. (1969); Milliman et al. (1972) noted that most oolites north of Georgia were deposited during the last regression of the sea during the late Pleistocene, while the oolites deposited off Florida were formed primarily during the Holocene transgression.

Substrate variations can also play an important role in determining calcareous constituents, although on modern shelves relict components frequently obscure this control. Muds and sands contain dominantly infaunal assemblages (Milliman, 1974). In the quieter, finer-grained sedimentary areas rich in organic matter, infaunal burrowers and detritus feeders are favored (Emery and Uchupi, 1972). The outer shelf sediments commonly contain higher carbonate percentages dominated by calcareous algae, ectoprocts and a few corals. The greater proportion of hard substrates on the outer shelf favors skeletal epifauna such as barnacles, ectoprocts, articulate brachiopods, corals, etc., and crustose coralline algae (Emery and Uchupi, 1972; Milliman, 1974).

On the shelf of the western Atlantic Ocean, such as off Long Island, New York, Holocene submergence coupled with a lack of influx of terrigenous sediment, resulted in the proliferation of benthic Foraminifera. A total of 91 species were identified, of these 67 were found to be living at the time of collection (Gevirtz et al., 1971). At a water depth of 35 fathoms the near-shore species of *Elphidium* has been concentrated, reflecting a former still-stand of sea level (McKinney and Friedman, 1970).

Off the New Jersey coast an aragonite-cemented shelly sandstone resembling a modern carbonate beachrock was brought to the surface from a depth of 43 fathoms (Allen et al., 1969; Friedman and Sanders, 1970; Friedman et al., 1971b). The shelly sandstone consisted of about equal parts of terrigenous particles and of whole and broken shells of bivalves. Well-preserved specimens of the encrusting bryozoan, *Cribulina punctata* (Hassall) were found on some molluskan shells. The cement consists of aragonite of two habits, fibrous and cryptocrystalline. Radiocarbon dates give an age of 4390 ± 120 years for the molluskan shells; cementation must have occurred more recently. This aragonite-cemented rock was sampled from a ledge which is part of a drowned shoreline and is an extensive regional feature. Underwater television has shown the presence of carbonate-cemented rock elsewhere along this drowned shoreline. Hence the widespread formation of aragonite cement may be responsible for carbonate marine and beach rock lithification in a temperate climatic zone (Allen et al., 1969; Friedman and Sanders, 1970; Friedman et al., 1971b).

Off southern Alaska ($56-58°$N) Hoskin and Nelson (1969) found carbonate sands dominated by barnacles, echinoderms and mollusks, with significant amounts of ectoprocts and minor amounts of ahermatypic corals, crustose coralline algae, articulate brachiopods and foraminifera. In places the ahermatypic coral *Allopora brochi* (Fisher) is abundant.

In the western English Channel, Boillet (1965) found mollusk- and ectoproct-dominated carbonate sediments on hard substrates composed of shells and pebbles. Shoreward, barnacles decreased and ectoprocts increased in abundance. Lees et al. (1968) found that in shallow nearshore Irish seas, generally less than 20 m deep, *Lithothamnium* is abundant on rocky substrates in association with mollusks, asteroids, fish and ophiuroids. On other Irish shelf areas, mollusks dominated with barnacles, echinoderms, and *Lithothamnium* being important secondary skeletal carbonate contributors.

[*Editors' Note:* Material has been omitted at this point.]

TABLE I

Latitudinal criteria for the common characteristics of carbonate sediment

	Tropical	Temperate	Polar
Overall abundance	large	small	small
Thickness of buildup	thick	thin	trace
Shell thickness	thick	moderate	thin
Rates of sedimentation	high	moderate	low
Diversity	high	intermediate	low
Equitability	low	moderate	high
Reefs	frequent	rare	absent
Muds	high	low	trace
Oolites	present	trace	absent
Grapestones	low	(?)	absent
Micrite envelopes	common	low	absent
Peloids	common	low	absent
Dolomite	rare	trace	absent
Low-Mg calcite	common	common	common
High-Mg calcite	common	less common	less common
Aragonite	common	less common	less common

TABLE II

Sensitivity matrix for evaluating the relative usefulness of various carbonate sediment parameters in order to distinguish the difference between a cold or warm environment.

This table is based on our present knowledge of recent shelf and beach sediments

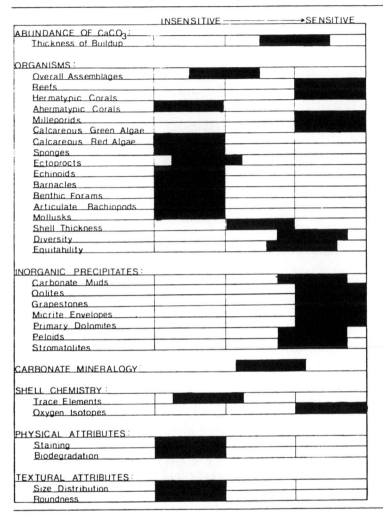

Summary of comparative criteria

In order to present the above-mentioned information concisely, we have constructed a matrix of the reviewed criteria for determining the latitudinal distribution of carbonate sediments (Table I). In addition, we have extracted most of the pertinent information on climatic differences of calcareous sediments in order to construct a 'sensitivity matrix' which indicates the relative usefulness of various carbonate sedimentological parameters (Table II).

We by no means assert that these tables of characteristics are complete and unchallengeable; and, we can find many exceptions ourselves. However, we do hope that this first attempt to characterize the usefulness of carbonate parameters in terms of climatic environments will open new, or at least different, avenues to the studies of paleoclimatology, paleoecology and environmental stratigraphy.

CONCLUSIONS

Due to the limited thrust of research emphasis, the current 'state of the art' gives us no single technique for differentiating cold- from warm-water calcareous sediments. This situation unfortunately holds true for both modern and ancient shelves and beaches. A pluralistic approach utilizing a combination of both positive and negative evidence must, therefore, be implemented. In spite of this dearth of information currently available, some general conclusions can be made:

(1) Because of the present tectonic framework and resulting greater continental land area and relief and because of the Pleistocene glaciation, the geology of the earth today does not provide an adequate model for comparison of warm- and cold-water carbonate deposition. Tectonics control the rate of terrigenous sedimentation which dilutes the carbonate fraction and ultimately determine whether a limestone forms.

(2) Although there is no direct cause-and-effect relationship, water temperature can be shown to influence the abundance of the carbonate fraction on present-day beaches and continental shelves. However, this parameter is less important than tectonism in determining limestone formation or the overall abundance of $CaCO_3$ in a sediment.

(3) At the present time continental shelves are not in equilibrium with their terrigenous sediment sources, i.e., the sediment is relict in most areas of the world. Thus, the only sediment forming in many shelf areas is essentially the carbonate fraction; this is true in both cold and warm waters.

(4) The abundance and thickness of carbonate build-ups is significantly

high in warm waters, relative to cold ones. Individual skeletons tend to be thicker and calcified at a faster rate in warm waters.

(5) There are no major higher taxonomic groups that are sensitive indicators of cold, shallow water, marine carbonate sedimentation. This is in contrast to warm waters where certain organisms, such as hermatypic reef corals and calcareous green algae, are good climatic indicators. Species diversity and equitability (abundance) are moderately sensitive, but they are also controlled by other ecological factors in addition to temperature.

(6) Biochemical and chemical carbonate precipitates, such as thick carbonate mud accumulations, micrite envelopes, calcareous stromatolites, syngenetic dolomites, evaporites, and oolites, are restricted to warm waters. Peloids are also restricted to warm waters.

(7) With respect to shell chemistry, mineralogy, physical and textural attributes of the carbonate fraction of beach and shelf sediments, only skeletal mineralogy and oxygen isotopes provide the best potential characteristics for recognizing cold-water marine carbonates. However, diagenetic events in the subsurface can void these characteristics.

ACKNOWLEDGEMENTS

This paper was originally presented at a symposium on 'Paleoclimatic Indicators in Sediments' that was organized by Rhodes W. Fairbridge and held at the 1976 Annual Meeting of the Society of Economic Paleontologists and Mineralogists. This project is partially a result of the joint Duke University—U.S. Geological Survey program in marine geology. We would like to thank the Duke University Oceanographic Program for cooperation in sample-collecting through the R.V. 'Eastward'. We would also like to thank the Woods Hole Oceanographic Institute for supplying many of the U.S. Atlantic shelf samples, through the joint W.H.O.I.—U.S. Geological Survey Atlantic continental marine program. Partial support from the National Science Foundation, Grant Numbers EAR76-84233 and EAR76-82433-A01 to S. Golubic and B. Cameron is acknowledged. The authors would like to thank John D. Milliman of W.H.O.I. for many informative discussions and G. Hoffman for the production of the manuscript.

REFERENCES

Adey, W.H. and MacIntyre, I.G., 1973. Crustose coralline algae: a reevaluation in the geological sciences. Geol. Soc. Am. Bull., 84: 883—904.

Alexandersson, T., 1972. Micritization of carbonate particles: processes of precipitation and dissolution in modern shallow-marine sediments. Bull. Geol. Inst. Univ. Uppsala, 3: 201—236.

Alexandersson, T., 1978. Distribution of submarine cements in a modern Caribbean fringing reef, Galeta Point, Panama: Discussion. J. Sediment. Petrol., 48: 665—668.

Allen, R.C., Gavish, E., Friedman, G.M. and Sanders, J.E., 1969. Aragonite-cemented sandstone from outer continental shelf off Delaware Bay: submarine lithification mechanism yields product resembling beachrock. J. Sediment. Petrol., 39: 136—149.

Bathurst, R.G.C., 1964. The replacement of aragonite by calcite in the molluscan shell wall. In: J. Imbrie and N. Newell (Editors), Approaches to Paleoecology. Wiley, New York, N.Y., pp. 357—376.

Bathurst, R.G.C., 1968. Precipitation of ooids and other aragonite fabrics of warm seas. In: G. Muller and G.M. Friedman (Editors), Recent Developments in Carbonate Sedimentology in Central Europe. Springer, New York, N.Y., pp. 1—10.

Bathurst, R.G.C., 1975. Carbonate Sediments and their Diagenesis. Elsevier, Amsterdam, 2nd ed., 620 pp.

Boillet, G., 1965. Organogenic gradients in the study of neritic deposits of biological origin: the example of the western English channel. Mar. Geol., 3: 359—367.

Briden, J.C. and Irving, E., 1964. Paleolatitude spectra of sedimentary paleoclimatic indicators. In: A.E.M. Nairn (Editor), Problems in Paleoclimatology. Wiley, London, pp. 199—224.

Cameron, B., 1979. Physical effects of quartzose algal mats and dune development — Plum Island spit, Massachusetts. In: D.G. Aubrey, Proceedings of a Workshop on Coastal Zone Research in Massachusetts (Nov. 27—28, 1978). W.H.O.I. Tech. Rep. 79-40, pp. 33—34.

Cayeux, L., 1970. Carbonate Rocks. Hafner, Darien, Conn., 493 pp.

Chave, K.E., 1954. Aspects of the biogeochemistry of magnesium. J. Geol., 62: 266—283, 587—599.

Chave, K.E., 1960. Carbonate skeletons to limestones: problems. Trans. N.Y. Acad. Sci., 23: 14—24.

Chave, K.E., 1967. Recent carbonate sediments — an unconventional view. J. Geol. Educ., 15: 200—204.

Clarke, F.W. and Wheeler, W.L., 1922. The inorganic constituents of marine invertebrates. U.S. Geol. Surv. Prof. Pap., 124: 62.

Dodd, J.R., 1967. Magnesium and strontium in calcareous skeletons: a review. J. Paleontol., 41: 1313—1329.

Emery, K.O., 1968. Relict sediments on continental shelves of the world. Am. Assoc. Pet. Geol. Bull., 52: 445—464.

Emery, K.O. and Uchupi, E., 1972. Western North Atlantic Ocean: topography, rocks, structure, water, life and sediments. Am. Assoc. Pet. Geol. Mem., 17: 532.

Fairbridge, R.W., 1964. The importance of limestone and its Ca/Mg content to paleoclimatology. In: A.E.M. Nairn (Editor), Problems in Paleoclimatology. Wiley, London, pp. 431—478.

Fairbridge, R.W., 1967. Carbonate rocks and paleoclimatology in the biogeochemical history of the planet. In G.V. Chilingar, H.J. Bissell and R.W. Fairbridge (Editors), Carbonate Rocks. Elsevier, Amsterdam, pp. 399—432.

Friedman, G.M., 1964. Early diagenesis and lithification in carbonate sediments. J. Sediment. Petrol., 34: 777—813.

Friedman, G.M., 1965. Occurrence and stability relationships of aragonite, high-magnesian calcite and low-magnesian calcite under deep sea conditions. Geol. Soc. Am. Bull., 76: 1191—1196.

Friedman, G.M., 1967. Obtaining environmental information: United States Patent Office, Patent No. 3,343,917, 3.

Friedman, G.M., 1968. Geology and geochemistry of reefs, carbonate sediments, and waters, Gulf of Aqaba (Elat), Red Sea. J. Sediment. Petrol., 38: 895—919.

Friedman, G.M., 1969. Trace elements as possible environmental indicators in carbonate sediments. In: G.M. Friedman (Editor), Depositional Environments in Carbonate Rocks. Soc. Econ. Paleontol. Mineral. Spec. Publ., 14: 193—198.

Friedman, G.M. and Sanders, J.E., 1967. Origin and occurrence of dolostones. In: G.V.

Chilingar, H.J. Bissell and R.W. Fairbridge (Editors), Carbonate Rocks. Elsevier, Amsterdam, pp. 267—348.

Friedman, G.M. and Sanders, J.E., 1970. Coincidence of high sea level with cold climate, and low sea level with warm climate: evidence from carbonate rocks. Geol. Soc. Am. Bull., 81: 2457—2458.

Friedman, G.M. and Sanders, J.E., 1978. Principles of Sedimentology. Wiley, New York, N.Y., 792 pp.

Friedman, G.M., Gebelein, C.D. and Sanders, J.E., 1971a. Micritic envelopes of carbonate grains are not exclusively of photosynthetic algal origin. Sedimentology, 16: 89—96.

Friedman, G.M., Sanders, J.E., Gavish, E. and Allen, R.C., 1971b. Marine lithification yields rock resembling beachrock. In: O.P. Bricker (Editor), Carbonate Cements. The John Hopkins Univ. Press, Baltimore, Md., 376 pp.

Garrels, R.M. and MacKenzie, F.T., 1971. Evolution of Sedimentary Rocks. W.W. Norton and Co., New York, N.Y., 397 pp.

Gevirtz, J.L., Park, R.A. and Friedman, G.M., 1971. Paleoecology of benthonic Foraminifera and associated micro-organisms of the continental shelf off Long Island, New York. J. Paleontol., 45: 153—177.

Giles, R.T. and Pilkey, O.H., 1965. Atlantic beach and dune sediment of the southern United States. J. Sediment. Petrol., 35: 900—910.

Ginsburg, R.N., 1959. Environmental relationships of grain size and constituent particles in some South Florida carbonate sediment. Am. Assoc. Pet. Geol. Bull., 40: 2384—2439.

Golubic, S., Perkins, R.D. and Lukas, K.J., 1975. Boring microorganisms and microborings in carbonate substrates. In: R.W. Frey (Editor), The Study of Trace Fossils. Springer, New York, N.Y., pp. 229—259.

Gordon, W.A., 1975. Distribution by latitude of phanerozoic evaporite deposits. J. Geol., 83: 671—684.

Goreau, T.F., 1961. On the relation of calcification to primary productivity in reef building organisms. In: The Biology of Hydra. Univ. of Miami Press, Miami, Fla., pp. 269—285.

Gunatilaka, A., 1977. Recent carbonate sedimentation in Connemara, Western Ireland. Estuarine Coastal Mar. Sci., 5: 609—629.

Hayes, M.O., 1967. Relationship between coastal climate and bottom sediment type on the inner continental shelf. Mar. Geol.; 5: 111—132.

Heckel, P.H., 1972. Recognition of ancient shallow marine environments. In: J.K. Rigby and W.K. Hamblin (Editors), Recognition of Ancient Sedimentary Environments. Soc. Econ. Paleontol. Mineral. Spec. Publ., 16: 226—286.

Hoskin, C.M. and Nelson, R.V., Jr., 1969. Modern marine carbonate sediment, Alexander Archipelago, Alaska. J. Sediment. Petrol., 39: 581—590.

Hoskin, C.M. and Nelson, R.V., Jr., 1971. Size modes in biogenic carbonate sediment, southeastern Alaska. J. Sediment. Petrol., 41: 1026—1037.

Hulsemann, J., 1967. The continental margin off the Atlantic coast of the United States: carbonate in sediments, Nova Scotia to Hudson Canyon. Sedimentology, 8: 121—145.

Keary, R., 1965. A note on the beach sands of the Cois Farraige coast. Ir. Nat. J., 15: 40—43.

Keary, R., 1967. Biogenic carbonate in beach sediments of the west coast of Ireland. Sci. Proc. R. Dublin Soc., Ser. A, 3: 75—85.

Khudoley, K.M., 1974. Circum-Pacific Mesozoic ammonoid distribution: relation to hypotheses of continental drift, polar wandering, and earth expansion. In: C.F. Kahle (Editor), Plate tectonics — assessments and reassessments. Am. Assoc. Pet. Geol. Mem., 23: 295—330.

Land, L.S., 1967. Diagenesis of skeletal carbonates. J. Sediment. Petrol., 37: 914—930.

LaPorte, L.F., 1968. Recent carbonate environments and their paleoecologic implications. In: E.J. Drake (Editor), Evolution and Environment — a Symposium. Yale Univ. Press, New Haven, Conn., pp. 229—258.

Lees, A., 1975. Possible influence of salinity and temperature on modern shelf carbonate sedimentation. Mar. Geol., 19: 159—198.

Lees, A. and Buller, A.T., 1971. Temperate water shallow marine carbonate sediments and their ancient equivalents. Prog. Abstr., 8th Int. Sedimentol. Congr., 58.

Lees, A. and Buller, A.T., 1972. Modern temperate water and warm water shelf carbonates contrasted. Mar. Geol., 13: M67—M73.

Lees, A., Buller, A.T. and Scott, J., 1968. Marine carbonate sedimentation in Connemara, Ireland. Reading University Geol. Rep. No. 2, 64.

Leonard, J.E., 1972. Dynamics of carbonate origin and accumulation in a cold water pocket beach system: Mt. Desert Island, Maine: Geol. Soc. Am., Abstr. Prog., 4: 27.

Leonard, J.E., 1973. The implication of coastal hydrodynamics for the accumulation of cold water carbonate sediments: Geol. Soc. Am. Abstr. Prog., 5, 188.

Leonard, J.E. and Cameron, B., 1979. Origin of a high latitude carbonate beach: Mt. Desert Island, Maine, U.S.A. Northeast. Geol., 1: 133—145.

Lewis, J.B., Axelsen, F., Goodbody, I., Page, C. and Chislett, G., 1968. Comparative growth rates of some reef corals in the Caribbean. Mar. Sci. Manuscript, Rep. 10, McGill Univ., Montreal, 26.

Lewy, Z., 1972. Recent and Senonian oncolites from Sinai and Southern Israel. Israel. J. Earth Sci., 21: 193—199.

Logan, B.W., 1961. *Cryptozoon* and associated stromatolites from the recent of Shark Bay, western Australia. J. Geol., 69: 517—533.

Lowenstam, H.A., 1954. Factors affecting the aragonite: calcite ratios in carbonate secreting maring organisms. J. Geol., 62: 284—321.

McKinney, T.F. and Friedman, G.M., 1970. Continental shelf sediments of Long Island, New York. J. Sediment. Petrol., 40: 213—248.

McMaster, R.L., Milliman, J.D. and Ashraf, A., 1971. Continental shelf and upper slope surface sediments off Portuguese Guinea, Guinea and Sierra Leone, West Africa. J. Sediment. Petrol., 41: 150—158.

Merrill, A.S., Emery, K.O. and Rubin, M., 1964. Ancient oyster shells on the Atlantic continental shelf. Science, 147: 398—400.

Meyerhoff, A.A. and Meyerhoff, H.A., 1974. Tests of plate tectonics. In: C.F. Kahle (Editor), Plate Tectonics — Assessments and Reassessments. Am. Assoc. Pet. Geol. Mem., 23: 43—145.

Milliman, J.D., 1971. The role of calcium carbonate in continental shelf sedimentation. In: D.J. Stanley (Editor), The New Concepts of Continental Margin Sedimentation — Supplement. American Geological Institute, Washington, D.C., p. 20.

Milliman, J.D., 1974. Marine Carbonates. Springer, New York, N.Y., 375 pp.

Milliman, J.D. and Emery, K.O., 1968. Sea levels during the past 35,000 years. Science, 162: 1121—1123.

Milliman, J.D., Pilkey, O.H. and Blackwelder, B.W., 1968. Carbonate sedimentation on the continental shelf, Cape Hatteras to Cape Romain. Southeast. Geol., 9: 1315—1334.

Milliman, J.D., Pilkey, O.H. and Ross, D.A., 1972. Sediments of the continental margin off the eastern United States. Geol. Soc. Am. Bull., 83: 1315—1334.

Molnia, B.F. and Pilkey, O.H., 1972. Origin and distribution of calcareous fines on the Carolina continental shelf. Sedimentology, 18: 293—310.

Muller, J. and Milliman, J.D., 1973. Relict carbonate-rich sediments on southwestern Grand Bank, Newfoundland. Can. J. Earth Sci., 10: 1744—1750.

Nelson, C.S., 1978. Temporate shelf carbonate sediments in the Cenozoic of New Zealand. Sedimentology, 25: 737—771.

Pettijohn, F.J., 1975. Sedimentary Rocks. Harper and Row, New York, N.Y., 3rd ed., 628 pp.

Pilkey, O.H., 1964. The size distribution and mineralogy of the carbonate fraction of United States south Atlantic shelf and upper slope sediments. Mar. Geol., 2: 121—136.

Pilkey, O.H., 1968. Sedimentation processes on the Atlantic southeastern United States continental shelf. Marit. Sediments., 4: 49—51.

Pilkey, O.H. and Blackwelder, B.W., 1968. Mineralogy of the sand size carbonate fraction of some recent marine terrigenous and carbonate sediments. J. Sediment. Petrol., 38: 799—810.

Pilkey, O.H. and Field, M.E., 1972. Onshore transportation of continental shelf sediment, Atlantic southeastern United States. In: D.J.P. Swift, D.B. Duane and O.H. Pilkey (Editors), Shelf Sediment Transport; Process and Pattern. Dowden, Hutchinson and Ross, Stroudsburg, Penn., 656 pp.

Pilkey, O.H., Blackwelder, B.W., Doyle, L.J., Estes, E. and Terlecky, P.M., 1969. Aspects of carbonate sedimentation on the Atlantic continental shelf off the southern United States. J. Sediment. Petrol., 39: 744—768.

Piotrowski, R., Pilkey, O.H. and Leonard, J.E., in prep. Distribution of carbonate sediments along the beaches of the U.S. Atlantic coast.

Raymond, P.E. and Hutchins, F., 1932. A calcareous beach at John O'Groats, Scotland. J. Sediment. Petrol., 2: 63—67.

Raymond, P.E. and Stetson, H.C., 1932. A calcareous beach on the coast of Maine. J. Sediment. Petrol., 2: 51—62.

Rodgers, J., 1957. The distribution of marine carbonate sediments: a review. Soc. Econ. Paleontol. Mineral. Spec. Publ., 5: 2—13.

Sanders, J.E. and Friedman, G.M., 1967. Origin and occurrence of limestones. In: G.V. Chilingar, H.J. Bissell and R.W. Fairbridge (Editors), Carbonate Rocks. Elsevier, Amsterdam, pp. 169—265.

Sanders, J.E. and Friedman, G.M., 1969. Position of regional carbonate/non carbonate boundary in nearshore sediments along a coast: possible climatic indicator. Geol. Soc. Am. Bull., 80: 1789—1796.

Schneider, J., 1977. Carbonate construction and decomposition by epilithic and endolithic micro-organisms in salt and freshwater. In: E. Flugel (Editor), Fossil Algae, Recent Results and Developments. Springer, New York, N.Y., pp. 248—260.

Schopf, T.J.M., 1970. Taxonomic diversity gradients on ectoprocts and bivalves and their geologic implications. Geol. Soc. Am. Bull., 81: 3765—3768.

Sneh, Amihai and Friedman, G.M., 1973. Recent and Senonian oncolites from Sinai and Southern Israel (discussion). Isr. J. Earth Sci., 22: 59—60.

Stehli, F.G. and Hower, T., 1961. Mineralogy and early diagenesis of carbonate sediments. J. Sediment. Petrol., 31: 358—371.

Stehli, F.G., McAlester, A.L. and Helsey, 1967. Taxonomic diversity of recent bivalves and some implications for geology. Geol. Soc. Am. Bull., 72: 455—466.

Stockman, K.W., Ginsburg, R.N. and Shinn, E.A., 1967. The production of lime mud by algae in south Florida. J. Sediment. Petrol., 37: 633—648.

Teichert, C., 1958. Cold- and deep-water coral banks. Am. Assoc. Pet. Geol. Bull., 42: 1064—1082.

Thorson, G., 1957. Bottom communities (sublittoral or shallow shelf). In: J.W. Hedgpeth (Editor), Treatise on Marine Ecology and Paleoecology, 1. Ecology. Geol. Soc. Am., Mem., 67: 461—534.

Titus, R. and Cameron, B., 1978. Endolithic algae, micrite envelopes, and micrite infillings of skeletal micropores as paleobathymetric indicators. Geol. Soc. Am., Abstr. Prog., 10: 89.

Valentine, J.W., 1968. Climatic regulations of species diversification and extinction. Geol. Soc. Am. Bull., 79: 273—276.

Wells, J.W., 1957. Corals. In: J.W. Hedgpeth (Editor), Treatise on Marine Geology and Paleoecology, 1. Ecology. Geol. Soc. Am. Mem., 67: 1087—1104.

Wigley, R.L., 1961. Bottom sediments of Georges Bank. J. Sediment. Petrol., 31: 165—188.

Zenger, D.H., 1972. Dolomitization and uniformitarianism. J. Geol. Educ., 20: 107—124.

AUTHOR CITATION INDEX

SUBJECT INDEX

About the Editors

AJIT BHATTACHARYYA is a reader in geology at Jadavpur University, Calcutta, India, and is currently a visiting post-doctoral research associate at the Department of Geology, Rensselaer Polytechnic Institute, Troy, New York. He holds the M.Sc. and Ph.D. degrees from Jadavpur University. He served as a junior research fellow of the Council of Scientific and Industrial Research in India for three years before joining the faculty of Jadavpur University in 1961. He has studied Proterozoic tidal-flat deposits in parts of Madhya Pradesh and Bihar, India, and has published a total of twenty-two papers. A member of the Society of Economic Paleontologists and Mineralogists and Sigma Gamma Epsilon, he visited the United States in 1977 on a UNESCO fellowship and worked in the laboratory of the Department of Geology, Rensselaer Polytechnic Institute, where he currently works.

GERALD M. FRIEDMAN is a professor of geology at Rensselaer Polytechnic Institute. He received the Ph.D. in geology from Columbia University in 1952 and the D.Sc. degree from the University of London in 1977. Dr. Friedman is national vice president of Sigma Gamma Epsilon and vice president of the Eastern Section of the American Association of Petroleum Geologists. He is a past president of the International Association of Sedimentologists and the Society of Economic Paleontologists and Mineralogists, past editor of the *Journal of Sedimentary Petrology*, former Distinguished Lecturer of the American Association of Petroleum Geologists, and past chairman of the geology section of the American Association for the Advancement of Science. A recipient of the Outstanding Paper Award from the *Journal of Sedimentary Petrology*, he is a member of the editorial boards of *Sedimentary Geology* and *Journal of Geology*.

Benchmark Papers
in Geology

Series Editor: Rhodes W. Fairbridge
Columbia University